Learning SAS® by Example

A Programmer's Guide

Second Edition

Ron Cody

sas.com/books

The correct bibliographic citation for this manual is as follows: Cody, Ron. 2018. *Learning SAS® by Example: A Programmer's Guide, Second Edition.* Cary, NC: SAS Institute Inc.

Learning SAS® by Example: A Programmer's Guide, Second Edition

ISBN 978-1-63526-659-7 (Paperback)
ISBN 978-1-63526-893-5 (Hardcover)
ISBN 978-1-63526-656-6 (EPUB)
ISBN 978-1-63526-657-3 (MOBI)
ISBN 978-1-63526-658-0 (PDF)

SAS Institute Inc., SAS Campus Drive, Cary, NC 27513-2414

July 2018

Contents

List of Programs

Chapter 1 Programs

Chapter 2 Programs

Chapter 3 Programs

Chapter 4 Programs

Chapter 5 Programs

Chapter 6 Programs

Chapter 7 Programs

Chapter 8 Programs

Chapter 9 Programs

Chapter 10 Programs

Chapter 11 Programs

Chapter 12 Programs

Chapter 13 Programs

Chapter 14 Programs

Chapter 15 Programs

Chapter 16 Programs

Chapter 17 Programs

Chapter 18 Programs

Chapter 19 Programs

Chapter 20 Programs

Chapter 21 Programs

Chapter 22 Programs

Chapter 23 Programs

Chapter 24 Programs

Chapter 25 Programs

Chapter 26 Programs

Chapter 27 Programs

Preface to the Second Edition

It's been almost 11 years since the first edition of this book was published, and I felt that it was time to bring it up-to-date. One major change is that all the SAS output is now displayed as HTML images, using the default style HTMLBlue. Not only does it look nicer than the old monospaced listing output, but the book figures are now images that can be viewed and sized on e-devices.

While we are on the topic of major changes, I should mention that a new chapter on Perl regular expressions was added, and the previous SAS/GRAPH chapter was replaced by a chapter on PROC SGPLOT. This procedure is easier to use than SAS/GRAPH, and (drum roll please) it is included with Base SAS.

People have come up to me at SAS conferences and told me that they learned SAS programming from the first edition of this book. That is really nice to hear. I hope that this edition will do the same for a whole new generation of programmers.

Ron Cody

Summer, 2018

About This Book

What Does This Book Cover?

This book teaches SAS programming from very basic concepts to more advanced topics. Because many programmers prefer examples over reference-type syntax, this book uses short examples to explain each topic. The second edition has brought this classic book about SAS programming up to the latest SAS version. There are new chapters that cover topics such as PROC SGPLOT (replacing the older chapter about SAS/GRAPH) and a completely new chapter about Perl regular expressions. This is a book that belongs on the shelf (or e-book reader) of every person who programs in SAS

Is This Book for You?

This book has been used by people with no programming experience who want to learn SAS as well as intermediate and even advanced SAS programmers who want to learn new techniques or see new ways to accomplish existing tasks.

What Are the Prerequisites for This Book?

There are no prerequisites for this book. It is for EVERYONE.

What's New in This Edition?

A new chapter about Perl regular expressions was added, and an old chapter about SAS/GRAPH was replaced by one describing PROC SGPLOT. This procedure can re-create all the output that was previously created by SAS/GRAPH but in a much simpler manner. All the programs in the second edition were examined to determine whether there was a newer, better way to accomplish the task. Finally, all the output is shown in the default HTML style.

What Should You Know about the Examples?

You can download every program and data set that is used in this book so that you can try the programs on your own—a valuable learning experience.

Software Used to Develop the Book's Content

The only SAS software that you need is Base SAS or the SAS University Edition. Because the latter is available to anyone as a free download, anyone can learn how to program using SAS even if he or she does not currently have access to SAS.

Example Code and Data

You can access the example code and data for this book by linking to its author page at support.sas.com/cody. If you are using the SAS University edition, you must copy the programs and data files to one of your shared folders. An example is `C:\SASUniversityEdition\Myfolders`.

SAS University Edition

 This book is compatible with SAS University Edition.

Where Are the Exercise Solutions?

Solutions to the odd-numbered problems are printed at the back of the book and are also included in the free download link on the author page. Professors can obtain copies of the solutions to the even-numbered problems, and self-learners can also request these solutions.

We Want to Hear from You

SAS Press books are written *by* SAS Users *for* SAS Users. We welcome your participation in their development and your feedback on SAS Press books that you are using. Please visit sas.com/books to do the following:

- Sign up to review a book.
- Recommend a topic.
- Request information about how to become a SAS Press author.
- Provide feedback on a book.

Do you have questions about a SAS Press book that you are reading? Contact the author through saspress@sas.com or https://support.sas.com/author_feedback.

SAS has many resources to help you find answers and expand your knowledge. If you need additional help, see our list of resources: sas.com/books.

About the Author

Ron Cody, EdD, is a retired professor from the Rutgers Robert Wood Johnson Medical School who now works as a national instructor for SAS and as an author of books on SAS and statistics. A SAS user since 1977, Ron's extensive knowledge and innovative style have made him a popular presenter at local, regional, and national SAS conferences. He has authored or co-authored numerous books, as well as countless articles in medical and scientific journals.

Learn more about this author by visiting his author page at http://support.sas.com/cody. There you can download free book excerpts, access example code and data, read the latest reviews, get updates, and more.

Acknowledgments

I hope you take the time to read this page because so many talented and hard-working people made this book possible, and I would like to thank and acknowledge their amazing work.

The idea for this book came from the editor of my last several books, Sian Roberts. Sian has moved on to a new role at SAS, and I am delighted to work with a new acquisitions and developmental editor, Lauree Shepard. She provided me with encouragement when needed and coordinated everything from technical reviews to copy editing and final assembly. There is truly more work in producing a book than just writing it.

One of my reviewers, Paul Grant, has reviewed almost every book I have written and he keeps coming back for more. Two other reviewers, Russ Tyndall and Charley Mullen, rounded out the team that reviewed the entire book. Thank you all so much. SAS also provided me with two experts, Kathryn McLawhorn and Leila McConnell, who reviewed sections of the book. Thank you both.

Vicki Leary had the difficult task of copy editing this second edition. It is the job of a copy editor to make it look like the author knows how to write. Before the book goes to press, I have read every chapter many, many times. But as hard as I try to catch every mistake or grammatical error, Vicki finds more.

Putting all the pieces (front matter, table of contents, the chapters themselves, and the index) together is quite a difficult and demanding job. Thanks so much to Denise Jones for producing the final product.

I have always felt that having an eye-popping and easily identifiable cover is really important and the artists at SAS are first rate, as you can see by this great cover created by Robert Harris.

Part 1: Getting Started

Chapter 1: What Is SAS?

1.1 Introduction

SAS is a collection of modules that are used to process and analyze data. It began in the late 60s and early 70s as a statistical package (the name *SAS* originally stood for Statistical Analysis System). However, unlike many competing statistical packages, SAS is also an extremely powerful, general-purpose programming language. SAS is the predominant software in the pharmaceutical industry and many Fortune 500 companies. In recent years, it has been enhanced to provide state-of-the-art data mining tools and programs for web development and analysis.

This book covers most of the basic data management and programming tools provided in Base SAS. Statistical procedures are not covered here. For a discussion of SAS statistical procedures, please see: Cody and Smith, *Applied Statistics and the Programming Language, 5th ed.* (Englewood Cliffs, NJ: Prentice Hall, 2005); Cody, *SAS Statistics by Example*, SAS Press (2011); Cody, *Biostatistics by Example Using SAS Studio,* SAS Press (2016).

The only way to really learn a programming language is to write lots of programs, make some errors, correct the errors, and then make some more. You can download all the programs and data files used in this book from this book's companion website at support.sas.com/cody. If you already have access to SAS at work or school, you are ready to go. If you are learning SAS on your own and do not have a copy of SAS to play with, you can obtain the free version of SAS (yes, I did say free) called the SAS University Edition. You can download the SAS University Edition by pointing your browser to www.sas.com/en_us/software/university-edition.html. You will be able to run any program in this book using the SAS University Edition.

1.2 Getting Data into SAS

SAS can read data from almost any source. Common sources of data are raw text files, Microsoft Office Excel spreadsheets, Access databases, and most of the common database systems such as DB2 and Oracle. Most of this book uses either text files or Excel spreadsheets as data sources. SAS also comes with a collection of data sets that you can use to practice your programming skills. You will find these data sets in the SASHELP library (more on this later).

1.3 A Sample SAS Program

Let's start out with a simple SAS program that reads data from a text file and produces some basic reports to give you an overview of the structure of SAS programs.

For this example, you have a text file with data on vegetable seeds. Each line of the file contains the following pieces of information (separated by spaces):

- Vegetable name
- Product code
- Days to germination
- Number of seeds
- Price

In SAS terminology, each piece of information is called a *variable*. (Other database systems, and sometimes SAS, use the term *column*.) A few sample lines from the file **`Veggies.txt`** are shown here:

```
Cucumber 50104-A 55 30    195
Cucumber 51789-A 56 30    225
Carrot   50179-A 68 1500 395
Carrot   50872-A 65 1500 225
Corn     57224-A 75 200  295
Corn     62471-A 80 200  395
Corn     57828-A 66 200  295
Eggplant 52233-A 70 30    225
```

In this example, each line of data produces what SAS calls an *observation* (also referred to as a *row* in other systems). A complete SAS program to read this data file and produce a list of the data, a frequency count showing the number of entries for each vegetable, the average price per seed, and the average number of days until germination is shown here:

Program 1.1: A Sample SAS Program

```
*SAS Program to read the Veggie.txt data file and to produce
 several reports;

options nonumber nodate;
data Veg;
   infile "C:\books\learning\Veggies.txt";
   input Name $ Code $ Days Number Price;
   CostPerSeed = Price / Number;
run;

title "List of the Raw Data";
proc print data=Veg;
run;

title "Frequency Distribution of Vegetable Names";
proc freq data=Veg;
   tables Name;
 run;

title "Average Cost of Seeds";
proc means data=Veg;
   var Price Days;
run;
```

At this point in the book, we won't explain every line of the program—we'll just give an overview.

SAS programs often contain DATA steps and PROC steps. *DATA steps* are parts of the program where you can read or write the data, manipulate the data, and perform calculations. *PROC* (short for procedure) *steps* are parts of your program where you ask SAS to run one or more of its procedures to produce reports, summarize the data, generate graphs, and much more. DATA steps begin with the word DATA and PROC steps begin with the word PROC. Most DATA and PROC steps end with a RUN statement (more on this later). SAS processes each DATA or PROC step completely and then goes on to the next step.

SAS also contains *global* statements that affect the entire SAS environment and remain in effect from one DATA or PROC step to another. In the program above, the OPTIONS and TITLE statements are examples of global statements. It is important to keep in mind that the actions of global statements remain in effect until they are changed by another global statement or until you end your SAS session.

All SAS programs, whether part of DATA or PROC steps, are made up of statements. Here is the rule: all SAS statements end with semicolons.

> **Note:** This is an important rule because if you leave out a semicolon where one is needed, the program may not run correctly, resulting in hard-to-interpret error messages.

Let's discuss some of the basic rules of SAS statements. First, they can begin in any column and can span several lines, if necessary. Because a semicolon determines the end of a SAS statement, you can place more than one statement on a single line (although this is not recommended as a matter of style).

To help make this clear, let's look at some of the statements in Program 1.1.

You could write the DATA step as shown in Program 1.2. Although this program is identical to the original, notice that it doesn't look organized, making it hard to read. Notice, also, that spacing is not critical either, though it is useful for legibility. It is a common practice to start each SAS statement on a new line and to indent each statement within a DATA or PROC step by several spaces (this author likes three spaces).

Program 1.2: An Alternative Version of Program 1.1

```
data Veg; infile "C:\books\learning\Veggies.txt";  input
Name $ Code $ Days Number
Price;   CostPerSeed =
Price /
Number;
run;
```

Another thing to notice about this program is that SAS is not case sensitive. Well, this is almost true. Of course references to external files must match the rules of your particular operating system. So, if you are running SAS under UNIX or Linux, file names will be case-sensitive. As you will see later, you get to name the variables in a SAS data set. The variable names in Program 1.1 are Name, Code, Days, Number, Price, and CostPerSeed. Although SAS doesn't care whether you write these names in uppercase, lowercase, or mixed case, it does "remember" the case of each variable the first time it encounters that variable and uses that form of the variable name when producing printed reports.

1.4 SAS Names

SAS names follow a simple naming rule: All SAS variable names and data set names can be no longer than 32 characters and must begin with a letter or the underscore (_) character. The remaining characters in the name may be letters, digits, or the underscore character. Characters such as dashes and spaces are not allowed. Here are some valid and invalid SAS names.

Valid SAS Names
Parts
LastName
First_Name
Ques5
Cost_per_Pound
DATE
time
X12Y34Z56

Invalid SAS Names	
8_is_enough	Begins with a number
Price per Pound	Contains blanks
Month-total	Contains an invalid character (-)
Num%	Contains an invalid character (%)

1.5 SAS Data Sets and SAS Data Types

We will talk a lot about SAS data sets throughout this book. For now, you need to know that when SAS reads data from anywhere (for example, raw data or spreadsheets), it stores the data in its own special form called a SAS data set. Only SAS can read and write SAS data sets. If you opened a SAS data set with another program (Microsoft Word, for example), it would not be a pretty sight—it would consist of some recognizable characters and many funny-looking graphics characters. In other words, it would look like nonsense. Even if SAS is reading data from an Oracle table or DB2, it is actually converting the data into SAS a data set format in the background.

The good news is that you don't ever have to worry about how SAS is storing its data or the structure of a SAS data set. However, it is important to understand that SAS data sets contain two parts: a descriptor portion and a data portion. Not only does SAS store the actual data values for you, it stores information about these values (things like storage lengths, labels, and formats). We'll discuss that more later.

SAS data sets have only two types of variables: character and numeric. This makes it much simpler to use and understand than some other programs that have many more data types (for example, integer, long integer, and logical). SAS determines a fixed storage length for every variable. Most SAS users never need to think about storage lengths for numerical values—they are stored in 8 bytes (about 14 or 15 significant digits, depending on your operating system) if you don't specify otherwise. The majority of SAS users will never have to change this default value (it can lead to complications and should only be considered by experienced SAS programmers). Each character value (data stored as letters, special characters, and digits) is assigned a fixed storage length explicitly by program statements or by various rules that SAS has about the length of character values.

1.6 The SAS Windowing Environment, SAS Enterprise Guide, and the SAS University Edition

Because SAS runs on many different platforms (mainframes, microcomputers running various Microsoft operating systems, UNIX, and Linux), the way you write and run programs will vary. You might use a general-purpose text editor on a mainframe to write a SAS program, submit it, and send the output back to a terminal or to a file. On PCs, you might use the SAS windowing environment, where you write your program in the Enhanced Editor (Editor window), see any error messages and comments about your program and the data in the Log window, and view your output in the Output window. Other ways to write and submit SAS programs are through a product called SAS Enterprise Guide, which is a front end to SAS that allows you to use a menu-driven system to write SAS programs and produce reports. One other alternative to the windowing environment or Enterprise Guide, is SAS Studio. The SAS University Edition (free SAS) uses SAS Studio as its entry into the SAS system.

There are many excellent books published by SAS that offer detailed instructions on how to run SAS programs on each specific platform and the appropriate access method into SAS. This book concentrates on how to write SAS programs. You will find that SAS programs, regardless of what computer or operating system you are using, look basically the same. Typically, the only changes you need to make to migrate a SAS program from one platform to another is the way you describe external data sources and where you store SAS programs and output.

1.7 Problems

Solutions to odd-numbered problems are located at the back of this book. Solutions to all problems are available to professors. If you are a professor, visit the book's companion website at support.sas.com/cody for information about how to obtain the solutions to all problems.

1. Identify which of the following variable names are valid SAS names:

```
Height
HeightInCentimeters
Height_in_centimeters
Wt-Kg
x123y456
76Trombones
MiXeDCasE
```

2. In the following list, classify each data set name as valid or invalid:

```
Clinic
clinic
work
hyphens-in-the-name
123GO
Demographics_2006
```

3. You have a data set consisting of Student_ID, English, History, Math, and Science_Scores on 10 students.

    a) The number of variables is __________
    b) The number of observations is __________

4. True or false:

    a) You can place more than one SAS statement on a single line.
    b) You can use several lines for a single SAS statement.
    c) SAS has three data types: character, numeric, and integer.
    d) OPTIONS and TITLE statements are considered global statements.

5. What is the default storage length for SAS numeric variables (in bytes)?

Chapter 2: Writing Your First SAS Program

2.1 A Simple Program to Read Raw Data and Produce a Report

Let's start out with a simple program to read data from a text file and produce some basic summaries. Then we'll go on to enhance the program.

The task: you have data values in a text file. These values represent Gender (M or F), Age, Height, and Weight. Each data value is separated from the next by one or more blanks. You want to produce two reports: one showing the frequencies for Gender (how many Ms and Fs); the other showing the average age, height, and weight for all the subjects.

Here is a listing of the raw data file **`Mydata.txt`** that you want to analyze:

```
M 50 68 155
F 23 60 101
M 65 72 220
F 35 65 133
M 15 71 166
```

Here is the program:

Program 2.1: Your First SAS Program

```
data Demographic;
  infile "C:\books\learning\Mydata.txt";
  input Gender $ Age Height Weight;
run;

title "Gender Frequencies";
proc freq data=Demographic;
   tables Gender;
run;
```

```
title "Summary Statistics";
proc means data=Demographic;
   var Age Height Weight;
run;
```

Notice that this program consists of one DATA step followed by two PROC steps. As we mentioned in Chapter 1, the DATA step begins with the word DATA. In this program, the name of the SAS data set being created is Demographic. The next line (the INFILE statement) tells SAS where the data values are coming from. In this example, the text file **Mydata.txt** is in the folder C:\books\learning on a Windows system.

If you decide to run some of the programs in this book, you can download all the programs and data files from the author website (support.sas.com/cody) and place them in a folder of your choice. For example, if you placed the text file **Mydata.txt** in a folder **C:\SASdata**, your INFILE statement would read:

```
infile "C:\SASdata\Mydata.txt";
```

If you are using the SAS University Edition, you may want to place all the data files in the folder **C:\SASUniversityEdition\Myfolders**, which is the default location you set up when you configured your virtual machine.

The INPUT statement shown here is one of four different methods that SAS has for reading raw data. This program uses the list input method, appropriate for data values separated by delimiters. The default data delimiter for SAS is a blank. SAS can also read data separated by any other delimiter (for example, commas or tabs) with a minor change to the INFILE statement. When you use the list input method for reading data, you need to list only the names you want to give each data value. SAS calls these *variable names*. As mentioned in Chapter 1, these names must conform to the SAS naming convention.

Notice the dollar sign ($) following the variable name Gender. The dollar sign following any variable name tells SAS that values for those variables are stored as character values. Without a dollar sign, SAS assumes values are numbers and should be stored as SAS numeric values.

Finally, the DATA step ends with a RUN statement. You will see later that, depending on what platform you are running your SAS program, RUN statements are not always necessary.

In Program 2.1 we placed a blank line between each step to make the program easier to read. Feel free to include blank lines whenever you wish to make the program more readable.

There are several TITLE statements in this program. You will see this statement in many of the SAS programs in this book. As you may have guessed, the text following the keyword TITLE (placed in single or double quotes, or even no quotes—as long as the title doesn't contain any single quotes) is printed at the top of each page of SAS output. Statements such as the TITLE statement are called *global* statements. The term *global* refers to the fact that the operations these statements perform are not tied to one single DATA or PROC step. They affect the entire SAS environment. In addition, the operations performed by these global statements remain in effect until they are changed. For example, if you have a single TITLE statement in the beginning of your program, that title will head every page of output from that point on until you write a new TITLE statement. It is a good practice to place a TITLE statement before every procedure that produces output to make it easy for someone to read and understand the information on the page. If you exit your SAS session, your titles are all reset and you need to submit new TITLE statements if you want them to appear.

In all the output displayed in this book, the global option NOPROCTITLE was in effect. Without this option, all output from every procedure would contain text such as "The MEANS Procedure" before it prints your own title statements. The way to set this option is to submit the line:

```
ODS NoProcTitle;
```

PROC FREQ is one of the many built-in SAS procedures. As the name implies, this procedure counts frequencies of data values.

To tell this procedure which variables you want to include in your frequency counts, you add an additional statement—the TABLES (or TABLE) statement. Following the word TABLES, you list those variables for which you want frequency counts. You could actually omit a TABLES statement but, if you did, PROC FREQ would compute frequencies for every variable in your data set (including all the numeric variables).

PROC MEANS is another built-in SAS procedure that computes means (averages) as well as some other statistics such as the minimum and maximum value of each variable. A VAR (short for variables) statement supplies PROC MEANS with a list of analysis variables (which must be numeric) for which you want to compute these statistics. Without a VAR statement, PROC MEANS computes statistics on every numeric variable in your data set.

Depending on whether you are using the SAS Display Manager on a Windows operating system, SAS Enterprise Guide, or SAS Studio (either on a standard version of SAS or the SAS University Edition, or even a mainframe computer), the actual mechanics of submitting your program may differ slightly. You can see screen shots for three different environments below:

Figure 2.1 shows a screen that runs SAS in the windowing environment on a Windows operating system: For most of the examples in this book, this is the system you will see. The programs that run under the other environments are very similar and you should not have any problems, regardless of which environment you are using.

Figure 2.1: View of the Enhanced Editor Window Using the SAS Windowing Environment

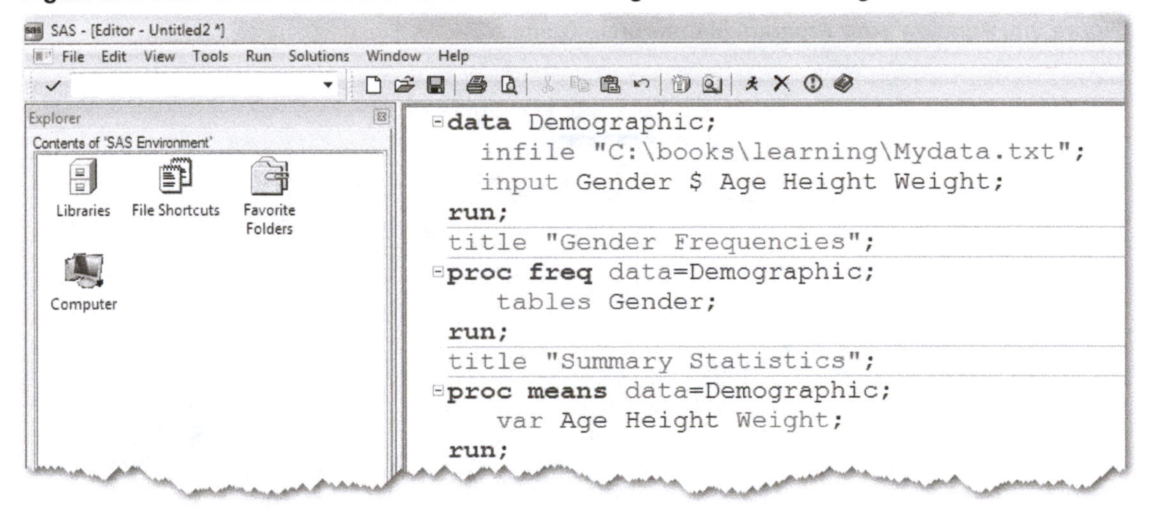

When you use the SAS windowing environment, you write your program in the Enhanced EDITOR window (shown in Figure 2.1). Other windows that you will see later are the LOG window (where you see a listing of your program, possible error messages, and information about data files that were read or written) and the OUTPUT window where you see your results.

To run this program, click the SUBMIT icon (see Figure 2.2).

Figure 2.2: SUBMIT Icon

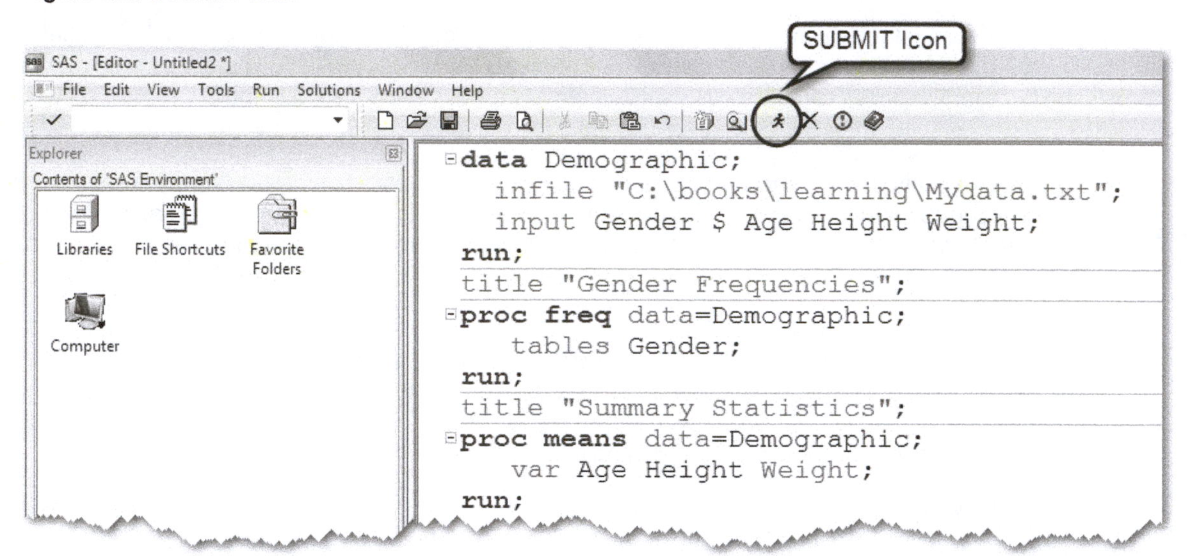

Before we show you the LOG and OUTPUT windows, here are screen shots (see Figure 2.3 and Figure 2.4) of the same program using SAS Enterprise Guide and SAS Studio (from the University Edition):

Figure 2.3: Running Your Program in Enterprise Guide

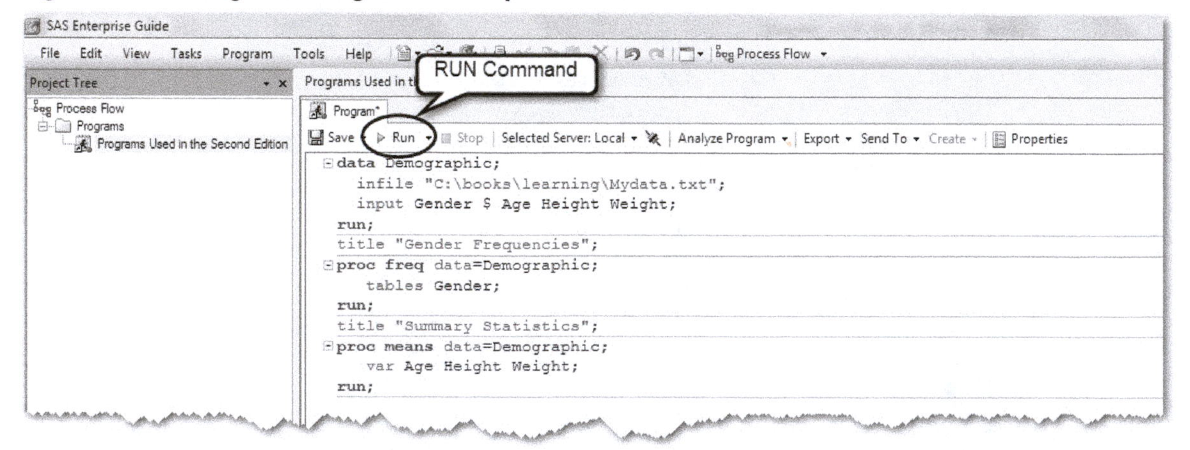

Figure 2.4: Running Your Program in SAS Studio (University Edition)

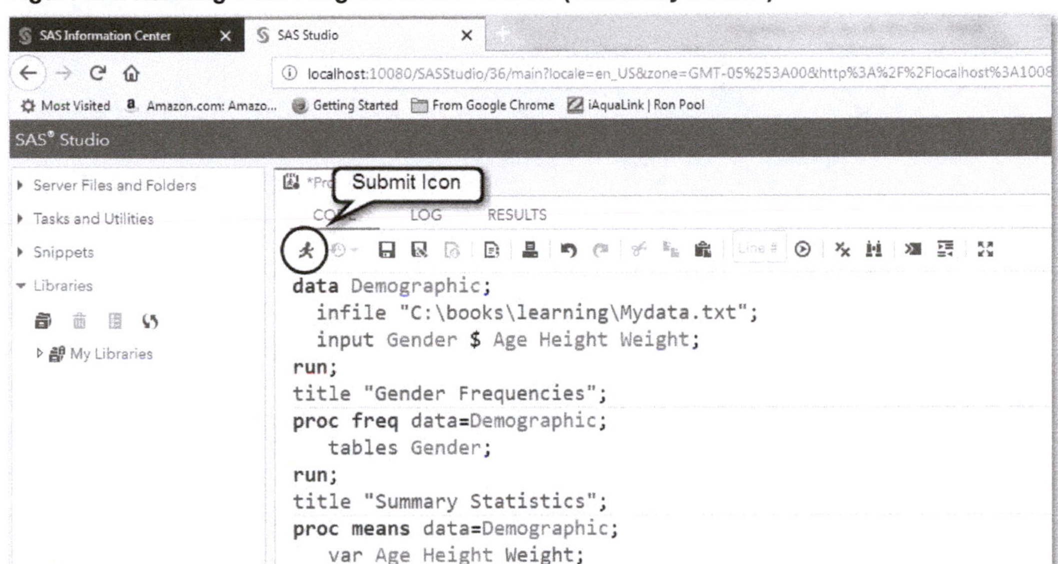

The programs are almost identical regardless of which SAS environment you are using. You might have noticed that the INFILE statement in the SAS Studio version is different from the other two programs. The "short answer" to this is that the SAS University Edition runs in a virtual environment and you need to direct your programs to find data on your disk in a slightly different manner. Please refer to *An Introduction to SAS University Edition* by this author for more information on how this works, or view the online information (PDFs and videos) supplied by SAS.

It's time to see what happens when you click the SUBMIT icon in the windowing environment example. Here is what you will see on your screen (see Figure 2.5):

Figure 2.5: Output from Program 2.1

Gender Frequencies

Gender	Frequency	Percent	Cumulative Frequency	Cumulative Percent
F	2	40.00	2	40.00
M	3	60.00	5	100.00

Summary Statistics

Variable	N	Mean	Std Dev	Minimum	Maximum
Age	5	37.6000000	20.2188031	15.0000000	65.0000000
Height	5	67.2000000	4.8682646	60.0000000	72.0000000
Weight	5	155.0000000	44.0056815	101.0000000	220.0000000

What you see here is the Output window. (The exact appearance of these windows will vary, depending on how you have set up SAS.) The top part of the output (produced by PROC FREQ) shows that there were two females and three males in the data set (the numbers listed under Frequency). The column labeled Percent shows the frequencies as a percent of all the non-missing data values in the data set. The last two columns display Cumulative Frequency and Cumulative Percentages. There were two females (representing 40% of the subjects) and two plus three or five males plus females, which are referred to as a cumulative frequency, (representing 100% of the subjects). The Cumulative Percent columns show the cumulative counts as percentages. You will see later how to eliminate these last two columns because they are seldom used.

Below the frequency display you see Summary Statistics for the three numeric variables (produced by PROC MEANS). N is the number of non-missing values, Mean is the arithmetic mean, Std Dev is the standard deviation, Minimum and Maximum are the smallest non-missing value and the largest value, respectively.

Notice the two titles correspond to the text you placed in the TITLE statement.

> **Note:** By default, SAS centers all output. For most of the output in this book, a system option called NOCENTER was used so that the output is left-justified. The statement (not shown here) Options NOCENTER was included at the beginning of every program.

You can switch among the three windows by clicking on the appropriate tab at the bottom of the screen. These tabs will be located in other places if you are using Enterprise Guide or SAS Studio, but you will have no trouble finding them. The tabs for the windowing environment are shown in Figure 2.6:

Figure 2.6: Tabs for Selecting the Editor, Log, or Output Windows in the Windowing Environment

Output - (Untitled) | Log - (Untitled) | Editor - Untitled1 * | Results Viewer - SAS ...

Figure 2.7 shows a complete listing of the Log window:

Figure 2.7: Inspecting the LOG Window

```
1      data Demographic;
2        infile "c:\books\learning\Mydata.txt";
3        input Gender $ Age Height Weight;
4      run;

NOTE: The infile "c:\books\learning\Mydata.txt" is:
      Filename=c:\books\learning\Mydata.txt,
      RECFM=V,LRECL=32767,File Size (bytes)=67,
      Last Modified=04Oct2017:14:29:45,
      Create Time=04Oct2017:14:29:45

NOTE: 5 records were read from the infile
      "c:\books\learning\Mydata.txt".
      The minimum record length was 11.
      The maximum record length was 12.
NOTE: The data set WORK.DEMOGRAPHIC has 5 observations and 4 variables.
NOTE: DATA statement used (Total process time):
      real time           0.01 seconds
      cpu time            0.01 seconds
```

```
5
6      title "Gender Frequencies";
7      proc freq data=demographic;
8         tables Gender;
9      run;

NOTE: Writing HTML Body file: sashtml.htm
NOTE: There were 5 observations read from the data set WORK.DEMOGRAPHIC.
NOTE: PROCEDURE FREQ used (Total process time):
      real time           0.43 seconds
      cpu time            0.37 seconds
```

```
10
11     title "Summary Statistics";
12     proc means data=demographic;
13        var Age Height Weight;
14     run;

NOTE: There were 5 observations read from the data set WORK.DEMOGRAPHIC.
NOTE: PROCEDURE MEANS used (Total process time):
      real time           0.01 seconds
      cpu time            0.01 seconds
```

Note: The Log window is very important. It is here that you see any error messages if you have made any mistakes in writing your program. In this example, there were no mistakes (a rarity for this author), so you see only the original program along with some information about the data file that was read and some timing information.

Let's spend a moment looking over the log. First, you see that the data came from the **`Mydata.txt`** file located in the **`C:\books\learning`** folder. Next, you see a note showing that five records (lines) of data were read and that the shortest line was 11 characters long and the longest was 13. The next note indicates that SAS created a data set called Work.Demographic. The Demographic part makes sense because that is the name you used in the DATA statement. The Work part is the way SAS tells you that this is a temporary data set—when you end the SAS session, this data set will self-destruct (and the secretary will disavow all knowledge of your actions). You will see later how to make SAS data sets permanent.

Also, as part of this note, you see that the Work.Demographic data set has five observations and four variables. The SAS term *observations* is analogous to rows in a table. The SAS term *variables* is analogous to columns in a table. In this example, each observation corresponds to the data collected on each subject and each variable corresponds to each item of information you collected on each subject.

The remaining notes show the real and CPU time used by SAS to process each procedure.

2.2 Enhancing the Program

At this point, it would be a good idea to access SAS somewhere, enter this program (you will probably want to change the name of the folder where you are storing your data file), and submit it.

Now, let's enhance the program so you can learn some more about how SAS works. For this version of the program, you will add a comment statement and compute a new variable based on the height and weight data. Here is the program:

Program 2. 2: Enhancing the Program

```
*Program name: Demog.sas stored in the C:\books\learning folder.

 Purpose: The program reads in data on height and weight
 (in inches and pounds, respectively) and computes a body
 mass index (BMI) for each subject.

 Programmer: Ron Cody
 Date Written: October 5, 2017;

data Demographic;
     infile "C:\books\learning\Mydata.txt";
     input Gender $ Age Height Weight;
     *Compute a body mass index (BMI);
     BMI = (Weight / 2.2) / (Height*.0254)**2;
run;
```

The statements beginning with an asterisk (*) are called comment statements. They enable you to include comments for yourself or others reading your program later. One way of writing a SAS comment is to start with an asterisk, write as many comment lines as you like, and end the statement (as you do all SAS statements) with a semicolon. Comments are not only useful for others trying to read and understand your program—they are useful to you as well. Just imagine trying to understand a section of a long program that you wrote a year ago and now need to correct or modify. Trust me—you will be glad you commented your program. You should usually include information about the file name used to store the program, the purpose of the program, and the date you wrote the program as well as the date and purpose of any changes you made to the program.

The statement that starts with BMI= is called an *assignment statement*. It is an instruction to perform the computation on the right-hand side of the equal sign and assign the resulting value to the variable named on the left. In this example, you are creating a new variable named BMI that is defined as a person's weight (in kilograms) divided by a person's Height (in meters) squared. BMI (body mass index) is a useful index of obesity. Medical researchers often use BMI when computing the health risks of various diseases (such as heart attacks).

This assignment statement uses three of the basic arithmetic operators used by SAS: the forward slash (/) for division, the asterisk (*) for multiplication, and the double asterisk (**) for exponentiation. This is a good time to mention the full set of arithmetic operators. They are as follows:

Operator	Description	Priority
+	Addition	Lowest
–	Subtraction	Lowest
*	Multiplication	Next Highest
/	Division	Next Highest
**	Exponentiation	Highest
–	Negation	Highest

The same rules you learned about the order of algebraic operations in school apply to SAS arithmetic operators. That is, multiplication and division occur before addition and subtraction. In the previous table, the two highest priority operations occur before all others; the next highest operations occur before the lowest. For example, the value of *x* in the following assignment statement is 14:

```
x = 2 + 3 * 4;
```

If you want to multiply the sum of 2 + 3 by 4, you need to use parentheses like this:

```
x = (2 + 3) * 4;
```

When you include parentheses in your expression, all operations within the parentheses are performed first. In this example, because parentheses surround the addition operation, the 2 and 3 are added together first and then multiplied by 4, yielding a value of 20.

As a further example of how the priority of arithmetic operators works, take a look at the expression here that uses each of the different operators:

```
x = 2**3 + 4 * -5;
```

Because exponentiation and negation occur first, you have the following equation:

```
x = 8 + 4 * -5;
```

This gives you:

```
8 + (-20) = -12
```

2.3 More on Comment Statements

Another way to add a comment to a SAS program is to start it with a slash star (/*) and end it with a star slash (*/). You may even embed comments of this type of comment within a SAS statement. For example, you could write:

```
input Gender $ Age /* age is in years */ Height Weight;
```

If you are using a mainframe computer, you may want to avoid starting your /* in column one because the operating system will interpret it as job control language (JCL) statement and terminate your SAS job.

Be sure that you do not nest the /* */ style comments. For example, you would get an error if you submitted Program 2.3. The first /* (shown in bold) would match the first */ (also shown in bold), leaving invalid SAS code to be processed.

Program 2.3: Incorrect Nesting of /* */ Style Comments

```
/* This comment contains a /* style */ comment embedded
   within another comment. Notice that the first star
   slash ends the comment and the remaining portion of
   the comment will cause a syntax error */
```

2.4 How SAS Works (a Look inside the "Black Box")

This is a good time to explain some of the inner workings of SAS as it processes a DATA step. Looking again at Program 2.2, let's "play computer." SAS processes DATA steps in two stages—a compile stage and an execution stage.

Here's how it works. SAS recognizes the keyword DATA and understands that it needs to process a DATA step. In the compile stage, it does some important housekeeping tasks. First, it prepares an area to store the SAS data set (Demographic). It checks the input file (described by the INFILE statement) and determines various attributes of this file (such as the length of each record). Next, it sets aside a place in memory called the *input buffer*, where it will place each record (line) of data as it is read from the input file. It then reads each line of the program, checks for invalid syntax, and determines the name of all the variables that are in the data set. Depending on your INPUT statement (or other SAS statements), SAS determines whether each variable is character or numeric and the storage length of each variable. This information is called the *descriptor portion* of the data set. In this compile stage, no data is read from the input file and no logical statements are evaluated. Each line is processed in order from the top to the bottom and left to right.

In this example, SAS sees the first four variables listed in the INPUT statement, decides that Gender is character (because of the dollar sign ($) following the name), and sets the storage length of each of these variables. Because no lengths are specified by the program, each variable is given a default length (8 bytes for the character and numeric variables). Eight bytes for a character variable means you can store values with up to eight characters. Eight bytes for numeric variables means that SAS can store numbers with approximately 14 or 15 significant figures (depending on the operating system). It is important to realize that the 8 bytes used to store numeric values does not limit you to numbers with eight digits. The information about each of the variables is stored in a reserved area of memory called the Program Data Vector (PDV for short). Think of the PDV as a set of post office boxes, with one box per variable, and information affixed to each box showing the variable name, type (character or numeric), and storage length. Some additional pieces of information are also stored for each variable. We'll discuss these later when we discuss more advanced programming techniques.

It helps to picture the PDV like this:

Gender **Character** **8 bytes**	**Age** **Numeric** **8 bytes**	**Height** **Numeric** **8 bytes**	**Weight** **Numeric** **8 bytes**

This shows that each variable has a name, a type, and a storage length. The second row of boxes is used to store the value for each of these variables.

Next, SAS sees the assignment statement defining a new variable called BMI. Because BMI is defined by an arithmetic operation, SAS decides that this variable is numeric, uses the default storage length for numerics (8 bytes), and adds it to the PDV.

Gender **Character** **8 bytes**	**Age** **Numeric** **8 bytes**	**Height** **Numeric** **8 bytes**	**Weight** **Numeric** **8 bytes**	**BMI** **Numeric** **8 bytes**

SAS has reached the bottom of the DATA step and the compile stage is complete. Now it begins the execution stage.

When you are reading text data from a file or variables defined by an assignment statement, SAS sets all the values in the PDV to a missing value. This happens before SAS reads in new line of data to ensure that there is a clean slate and that no values are left over from a previous operation. SAS uses blanks to represent missing character values and periods to represent missing numeric values. Therefore, you can now picture the PDV like this:

Gender **Character** **8 bytes**	**Age** **Numeric** **8 bytes**	**Height** **Numeric** **8 bytes**	**Weight** **Numeric** **8 bytes**	**BMI** **Numeric** **8 bytes**
	.	.	.	.

The first line of data from the input file is copied to the input buffer.

M	50	68	155

An internal pointer that keeps track of the current record in the input file now moves to the next line.

In this example, the values in the text file are separated by one or more blanks. This arrangement of data values is called *delimited data* and the method that SAS uses to read this type of data is called *list input*.

SAS expects blanks as the default delimiter but, as you will see later, you can tell SAS if your file contains other delimiters (such as commas) between the data values.

SAS reads each value until it reaches a delimiter and then moves along until it finds the next value. The values in the input buffer are now copied to the PDV as follows:

Gender Character 8 bytes	Age Numeric 8 bytes	Height Numeric 8 bytes	Weight Numeric 8 bytes	BMI Numeric 8 bytes
M	50	68	155	.

Next, BMI is evaluated by substituting the values in the PDV for Height and Weight and evaluating the equation. This value is then added to the PDV:

Gender Character 8 bytes	Age Numeric 8 bytes	Height Numeric 8 bytes	Weight Numeric 8 bytes	BMI Numeric 8 bytes
M	50	68	155	23.616947202

SAS has reached the bottom of the DATA step (because it sees the RUN statement—an explicit step boundary).

> Note that SAS would sense the end of the DATA step without a RUN statement if the next line were a DATA or PROC statement (an implicit step boundary). As a matter of style, it is preferable to end each DATA or PROC step with a RUN statement.

At this point the values in the PDV are written to the SAS data set (Demographic), forming the first observation. There is, by default, an implied OUTPUT statement at the bottom of each DATA step. SAS returns back to the top of the DATA step (the line following the DATA statement) and sees that there are more lines of data to read (when it executes the INPUT statement). It repeats the process of setting values in the PDV to missing, reading new data values, computing the BMI, and outputting observations to the SAS data set. This continues until the INPUT statement reads the end-of-file marker in the input file. You can think of a DATA step as a loop that continues until all data values have been read.

At this time, you may find this discussion somewhat tedious. However, as you learn more advanced programming techniques, you should review this discussion—it can really help you understand the more advanced and subtle features of SAS programming.

2.5 Problems

Solutions to odd-numbered problems are located at the back of this book. Solutions to all problems are available to professors or by permission of SAS Press. If you are a professor, visit the book's companion website at support.sas.com/cody for information about how to obtain the solutions to all problems.

1. You have a text file called **Stocks.txt** containing a stock symbol, a price, and the number of shares. Here are some sample lines of data:

```
AMGN 67.66 100
DELL 24.60 200
GE 34.50 100
HPQ 32.32 120
IBM 82.25 50
MOT 30.24 100
```

   a. Using this raw data file, create a temporary SAS data set (Portfolio). Choose your own variable names for the stock symbol, price, and number of shares. In addition, create a new variable (call it Value) equal to the stock price times the number of shares. Include a comment in your program describing the purpose of the program, your name, and the date the program was written.

   b. Write the appropriate statements to compute the average price and the average number of shares of your stocks.

2. Given the program here, add the necessary statements to compute four new variables:

   a. Weight in kilograms (1 kg = 2.2 pounds). Name this variable WtKg.

   c. Height in centimeters (1 inch = 2.54 cm). Name this variable HtCm.

   d. Average blood pressure (call it AveBP) equal to the diastolic blood pressure plus one-third the difference of the systolic blood pressure minus the diastolic blood pressure.

   e. A variable (call it HtPolynomial) equal to 2 times the height squared plus 1.5 times the height cubed.

Here is the program for you to modify:

```
data Prob2;
   input ID $
         Height /* in inches */
         Weight /* in pounds */
         SBP    /* systolic BP  */
         DBP    /* diastolic BP */;

< place your statements here >

datalines;
001 68 150 110 70
002 73 240 150 90
003 62 101 120 80
;

title "Listing of Prob2";
proc print data=Prob2;
run;
```

Note: This program uses a DATALINES statement, which enables you to include the input data directly in the program. You can read more about this statement in the next chapter.

3. You are given an equation to predict electromagnetic field (EMF) strength, as follows:

   $$EMF = 1.45 \times V + (R/E) \times V^3 - 125.$$

   If your SAS data set contains variables called *V*, *R*, and *E*, write a SAS assignment statement to compute the EMF strength.

4. What is wrong with this program?

```
001   data New-Data;
002      infile C:\books\learning\Prob4data.txt;
003      input x1 x2
004      y1 = 3(x1) + 2(x2);
005      y2 = x1 / x2;
006      New_Variable_from_X1_and_X2 = X1 + X2 - 37;
007   run;
```

   Note: Line numbers are for reference only; they are not part of the program.

5. What is wrong with this program?

```
001 data XYZ;
002    infile "C:\books\learning\DataXYZ.txt";
003    input Gender X Y Z;
004    Sum = X + y + Z;
005 run;
```

   The File **`C:\books\learning\DataXYZ.txt`** looks as follows:

```
Male 1 2 3
Female 4 5 6
Male 7 8 9
```

Part 2: DATA Step Processing

Chapter 3: Reading Raw Data from External Files

3.1 Introduction

One way to provide SAS with data is to have SAS read the data from a text file and create a SAS data set. Some SAS users already have data in SAS data sets. If this is your case, you can skip this chapter!

SAS has different ways of reading data from text files and, depending on how the data values are arranged, you can choose an input method that is most convenient. You have already seen one method, called *list input*, that was used in the introductory program in Chapter 2. This chapter discusses list input as well as two other methods that are appropriate for data arranged in fixed columns.

Some of the more advanced aspects of reading raw data are covered in Chapter 21.

3.2 Reading Data Values Separated by Blanks

One of the easiest methods of reading data is called list input. By default, SAS assumes that data values are separated by one or more blanks.

Task: you have a raw data file called **Mydata.txt** stored in your **C:\books\learning** folder. It is shown here:

```
M 50 68 155
F 23 60 101
M 65 72 220
F 35 65 133
M 15 71 166
```

These values represent gender, age, height (in inches), and weight (in pounds). Notice that this file meets the criteria for list input—each data value is separated from the next by one or more blanks. Program 3.1 reads data from this file and creates a SAS data set.

Program 3.1: Demonstrating List Input with Blanks as Delimiters

```
data Demographics;
  infile 'C:\books\learning\Mydata.txt';
  input Gender $ Age Height Weight;
run;

title "Listing of data set Demographics";
proc print data=Demographics;
run;
```

The INFILE statement tells SAS where to find the data. The INPUT statement contains the variable names you want to associate with each data value. The order of these names matches the order of the values in the file. The dollar sign ($) following Gender tells SAS that Gender is a character variable.

To see that this program works properly, we added a PROC PRINT step to list the observations in the SAS data set (details on PROC PRINT can be found in Chapter 14).

Here is the output:

Figure 3.1: Output from Program 3.1

Listing of data set Demographics

Obs	Gender	Age	Height	Weight
1	M	50	68	155
2	F	23	60	101
3	M	65	72	220
4	F	35	65	133
5	M	15	71	166

Each column represents a variable in the data set and each row represents the data on a single person (an observation). The first column, labeled Obs (short for observation), is generated by PROC PRINT. The values in this column go from 1 to the number of observations in the data set. The order of rows in this list reflects the order that the observations were read from the input data and created in the DATA step. If you change the order of the observations or add new observations to the data set, the numbers in the Obs column may change.

The order of the variables (columns) reflects the order that the variables were encountered in the DATA step.

3.3 Specifying Missing Values with List Input

What would happen if you didn't have a value for Age for the second subject in your file? Your data file would look like this (with a missing value in line 2):

```
M 50 68 155
F    60 101
M 65 72 220
F 35 65 133
M 15 71 166
```

It should be obvious that this will cause errors. SAS reads the value 60 for the Age and 101 for the Height. Because there are no more values on the second line of data, SAS goes to the next line and attempts to read the M as a Height value (and causes a data error message in the log). Clearly, you need a way to tell SAS that there is a missing value for Age in the second line. One way to do this is to use a period to represent the missing value, like this in **Mydata.txt**:

```
M 50 68 155
F .  60 101
M 65 72 220
F 35 65 133
M 15 71 166
```

You must separate the period from the values around it by one or more spaces because a space is the default delimiter character. SAS now assigns a missing value for Age for the second subject. By the way, a missing value is not the same as a 0. This is important because if you asked SAS to compute the mean (average) Age for all the subjects, it would average only the non-missing values.

You can use a period to represent a missing character or numeric value when you use list input.

3.4 Reading Data Values Separated by Commas (CSV Files)

A common way to store data on Windows and UNIX platforms is in comma-separated values (CSV) files. These files use commas instead of blanks as data delimiters. They may or may not enclose character values in quotes. The file **Mydata.csv** contains the same values as the file **Mydata.txt**. It is shown here.

File C:\books\learning\Mydata.csv

```
"M",50,68,155
"F",23,60,101
"M",65,72,220
"F",35,65,133
"M",15,71,166
```

Program 3.2 (below) reads this file and creates a new data set called Demographics.

Program 3.2: Reading Data From a Comma-Separated Values (Csv) File

```
data Demographics;
   infile 'C:\books\learning\Mydata.csv' dsd;
   input Gender $ Age Height Weight;
run;
```

Notice the INFILE statement in this example. The DSD (delimiter-sensitive data) following the file name is an INFILE option. It performs several functions. First, it changes the default delimiter from a blank to a comma. Next, if there are two delimiters in a row, it assumes there is a missing value between. Finally, if character values are placed in quotes (single or double quotes), the quotes are stripped from the value. That's a lot of mileage for just three letters!

The INPUT statement is identical to Program 3.1 as is the resulting SAS data set.

3.5 Using an Alternative Method to Specify an External File

The INFILE statement in Program 3.2 used the actual file name (placed in quotes) to specify your raw data file. An alternative method is to use a separate FILENAME statement to identify the file and to use this reference (called a fileref) in your INFILE statement instead of the actual file name. Program 3.3 is identical to Program 3.2 except for the way it references the external file.

Program 3.3: Using a Filename Statement to Identify an External File

```
filename Preston 'C:\books\learning\Mydata.csv';

data Demographics;
   infile Preston dsd;
   input Gender $ Age Height Weight;
run;
```

The name following the FILENAME statement (Preston, in this example) is an alias for the actual file name. For certain operating environments, the fileref can be created outside of SAS (for example, in a DD statement in JCL on a mainframe). Notice also that the fileref (Preston) in the INFILE statement is not placed in quotes. This is how SAS knows that Preston is not the name of a file but rather a reference to it.

3.6 Reading Data Values Separated by Delimiters Other Than Blanks or Commas

Remember that the default data delimiter for list input is a blank. Using the INFILE option DSD changes the default to a comma. What if you have a file with other delimiters, such as tabs or colons? No problem! You only need to add the DLM= option to the INFILE statement. For example, the following lines of data use colons as delimiters.

Example of a file using colon delimiters:

```
M:50:68:155
F:23:60:101
M:65:72:220
F:35:65:133
M:15:71:166
```

To read this file, you could use this INFILE statement:

```
infile 'file-description' dlm=':';
```

You can spell out the name of the DELIMITER= option instead of using the abbreviation DLM= if you like, for example:

```
infile 'file-description' delimiter=':';
```

You can use the DSD and DLM= options together. This combination of options performs all the actions requested by the DSD option (see Section 3.4) but overrides the default DSD delimiter (comma) with a delimiter of your choice.

```
infile 'file-description' dsd dlm=':';
```

Tabs present a particularly interesting problem. What character do you place between the quotes on the DLM= option? You cannot click the TAB key. Instead, you need to represent the tab by its hexadecimal equivalent. For ASCII files (the coding method used on Windows platforms and UNIX operating systems—it stands for American Standard Code for Information Interchange), you would use the following:

```
infile 'file-description' dlm='09'x;
```

For EBCDIC files (used on most mainframe computers—it stands for Extended Binary-Coded Decimal Interchange Code), you would use the following statement:

```
infile 'file-description' dlm='05'x;
```

Note: These two values are called hexadecimal constants. If you know (or look up) the hexadecimal value of any character, you can represent it in a SAS statement by placing the hexadecimal value in single or double quotes and following the value immediately (no space) by an upper- or lowercase x.

3.7 Placing Data Lines Directly in Your Program (the DATALINES Statement)

Suppose you want to write a short test program in SAS. Instead of having to place your data in an external file, you can place your lines of data directly in your SAS program by using a DATALINES statement. For example, if you want to read data from the text file **`Mydata.txt`** (blank delimited data with values for Gender, Age, Height, and Weight), but you don't want to go to the trouble of writing the external file, you could use Program 3.4.

Program 3.4: Demonstrating the DATALINES Statement

```
data demographic;
   input Gender $ Age Height Weight;
datalines;
M 50 68 155
F 23 60 101
M 65 72 220
F 35 65 133
M 15 71 166
;
```

As you can see from this example, the INFILE statement was removed and a DATALINES statement was added. Following DATALINES are your lines of data. Finally, a semicolon is used to end the DATA step. (Note: You may either use a single semicolon or a RUN statement to end the DATA step.) The lines of data must be the last element in the DATA step—any other statements must come before the lines of data.

While you would probably not use DATALINES in a real application, it is extremely useful when you want to write short test programs.

As a historical note, the DATALINES statement used to be called the CARDS statement. If you don't know what a computer card is, ask an old person. By the way, you can still use the word CARDS in place of DATALINES if you want.

3.8 Specifying INFILE Options with the DATALINES Statement

What if you use DATALINES and want to use one or more of the INFILE options, such as DLM= or DSD? You can use many of the INFILE options with DATALINES by using a reserved file reference called DATALINES. For example, if you wanted to run Program 3.2 without an external data file, you could use Program 3.5.

Program 3.5: Using INFILE Options with DATALINES

```
data Demographics;
   infile datalines dsd;
   input Gender $ Age Height Weight;
datalines;
"M",50,68,155
"F",23,60,101
"M",65,72,220
"F",35,65,133
"M",15,71,166
;
```

3.9 Reading Raw Data from Fixed Columns—Method 1: Column Input

Many raw data files store specific information in fixed columns. This has several advantages over data values separated by delimiters. First, you don't have to worry about missing values. If you do not have a value, you can leave the appropriate columns blank. Next, when you write your INPUT statement, you can choose which variables to read and in what order to read them.

The simplest method for reading data in fixed columns is called column input. This method of input can read character data and standard numeric values. By standard numeric values, we mean positive or negative numbers as well as numbers in exponential form (for example, 3.4E3 means 3.4 times 10 to the 3rd power). This form of input cannot handle values with commas or dollar signs. You can only read dates as character values with this form of input as well. Now for an example.

You have a raw data file called **`Bank.txt`** in a folder called C:\books\learning on your Windows computer. A data description for this file follows.

Variable	Description	Starting Column	Ending Column	Data Type
Subj	Subject Number	1	3	Character
DOB	Date of Birth	4	13	Character
Gender	Gender	14	14	Character
Balance	Bank Account Balance	15	21	Numeric

File C:\books\learning\Bank.txt

```
         1         2
1234567890123456789012345 ← Columns (not part of the file)
-------------------------
00110/21/1955M    1145
00211/18/2001F   18722
00305/07/1944M 123.45
00407/25/1945F -12345
```

Program 3.6 is a SAS program that reads data values from this file.

Program 3.6: Demonstrating Column Input

```
data Financial;
   infile 'C:\books\learning\Bank.txt';
   input Subj     $   1-3
         DOB      $  4-13
         Gender   $    14
         Balance    15-21;
run;

title "Listing of Financial";
proc print data=Financial;
run;
```

You specify a variable name, a dollar sign if the variable is a character value, the starting column, and the ending column (if the value takes more than one column). In this program, the number of columns you specify for each character variable determines the number of bytes SAS uses to store these values; for numeric variables, SAS will always use 8 bytes to store these values, regardless of how many columns you specify in your INPUT statement. (There are advanced techniques to change the storage length for numeric variables—and these techniques should be used only when you need to save storage space and you understand the possible problems that can result.)

Notice that this program uses a separate line for each variable. This is not necessary, but it makes the program more readable. You could have written the program like this:

```
data Financial;
   infile 'C:\books\learning\Bank.txt';
   input Subj $ 1-3 DOB $ 4-13 Gender $ 14 Balance 15-21;
run;
```

It just doesn't look as nice and is harder to read. This is a good time to recommend that you get into good habits in writing your SAS programs. It is amazing how much easier it is to read and understand a program where some care is taken in its appearance.

The resulting listing is:

Figure 3.2: Output from Program 3.6

Listing of Financial

Obs	Subj	DOB	Gender	Balance
1	001	10/21/1955	M	1145.00
2	002	11/18/2001	F	18722.00
3	003	05/07/1944	M	123.45
4	004	07/25/1945	F	-12345.00

It is important to remember that the date of birth (DOB) is a character value in this data set. To create a more useful, numerical SAS date, you need to use formatted input, the next type of input to be described.

3.10 Reading Raw Data from Fixed Columns—Method 2: Formatted Input

Formatted input also reads data from fixed columns. It can read both character and standard numeric data as well as nonstandard numerical values, such as numbers with dollar signs and commas, and dates in a variety of formats. Formatted input is the most common and powerful of all the input methods. Anytime you have nonstandard data in fixed columns, you should consider using formatted input to read the file.

Let's start with the same raw data file (**`Bank.txt`**) that was used in Program 3.6. First examine the program, and then read the explanation.

Program 3.7: Demonstrating Formatted Input

```
data Financial;
   infile 'C:\books\learning\Bank.txt';
   input @1  Subj         $3.
         @4  DOB    mmddyy10.
         @14 Gender       $1.
         @15 Balance       7.;
run;

title "Listing of Financial";
proc print data=Financial;
run;
```

The @ (at) signs in the INPUT statement are called column pointers—and they do just that. For example, @4 says to SAS, go to column 4. Following the variable names are SAS informats. Informats are built-in instructions that tell SAS how to read a data value. The choice of which informat to use is dictated by the data.

Two of the most basic informats are w.d and $w. The w.d informat reads standard numeric values. The w tells SAS how many columns to read. The optional d tells SAS that there is an implied decimal point in the value. For example, if you have the number 123 and you read it with a 3.0 informat, SAS stores the value 123.0. If you read the same number with a 3.1 informat, SAS stores the value 12.3. If the number you are reading already has a decimal point in it (this counts as one of the columns to be read), SAS ignores the d portion of the informat. So, if you read the value 1.23 with a 4.1 informat, SAS stores a value of 1.23.

The $w. informat tells SAS to read w columns of character data. In this program, Subj is read as character data and takes up three columns; values of Gender take up a single column.

Now it's time to read the date. The mmddyy10. informat tells SAS that the date you are reading is in the mm/dd/yyyy form. SAS reads the date and converts the value into a SAS date. SAS stores dates as numeric values equal to the number of days from January 1, 1960.

If you read the value 01/01/1960 with the mmddyy10. informat, SAS stores a value of 0.

The date 01/02/1960 read with the same informat would result in a value of 1, and so forth. SAS knows all about leap years and correctly converts any date from 1582 to way into the future (1582 is the year Pope Gregory started the Gregorian calendar—dates before this are not defined in SAS).

So, getting back to our example, since date values are in the mm/dd/yyyy form and start in column 4, you use @4 to move the column pointer to column 4 and the mmddyy10. informat to tell SAS to read the next 10 columns as a date in this form. SAS then computes the number of days from January 1, 1960, corresponding to each of the date values. Let's see the results:

Figure 3.3: Output from Program 3.7

Listing of Financial

Obs	Subj	DOB	Gender	Balance
1	001	-1533	M	1145.00
2	002	15297	F	18722.00
3	003	-5717	M	123.45
4	004	-5273	F	-12345.00

Well, the dates (variable DOB) look rather strange. What you are seeing are the actual values SAS is storing for each DOB (the number of days from January 1, 1960).

You need a way to display these dates in a more traditional form, such as the way the dates were displayed in the raw data file (10/21/1955, in the first observation) or in some other form (such as 10Oct1955). While you are at it, why not add dollar signs and commas to the Balance figures?

You can accomplish both of these tasks by associating a format with each of these two variables. There are many built-in formats in SAS that allow you to display dates and financial values in easily readable ways. You associate these formats with the appropriate variables in a FORMAT statement. Program 3.8 shows how to add a FORMAT statement to PROC PRINT.

Program 3.8: Demonstrating a FORMAT Statement

```
title "Listing of Financial";
proc print data=Financial;
   format DOB     mmddyy10.
          Balance dollar11.2;
run;
```

Here you are using the mmddyy10. format to print the DOB values and the dollar11.2 format to print the Balance values. Notice the period in each of the formats. All SAS formats need to end either in a period or in a period followed by a number. This is how SAS distinguishes between the names of variables or data sets and the names of formats. The 11.2 following the dollar format says to allow up to 11 columns to print the Balance values (including the dollar sign, the decimal point, and possibly a comma or a minus sign). The 2 following the period says to include two decimal places after the decimal point. Here is the revised output:

Figure 3.4: Output from Program 3.8

Listing of Financial

Obs	Subj	DOB	Gender	Balance
1	001	10/21/1955	M	$1,145.00
2	002	11/18/2001	F	$18,722.00
3	003	05/07/1944	M	$123.45
4	004	07/25/1945	F	$-12,345.00

It is important to remember that the formats only affect the way these values appear in printed output—the internal values are not changed.

To be sure that you understand what formats do, let's repeat Program 3.8 and use another format for date of birth (DOB).

Program 3.9: Rerunning Program 3.8 with a Different Format

```
title "Listing of Financial";
proc print data=Financial;
   format DOB     date9.
          Balance dollar11.2;
run;
```

This produces the resulting output:

Figure 3.5: Output from Program 3.9

Listing of Financial

Obs	Subj	DOB	Gender	Balance
1	001	21OCT1955	M	$1,145.00
2	002	18NOV2001	F	$18,722.00
3	003	07MAY1944	M	$123.45
4	004	25JUL1945	F	$-12,345.00

The date9. format prints dates as a two-digit day of the month, a three-character month abbreviation, and a four-digit year. This format helps avoid confusion between the month-day-year and day-month-year formats used in the United States and Europe, respectively.

Notice also that the dollar11.2 format makes the Balance figures much easier to read. This is a good place to mention that the commaw.d format is useful for displaying large numbers with commas, where you don't need or want dollar signs.

3.11 Using a FORMAT Statement in a DATA Step versus in a Procedure

Program 3.8 and Program 3.9 demonstrated using a FORMAT statement in a procedure. Placing a FORMAT statement here associates the formats and variables only for that procedure. It is usually more useful to place your FORMAT statement in the DATA step. When you do this, there is a permanent association of the formats and variables in the data set. You can override any permanent format by placing a FORMAT statement in a particular procedure where you would like a different format. You will usually want to place all of your date formats in a DATA step because no one wants to see unformatted SAS dates. You can also remove a format from a variable by issuing a FORMAT statement for one or more variables and not specify a format. For example, if a variable called Age was formatted in a DATA step and you wanted to see unformatted values in a listing, you could write the following FORMAT statement:

```
format Age;
```

3.12 Using Informats with List Input

Suppose you have a blank- or comma-delimited file containing dates and character values longer than 8 bytes (or other values that require an informat). One way to provide informats with list input is to follow each variable name in your INPUT statement with a colon, followed by the appropriate informat. To see how this works, suppose you want to read the CSV file **`List.csv`**:

```
"001","Christopher Mullens",11/12/1955,"$45,200"
"002","Michelle Kwo",9/12/1955,"$78,123"
"003","Roger W. McDonald",1/1/1960,"$107,200"
```

Variables in this file represent a subject number (Subj), Name, date of birth (DOB), and yearly salary (Salary). You need to supply informats for Name (length is greater than 8 bytes), DOB (you need a date informat here), and Salary (this is a nonstandard numeric value—with a dollar sign and commas). Program 3.10 shows one way to supply the appropriate informats for these variables.

Program 3.10: Using Informats with List Input

```
data List_Example;
   infile 'C:\books\learning\List.csv' dsd;
   input Subj   :        $3.
         Name   :       $20.
         DOB    : mmddyy10.
         Salary :  dollar8.;
   format DOB date9. Salary dollar8.;
run;
```

You see here that there is a colon preceding each informat. This colon (called an informat modifier) tells SAS to use the informat supplied but to stop reading the value for this variable when a delimiter is encountered. Do not forget the colons because without them SAS may read past a delimiter to satisfy the width specified in the informat.

This program would also work if the informat for Subj were omitted and the variable name was followed by a dollar sign (to signify that Subj is a character variable). However, the Subj variable would then be stored in 8 bytes (the default length for character variables with list input). By providing the $3. informat, you tell SAS to use 3 bytes to store this variable.

3.13 Supplying an INFORMAT Statement with List Input

Another way to supply informats when using list input is to use an INFORMAT statement before the INPUT statement. Following the keyword INFORMAT, you list each variable and the informat you want to use to read each variable. You may also use a single informat for several variables if you follow a list of variables by a single informat.

To see how this works, see Program 3.11, that uses an INFORMAT statement.

Program 3.11: Supplying an INFORMAT Statement with List Input

```
data List_Example;
   informat Subj        $3.
            Name       $20.
            DOB   mmddyy10.
            Salary dollar8.;
   infile 'C:\books\learning\List.csv' dsd;
   input Subj
         Name
         DOB
         Salary;
   format DOB date9. Salary dollar8.;
run;
```

This program uses an INFORMAT statement to associate an informat to each of the variables. When choosing informats for your variables, be sure to make the length long enough to accommodate the longest data value you will encounter. Notice that the INPUT statement does not require anything other than the variable names because each variable already has an assigned informat. A listing from PROC PRINT confirms that all is well:

Figure 3.6: Output from Program 3.11

Listing of Data Set List_Example

Obs	Subj	Name	DOB	Salary
1	001	Christopher Mullens	12NOV1955	$45,200
2	002	Michelle Kwo	12SEP1955	$78,123
3	003	Roger W. McDonald	01JAN1960	$107,200

3.14 Using List Input with Embedded Delimiters

What if the data in the previous CSV file was placed in a text file where blanks were used as delimiters instead of commas and there were no quotes around each character value? Here's what the file **List.txt** would look like:

```
001 Christopher Mullens 11/12/1955 $45,200
002 Michelle Kwo 9/12/1955 $78,123
003 Roger W. McDonald 1/1/1960 $107,200
```

Houston, we have a problem! If you try to read this file with list input, the blank(s) in the Name field will trigger the end of the variable. SAS, in its infinite wisdom, came up with a novel solution—the ampersand (&) informat modifier. The ampersand, like the colon, says to use the supplied informat, but the delimiter is now two or more blanks instead of just one. So, if you use an ampersand modifier to read the **List.txt** file here, you need to use the ampersand modifier following Name. You also need to have two or more spaces between the end of the name and the date of birth. Here is the modified **List.txt** file:

```
001 Christopher Mullens    11/12/1955 $45,200
002 Michelle Kwo           9/12/1955  $78,123
003 Roger W. McDonald      1/1/1960   $107,200
```

And here is the program using the ampersand modifier:

Program 3.12: Demonstrating the Ampersand Modifier for List Input

```
data list_example;
   infile 'C:\books\learning\list.txt';
   input Subj    :        $3.
         Name    &       $20.
         DOB     : mmddyy10.
         Salary :   dollar8.;
   format DOB date9. Salary dollar8.;
run;
```

The INPUT statement is one of the most powerful and versatile SAS statements. Please refer to Chapter 25 to learn even more about the ability of SAS to read raw data.

3.15 Problems

Solutions to odd-numbered problems are located at the back of this book. Solutions to all problems are available to professors. If you are a professor, visit the book's companion website at support.sas.com/cody for information about how to obtain the solutions to all problems.

1. You have a text file called **Scores.txt** containing information on gender (M or F) and four test scores (English, history, math, and science). Each data value is separated from the others by one or more blanks. Here is a listing of the data file **Scores.txt**:

```
M   80   82 85 88
F   94   92 88 96
M   96   88 89 92
F   95    . 92 92
```

   a. Write a DATA step to read in these values. Choose your own variable names. Be sure that the value for Gender is stored in 1 byte and that the four test scores are numeric.

   b. Include an assignment statement computing the average of the four test scores.

   c. Write the appropriate PROC PRINT statements to list the contents of this data set.

2. You are given a CSV (comma-separated values) file called **`Political.csv`** containing state, political party, and age. A listing of the file **`Political.txt`** is shown here:

```
"NJ",Ind,55
"CO",Dem,45
"NY",Rep,23
"FL",Dem,66
"NJ",Rep,34
```

   a. Write a SAS program to create a temporary SAS data set called Vote. Use the variable names State, Party, and Age. Age should be stored as a numeric variable; State and Party should be stored as character variables.
   b. Include a procedure to list the observations in this data set.
   c. Include a procedure to compute frequencies for Party.

3. You are given a text file where dollar signs were used as delimiters. To indicate missing values, two dollars signs were entered. Values in this file represent last name, employee number, and annual salary.

   Here is a listing of the file **`Company.txt`**:

```
Roberts$M234$45000
Chien$M74777$$
Walters$$75000
Rogers$F7272$78131
```

   Using this data file as input, create a temporary SAS data set called Company with the variables LastName (character), EmpNo (character), and Salary (numeric).

4. Repeat Problem 2 using a FILENAME statement to create a fileref instead of using the file name on the INFILE statements.

5. You want to create a program that uses a DATALINES statement to read in values for X and Y. In the DATA step, you want to create a new variable, Z, equal to $100 + 50X + 2X^2 - 25Y + Y^2$. Use the following (X,Y) data pairs: (1,2), (3,6), (5,9), and (9,11).

6. You have a text file called **`Bankdata.txt`** with data values arranged as follows:

Variable	Description	Starting Column	Ending Column	Data Type
Name	Name	1	15	Char
Acct	Account number	16	20	Char
Balance	Acct balance	21	26	Num
Rate	Interest rate	27	30	Num

   Create a temporary SAS data set called Bank using this data file. Use column input to specify the location of each value. Include in this data set a variable called Interest computed by multiplying Balance by Rate. List the contents of this data set using PROC PRINT.

   Here is a listing of the text file:

```
Philip Jones   V1234   4322.32
Nathan Philips V1399  15202.45
Shu Lu         W8892 451233.45
Betty Boop     V7677  50002.78
```

7. You have a text file called **Geocaching.txt** with data values arranged as follows:

Variable	Description	Starting Column	Ending Column	Data Type
Name	Cache name	1	20	Char
LongDeg	Longitude degrees	21	22	Num
LongMin	Longitude minutes	23	28	Num
LatDeg	Latitude degrees	29	30	Num
LatMin	Latitude minutes	31	36	Num

Here is a listing of the file:

```
Higgensville Hike   4030.2937446.539
Really Roaring      4027.4047442.147
Cushetunk Climb     4037.0247448.014
Uplands Trek        4030.9907452.794
```

Create a temporary SAS data set called Cache using this data file. Use column input to read the data values.

To learn about geocaching (treasure hunting with a hand-held GPS), go to www.geocaching.com. The author and his wife use the geocaching name "Jan and the Man." Check it out.

8. Repeat Problem 6 using formatted input to read the data values instead of column input.

9. Repeat Problem 7 using formatted input to read the data values instead of column input.

10. You are given a text file called **Stockprices.txt** containing information on the purchase and sale of stocks. The data layout is as follows:

Variable	Description	Starting Column	Length	Type
Stock	Stock symbol	1	4	Char
PurDate	Purchase date	5	10	*mm/dd/yyyy*
PurPrice	Purchase price	15	6	Dollar signs and commas
Number	Number of shares	21	4	Num
SellDate	Selling date	25	10	*mm/dd/yyyy*
SellPrice	Selling price	35	6	Dollar signs and commas

A listing of the data file is:

```
IBM  5/21/2006 $80.0 10007/20/2006 $88.5
CSCO04/05/2005 $17.5 20009/21/2005 $23.6
MOT 03/01/2004 $14.7 50010/10/2006 $19.9
XMSR04/15/2006 $28.4 20004/15/2007 $12.7
BBY 02/15/2005 $45.2 10009/09/2006 $56.8
```

Create a SAS data set (call it Stocks) by reading the data from this file. Use formatted input. Compute several new variables as follows:

Variable	Description	Computation
TotalPur	Total purchase price	Number times PurPrice
TotalSell	Total selling price	Number times SellPrice
Profit	Profit	TotalSell minus TotalPur

Print out the contents of this data set using PROC PRINT.

11. You have a CSV file called **`employee.csv`**. This file contains the following information:

Variable	Description	Desired Informat
ID	Employee ID	$3.
Name	Employee name	$20.
Depart	Department	$8.
DateHire	Hire date	MMDDYY10.
Salary	Yearly salary	DOLLAR8.

Use list input to read data from this file. You will need an informat to read most of these values correctly (i.e., DateHire needs a date informat). You can do this in either of two ways. First is to include an INFORMAT statement to associate each variable with the appropriate informat. The other is to use the colon modifier and supply the informats directly in the INPUT statement. Create a temporary SAS data set (Employ) from this data file. Use PROC PRINT to list the observations in your data set and the appropriate procedure to compute frequencies for the variable Depart.

A listing of the raw data file **`Employee.csv`** is:

```
123,"Harold Wilson",Acct,01/15/1989,$78123.
128,"Julia Child",Food,08/29/1988,$89123
007,"James Bond",Security,02/01/2000,$82100
828,"Roger Doger",Acct,08/15/1999,$39100
900,"Earl Davenport",Food,09/09/1989,$45399
906,"James Swindler",Acct,12/21/1978,$78200
```

Chapter 4: Creating Permanent SAS Data Sets

4.1 Introduction

SAS procedures cannot read raw data files or spreadsheets directly. One way or another, they need the data in SAS data sets. Remember that SAS DATA steps can create SAS data sets. You can also have SAS convert data from other sources, such as Microsoft Office Excel or Access, Oracle, or DB2. This conversion process can be automated by using automated Import facilities in the Display Manager, SAS Enterprise Guide, or SAS Studio (in the University Edition), or by using data access *engines*, which automatically convert the data into a form SAS can process.

This chapter describes how to make your SAS data set permanent and how to determine the contents of a SAS data set.

4.2 SAS Libraries—The LIBNAME Statement

When you write a DATA statement such as

```
data Test;
```

SAS creates a temporary SAS data set called Test. When you close your SAS session, this data set disappears. SAS data set names actually have two-part names in the form:

```
libref.data-set-name
```

The part of the name before the period is called a libref (short for library reference), and this tells SAS where to store (or retrieve) the data set. The part of the name after the period identifies the name you want to give the data set.

Up until now, all the programming examples in this book used a data set name without a period. When you use a name like Test in the DATA statement, SAS uses a default libref called Work that SAS creates automatically every time you open a SAS session. For example, if you write a DATA statement such as

```
data Test;
```

SAS adds the default libref Work, so this DATA statement is equivalent to

```
data Work.Test;
```

All that is required to make your SAS data sets permanent is to create your own libref using a LIBNAME statement and use that libref in the two-level SAS data set name.

Suppose you want to create a permanent SAS data set called Test_Scores in your **`C:\books\learning`** folder. You could use following program.

Program 4.1: Creating a Permanent SAS Data Set

```
libname Mozart 'C:\books\learning';

data Mozart.Test_Scores;
   length ID $ 3 Name $ 15;
   input ID $ Score1-Score3 Name $;
datalines;
1 90 95 98 Jan
2 78 77 99 Preston
3 88 91 92 Russell
;
```

The LIBNAME statement starts with the keyword LIBNAME and then specifies the name of the library (called a libref), followed by the directory or folder where you want to store your permanent SAS data sets. The libref you use must not be more than 8 characters in length and must be a valid SAS name.

When you run this program, data set Test_Scores becomes a permanent SAS data set in the **`C:\books\learning`** folder. It is important to remember that any libref that you create exists only for your current SAS session. If you open a new SAS session, you need to reissue that LIBNAME statement. A good way to think of a libref is as an alias for the name of the folder (on Windows or UNIX platforms). On mainframe computers, a SAS library is actually a single file that can hold multiple SAS data sets.

If you run Program 4.1 on a Windows platform, the SAS data set will be stored as the SAS data set called Test_Scores, and it will be stored as the file Test_Scores.sas7bdat in the **`C:\books\learning`** folder. The file extension stands for SAS binary data version 7. You may wonder why there is a 7 rather than a 9 in the file extension when this data set was created using SAS9. Since the structure of SAS data sets has not changed since SAS 7, SAS has maintained the same file extension it used in SAS 7.

4.3 Why Create Permanent SAS Data Sets?

If your data sets are small, you may choose to create them each time you start a SAS session. However, it takes computer resources to create SAS data sets and it makes more sense to make your data sets permanent if you plan to use them more than once, especially if they are large.

4.4 Examining the Descriptor Portion of a SAS Data Set Using PROC CONTENTS

A SAS data set consists of two parts: a descriptor portion and a data portion. One way to examine the descriptor portion of a SAS data set is by using PROC CONTENTS. If you want to see the contents of the descriptor portion of the Test_Scores data set, submit the following program:

Program 4.2: Using PROC CONTENTS to Examine the Descriptor Portion of a SAS Data Set

```
title "The Descriptor Portion of Data Set TEST_SCORES";
proc contents data=Mozart.Test_Scores;
run;
```

The resulting output is shown next:

Figure 4.1: Output from Program 4.2

The Descriptor Portion of Data Set Test_Score

Data Set Name	MOZART.TEST_SCORES	Observations	3
Member Type	DATA	Variables	5
Engine	V9	Indexes	0
Created	10/11/2017 09:48:00	Observation Length	48
Last Modified	10/11/2017 09:48:00	Deleted Observations	0
Protection		Compressed	NO
Data Set Type		Sorted	NO
Label			
Data Representation	WINDOWS_64		
Encoding	wlatin1 Western (Windows)		

Alphabetic List of Variables and Attributes			
#	Variable	Type	Len
1	ID	Char	3
2	Name	Char	15
3	Score1	Num	8
4	Score2	Num	8
5	Score3	Num	8

Engine/Host Dependent Information	
Data Set Page Size	65536
Number of Data Set Pages	1
First Data Page	1
Max Obs per Page	1361
Obs in First Data Page	3
Number of Data Set Repairs	0
ExtendObsCounter	YES
Filename	c:\books\learning\test_scores.sas7bdat
Release Created	9.0401M2
Host Created	X64_7HOME

This output displays information about the data set, such as the number of variables, the number of observations, and the creation and modification dates. It also displays information about the SAS version used to create the data set. In addition, it displays information on each of the variables in the data set—the variable name, type, and storage length.

A quick look at this output shows that the data set Test_Scores has 3 observations and 5 variables. The list of variables is in alphabetical order. It shows that ID and Name are character variables stored in 3 and 15 bytes, respectively; the three Score variables are numeric and are stored in 8 bytes each.

The title at the top of the page is created by using a TITLE statement as shown in Program 4.2. You may use either single or double quotes to enclose your title. If the title contains any single quotation marks (or apostrophes), you should use double quotation marks.

Note: If your title does not contain any apostrophes or macro variable names (see Chapter 25), you can actually omit the quotation marks altogether. However, as a matter of style, you may want to use quotation marks on all your TITLE statements.

A more useful way to list variable information is to list them in the order the variables are stored in the SAS data set, rather than alphabetically. To create such a list, use the VARNUM option of PROC CONTENTS, like this:

Program 4.3: Demonstrating the VARNUM option of PROC CONTENTS

```
title "The Descriptor Portion of Data Set Test_Scores";
proc contents data=Mozart.Test_Scores varnum;
run;
```

Output from this program is identical to the previous output except that the variable list is now in the same order as the variables in the data set. This portion of the output is shown below to demonstrate the effect of this option:

Figure 4.2: Demonstrating the VARNUM Option of PROC CONTENTS

Variables in Creation Order			
#	Variable	Type	Len
1	ID	Char	3
2	Name	Char	15
3	Score1	Num	8
4	Score2	Num	8
5	Score3	Num	8

It is important to remember that if you have just opened a new SAS session, you must reissue a LIBNAME statement if you want to access a previously created SAS data set or to create a new one. You may use any library name (libref) you want each time you open a SAS session, although in practice you usually use the same library reference each time.

For example, if you open up a new SAS session, you can submit the following statements to obtain information on the Test_Scores data set:

Program 4.4: Using a LIBNAME in a New SAS Session

```
libname Proj99 'C:\books\learning';

title "Descriptor Portion of Data Set Test_Scores";
proc contents data=Proj99.Test_Scores varnum;
run;
```

Note: One other useful PROC CONTENTS option is POSITION. This option produces both an alphabetical and ordered list of the variables in your SAS data set, which is this author's favorite option.

4.5 Listing All the SAS Data Sets in a SAS Library Using PROC CONTENTS

You can use PROC CONTENTS to list the names of all the SAS data sets in a SAS library (folder). To do this, use the following program:

Program 4.5: Using PROC CONTENTS to List the Names of all the SAS Data Sets in a SAS Library

```
title "Listing All the SAS Data Sets in a Library";
proc contents data=Mozart._all_ nods;
run;
```

The keyword _ALL_ is used in place of a data set name. The NODS option gives you the name of the SAS data sets only, omitting the detail listing for each data set. A partial listing is shown next:

Figure 4.3: Partial Listing from Program 4.5

Listing All the SAS Data Sets in a Library

Directory	
Libref	MOZART
Engine	V9
Physical Name	c:\books\learning
Filename	c:\books\learning

#	Name	Member Type	File Size	Last Modified
1	ADDRESS	DATA	13312	03/23/2009 23:29:34
2	ASSIGN	DATA	5120	03/23/2009 23:29:38
3	BICYCLES	DATA	9216	03/23/2009 23:29:36
4	[illegible]	DATA	[illegible]	[illegible]

4.6 Viewing the Descriptor Portion of a SAS Data Set Using a Point-and-Click Approach

If you are running SAS using any of the three windowing environments, you can use simple point-and-click methods to view either the descriptor or data portion of a SAS data set. The screen shots that are displayed below were made using the SAS windowing environment on a Windows platform. Similar techniques can be used with Enterprise Guide or SAS Studio.

First, click the **Explorer** tab on the bottom left of your editor window:

Figure 4.4: Using a Point-and-Click Approach to Investigate Your SAS Data Set

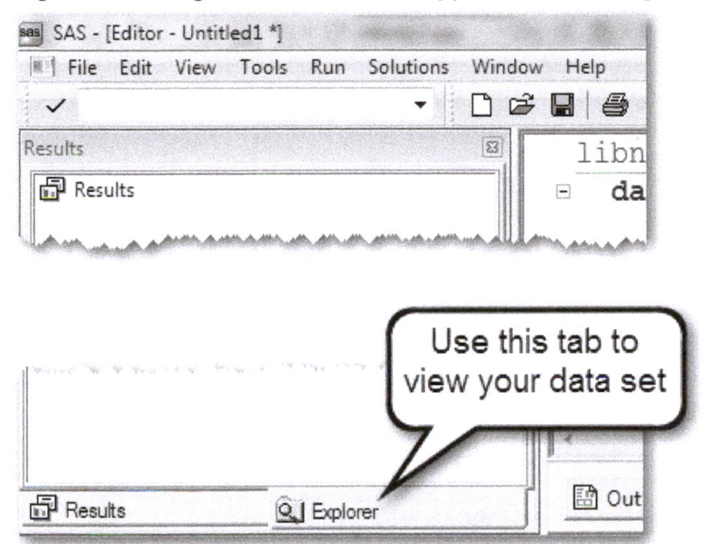

This brings up the following window:

Figure 4.5: The LIBRARY Icon

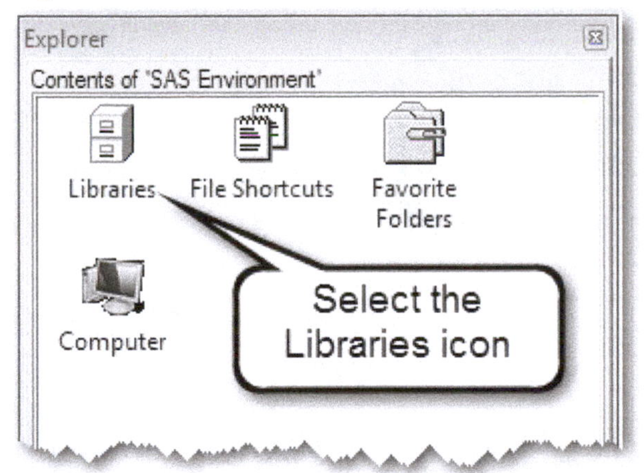

The Libraries icon shows the built-in libraries plus any libraries you have created using LIBNAME statements. Double-click the Libraries icon.

Figure 4.6: Select the Library You Want to Inspect

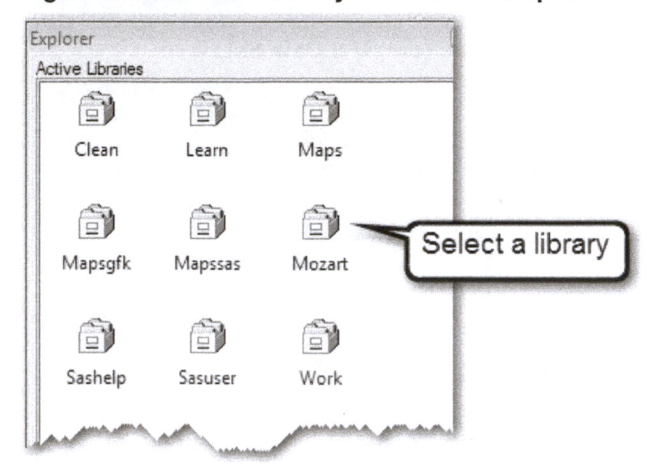

Selecting a library enables you to see all the SAS data sets stored there.

Figure 4.7: Partial List of SAS Data Sets in the Mozart Library

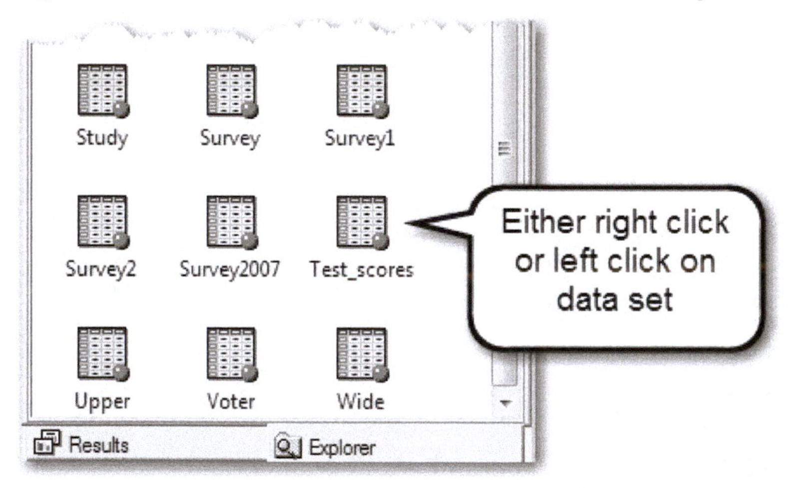

A right-click on the data set icon brings up a menu that includes a choice to see the variables (columns) in the data set and their attributes. A left-click opens the SAS Viewer, showing the data portion of your data set. If you right-click and select View Columns, you will see the following:

Figure 4.8: Selecting View Columns to See Your Variables and Attributes

General | Details | Columns | Indexes | Integrity | Passwords

Find column name:

Column Name	Type	Length	Format	Informat	Label
ID	Text	3			
Name	Text	15			
Score1	Num...	8			
Score2	Num...	8			
Score3	Num...	8			

This shows the same information that you can obtain by running PROC CONTENTS. The order of the variables in the list is the same as the order you will see when using the VARNUM option.

If you left-click the data set icon, you will see your data set in the SAS Viewer. It looks like this:

Figure 4.9: Using the SAS Viewer to List Your SAS Data Set

	ID	Name	Score1	Score2	Score3
1	1	Jan	90	95	98
2	2	Preston	78	77	99
3	3	Russell	88	91	92

Note: Be sure to close this window if you plan to make changes to the data set. If the viewer is still open, you will be prevented from making any changes to the data set.

Similar resources are available in Enterprise Guide and SAS Studio.

4.7 Viewing the Data Portion of a SAS Data Set Using PROC PRINT

As you have seen in several programs so far in this book, PROC PRINT can be used to list the data in a SAS data set. Although there are a number of options to control how this listing appears, you can use it with all the defaults to get a quick listing of your data set. Here is the code to list the data portion of data set Test_Scores:

Program 4.6: Using PROC PRINT to List the Data Portion of a SAS Data Set

```
title "Listing of Test_Scores";
proc print data=Mozart.Test_Scores;
run;
```

This code generates the following output:

Figure 4.10: Output from Program 4.6

Listing of Test_Scores

Obs	ID	Name	Score1	Score2	Score3
1	1	Jan	90	95	98
2	2	Preston	78	77	99
3	3	Russell	88	91	92

This listing displays all the variables and all the observations in the Test_Scores data set.

Program 4.6 is an example of a procedure that uses all the default actions. That is, you did not specify any details such as which variables to print or other controllable aspects of this procedure. Chapter 14 describes how to add options and statements to PROC PRINT to customize your report.

4.8 Using a SAS Data Set as Input to a DATA Step

Besides raw data files, SAS data sets can also be used as input to a DATA step. As an example, you might want to use the information in an existing SAS data set to compute new variables.

Consider the data set Test_Scores (stored in the `C:\books\learning` folder). This data set contains the variables ID, Name, and Score1–Score3 (three test scores). Suppose you want to compute an average score for each subject in this data set. Program 4.7 performs this task:

Program 4.7: Using Observations from a SAS Data Set as Input to a New SAS Data Set

```
libname Learn 'C:\books\learning';

data New;
   set Learn.Test_Scores;
   AveScore = mean(of Score1-Score3);
run;

title "Listing of Data Set New";
proc print data=New;
   var ID Score1-Score3 AveScore;
run;
```

The key to this program is the SET statement. You can think of a SET statement as an INPUT statement except you are reading observations from a SAS data set instead of lines from a raw data file. There is a difference, however. Each time you read a line of data from a raw data file, the variables being read from the raw data file or created by assignment statements in the DATA step are initialized to a missing value during each iteration of the DATA step. Variables that are read from SAS data sets are not set to missing values during each iteration of the DATA step—they are said to be *retained*. In Program 4.7, the variables ID, Name, and Score1–Score3 are retained; the variable AveScore is not. This fact is not a concern to us here, but it can be used to advantage in more advanced programs.

The assignment statement that creates the AveScore variable uses the MEAN function to compute the mean of the three Score variables. You can read more about the MEAN function in Chapter 11. For now, you should notice that the variable list Score1–Score3 is preceded by the word *of*. This is typical of many SAS statistical functions that can take a variable list as an argument. Without the word *of*, the MEAN function would return the difference of Score1 and Score3 (that is, the dash would be interpreted as a minus sign).

Here is a listing of data set New:

Figure 4.11: Output from Program 4.7

Listing of Data Set New

Obs	ID	Score1	Score2	Score3	AveScore
1	1	90	95	98	94.3333
2	2	78	77	99	84.6667
3	3	88	91	92	90.3333

4.9 DATA _NULL_: A Data Set That Isn't

There are many applications where you want to process observations in a SAS data set, perhaps to print out data errors or to produce a report, and you don't need to create a new data set.

You can use the data set name _NULL_ for these applications. The reserved data set name _NULL_ tells SAS not to create a data set. It enables you to process observations from raw data or an existing data set without the overhead of creating a new data set. Here is an example.

You have a permanent SAS data set (Test_Scores) and you want to create a list of all the IDs of students who achieved a score of 95 or higher on any of the tests. You could create a new SAS data set and use PROC PRINT to list these students or you could do it more efficiently with a DATA _NULL_ step, like this:

Program 4.8: Demonstrating a DATA _NULL_ Step

```
title "Scores Greater Than or Equal to 95";
data _null_;
   set Learn.Test_Scores;
   if Score1 ge 95 or Score2 ge 95 or Score3 ge 95 then
      put ID= Score1= Score2= Score3=;
run;
```

The IF statement checks if any of the three test scores is greater than or equal to 95. If so, the PUT statement writes out the values of ID and the three test scores. A PUT statement writes text to a location of your choice: an external text file, the SAS log, or the OUTPUT window. In Program 4.8, an output location is not specified so the default location, the SAS log, is used. Here is a listing of the SAS log after running this program:

Figure 4.12: Listing of the SAS Log After Running Program 4.8

```
15        ;
16     title "Scores Greater Than or Equal to 95";
17       data _null_;
18          set Learn.Test_Scores;
19          if Score1 ge 95 or Score2 ge 95 or Score3 ge 95 then
20             put ID= Score1= Score2= Score3=;
21       run;

ID=1 Score1=90 Score2=95 Score3=98
ID=2 Score1=78 Score2=77 Score3=99
NOTE: There were 3 observations read from the data set
      LEARN.TEST_SCORES.
NOTE: DATA statement used (Total process time):
      real time           0.00 seconds
      cpu time            0.00 seconds
```

Placing PUT statements in a DATA step is an excellent way to help debug SAS programs. You can examine the values of your variables at any place in the DATA step. (You can also use the SAS debugger, available on the PC platform for this purpose.)

If you want to send the output to a file called **`C:\books\learning\HighScores.txt`**, you would need to place a FILE statement before the PUT statement, as follows:

```
file 'C:\books\learning\HighScores.txt';
```

A FILE statement is somewhat like an INFILE statement—that is, it works in concert with a PUT statement, telling SAS the destination of the text you are outputting.

If you want the results of the PUT statement to be written to the output device (on a PC, this would be the OUTPUT window), you can use the reserved file reference PRINT, like this:

```
file print;
```

DATA _NULL_ steps are sometimes used to create custom reports. As a matter of fact, this type of report is referred to as *DATA _NULL_ reporting*. To control how SAS writes this output, you can use pointers and formats to specify exactly what columns to write to and how the values are to be formatted. With more powerful procedures such as PROC REPORT, writing your own custom report using a DATA _NULL_ DATA step is rarely needed.

4.10 Problems

Solutions to odd-numbered problems are located at the back of this book. Solutions to all problems are available to professors. If you are a professor, visit the book's companion website at support.sas.com/cody for information about how to obtain the solutions to all problems.

1. Run the program here to create a permanent SAS data set called Perm. You will need to modify the program to specify a folder where you want to place this data set. Run PROC CONTENTS on this data set.

```
libname learn 'C:\your-folder-name';

   data learn.Perm;
      input ID : $3. Gender : $1. DOB : mmddyy10.
            Height Weight;
      label DOB = 'Date of Birth'
            Height = 'Height in inches'
            Weight = 'Weight in pounds';
      format DOB date9.;
   datalines;
   001 M 10/21/1946 68 150
   002 F 5/26/1950 63 122
   003 M 5/11/1981 72 175
   004 M 7/4/1983 70 128
   005 F 12/25/2005 30 40
   ;
```

2. Run PROC PRINT on the data set you created in Problem 1. Use the SAS VIEWTABLE window (if available on your system) to open this data set and compare the headings in the window to the column headings from your PROC PRINT. What is the difference?

3. Run this program to create a permanent SAS data set called Survey2018. Close your SAS session, open up a new session, and write the statements necessary to compute the mean of age.

```
 *Write your LIBNAME statement here;
 data <fill in your data set name here> ;
    input Age Gender $ (Ques1-Ques5)($1.);
    /* See Chapter 21, Section 14 for a discussion
       of variable lists and format lists used above */
 datalines;
 23 M 15243
 30 F 11123
 42 M 23555
 48 F 55541
 55 F 42232
 62 F 33333
 68 M 44122
;

*Write this code after you closed and reopened your SAS session;
*Write your libname statement here;
proc means data= - insert the correct data set name -;
   var Age;
run;
```

Chapter 5: Creating Labels and Formats

5.1 Adding Labels to Your Variables

If you are using SAS to produce listings and reports for others, you will want to make the output more readable and attractive. SAS formats and labels help you do this. They also help you to remember what each variable represents.

Many SAS procedures use variable labels to improve readability. You can create labels either in a DATA or PROC step. As an example, you can add labels to the variables in the Test_Scores data set like this:

Program 5.1: Adding Labels to Variables in a SAS Data Set

```
libname Learn 'C:\books\learning';

data Learn.Test_Scores;
   length ID $ 3 Name $ 15;
   input ID $ Score1-Score3 Name $;
   label ID = 'Student ID'
         Score1 = 'Math Score'
         Score2 = 'Science Score'
         Score3 = 'English Score';
datalines;
1 90 95 98 Jan
2 78 77 99 Preston
3 88 91 92 Russell
;

title "Descriptive Statistics for Student Scores";
proc means data=Learn.Test_Scores;
run;
```

Labels are created with a LABEL statement. Following the keyword LABEL, you enter a variable name, followed by an equal sign, followed by your label, placed in single or double quotes. Labels can be up to 256 characters long (255 on UNIX platforms). You may continue with variable names and labels for as many variables as you want. Just make sure that you complete the LABEL statement with a semicolon.

When you run certain SAS procedures, these labels are printed along with the variable names.

Here is output from the program above:

Figure 5.1: Output from Program 5.1

Descriptive Statistics for Student Scores

Variable	Label	N	Mean	Std Dev	Minimum	Maximum
Score1	Math Score	3	85.3333333	6.4291005	78.0000000	90.0000000
Score2	Science Score	3	87.6666667	9.4516313	77.0000000	95.0000000
Score3	English Score	3	96.3333333	3.7859389	92.0000000	99.0000000

Notice how the labels improve the readability of this output.

If you include your LABEL statement in the DATA step, the labels remain associated with the respective variables; if you include your LABEL statement in a PROC step, the labels are used only for that procedure. This is because the label created in a DATA step is stored in the descriptor portion of the SAS data set.

5.2 Using Formats to Enhance Your Output

SAS provides built-in formats to improve the appearance of printed output. For example, you can print financial data with dollar signs or add commas to large numbers. You saw an example of this in program 3.8.

You can also create your own user-defined formats. For example, if you have a variable called Gender with values of `F` and `M`, you can format these values so that they print as Male and Female. If you have a variable representing age, you can use formats to display the values as age groups instead of actual ages. You can have one format for each variable or use one format for a group of variables.

You create user-defined formats with PROC FORMAT; you associate your formats (or SAS built-in formats) with one or more variables in a FORMAT statement. A SAS data set called Survey shows how formats can be used. Here is a listing of this data set, without any formats:

Figure 5.2: Listing of Data Set Survey

Listing of Data Set Survey

ID	Gender	Age	Salary	Ques1	Ques2	Ques3	Ques4	Ques5
001	M	23	28000	1	2	1	2	3
002	F	55	76123	4	5	2	1	1
003	M	38	36500	2	2	2	2	1
004	F	67	128000	5	3	2	2	4
005	M	22	23060	3	3	3	4	2
006	M	63	90000	2	3	5	4	3
007	F	45	76100	5	3	4	3	3

Let's see how formats can improve the readability of this listing:

Program 5.2: Using PROC FORMAT to Create User-Defined Formats

```
proc format;
   value $Gender 'M' = 'Male'
                 'F' = 'Female'
                 ' ' = 'Not entered'
               other = 'Miscoded';

   value Age low-29  = 'Less than 30'
             30-50   = '30 to 50'
             51-high = '51+';

   value $Likert '1' = 'Str Disagree'
                 '2' = 'Disagree'
                 '3' = 'No Opinion'
                 '4' = 'Agree'
                 '5' = 'Str Agree';
run;
```

You should notice several things about this procedure. First, you use a VALUE statement to create each user-defined format. Notice that when you are creating the format in a VALUE statement, you do not include a period following the format name.

Next, formats used with character variables start with a dollar sign. Following the format name are either unique values and/or ranges of values, an equal sign, and then the text you want to associate with each value or range of values. Rules concerning format names are the similar to those for SAS variable names with the exception that format names cannot end in a digit and the maximum length of a format name is 31 characters. For those curious readers, the length of 31 instead of 32 comes from the fact that you add a period at the end of a format name when you use it in a FORMAT statement in a DATA or PROC step. (SAS versions prior to SAS 9 allowed only 8 character format names.)

The first format to be defined is $GENDER. Format names do not need to be related to a variable name—calling this format $GENDER makes it easier to remember that you will use it later to alter how the Gender values will be printed in SAS output.

Values for Gender are stored as **M** and **F**. Associating the $GENDER format with the variable Gender results in **M** displaying as **Male**, **F** displaying as **Female**, and missing values displayed as `Not entered`. The keyword **other** in the VALUE statement causes the text **Miscoded** to be printed for any characters besides **M**, **F**, or a missing value.

The format AGE is used to group ages into three categories. Notice that it is OK to use the same name for a format and a variable. (SAS knows that a name containing a period is a format.) If you apply this format to the variable Age, the age groups are printed instead of the actual ages. Remember that the internal values of SAS variables are not changed because they have been associated with a format. The format only affects how values print or, in some cases, how SAS procedures process a variable. (For example, PROC FREQ computes frequencies of formatted values rather than raw values; PROC MEANS uses formatted values for variables listed in the CLASS statement, and so forth.)

In the AGE format, the keywords LOW and HIGH refer to the lowest nonmissing value and the highest value, respectively.

> **Note:** The keyword LOW when used with character formats includes missing values.

The last format, $LIKERT, is used to substitute the appropriate text for the digits **1 (strongly disagree)** to **5 (strongly agree)**.

> **Note:** The name Likert was chosen because that is the name that psychometricians use for responses to questions that range from strongly disagree to strongly agree.

Let's first see what happens if you place a format statement in PROC PRINT, as follows:

Program 5.3: Adding a FORMAT Statement in PROC PRINT

```
title "Data Set SURVEY with Formatted Values";
proc print data=Learn.Survey;
   id ID;
   var Gender Age Salary Ques1-Ques5;
   format Gender      $Gender.
          Age         Age.
          Ques1-Ques5 $Likert.
          Salary      Dollar11.2;
run;
```

Here the formats $GENDER and AGE are used to format the variables Gender and Age, respectively. The format $LIKERT formats the five variables Ques1 through Ques5. Notice that each format is followed by a period, just the same as built-in SAS formats.

The format for Salary, DOLLAR11.2, is a SAS format. The name *dollar* indicates that you want to use the dollar format (which adds a dollar sign and commas to the value); the number 11 tells SAS to print a value using 11 columns; the 2 following the decimal point tells SAS that you want to print two digits to the right of the decimal point. Take note that the 11 columns include the dollar sign, the commas, and the decimal point, in addition to the digits. The largest value for Salary using the DOLLAR11.2 format would be:

```
$999,999.99
```

It is a good idea to make the total width a bit larger than you think you need, just in case your data contains a larger number than you expect. SAS has a set of rules that will allow numbers to print when you have not allocated enough columns. However, it is better to ensure you have enough columns and not be concerned with what happens when the format is too small.

Before we show you the output, notice the ID statement in Program 5.3. When you include an ID statement in PROC PRINT, the variable (or variables) you list show up in the first column (or columns) of your report, replacing the Obs column that SAS usually displays in the first column. If you list a variable in an ID statement, don't also list it in the VAR statement. If you do, it appears twice on the listing. If you have an ID variable such as Subject or ID, it is recommended that you use an ID statement.

Here is the listing:

Figure 5.3: Output from Program 5.3

Data Set SURVEY with Formatted Values

ID	Gender	Age	Salary	Ques1	Ques2	Ques3	Ques4	Ques5
001	Male	Less than 30	$28,000.00	Str Disagree	Disagree	Str Disagree	Disagree	No Opinion
002	Female	51+	$76,123.00	Agree	Str Agree	Disagree	Str Disagree	Str Disagree
003	Male	30 to 50	$36,500.00	Disagree	Disagree	Disagree	Disagree	Str Disagree
004	Female	51+	$128,000.00	Str Agree	No Opinion	Disagree	Disagree	Agree
005	Male	Less than 30	$23,060.00	No Opinion	No Opinion	No Opinion	Agree	Disagree
006	Male	51+	$90,000.00	Disagree	No Opinion	Str Agree	Agree	No Opinion
007	Female	30 to 50	$76,100.00	Str Agree	No Opinion	Agree	No Opinion	No Opinion

5.3 Regrouping Values Using Formats

You can use formats to group various values together. For example, suppose you want to see the survey results, but instead of looking at the five possible responses for Questions 1 through 5, you want to group the values `1` and `2` (strongly disagree and disagree) together and the values `4` and `5` (agree and strongly agree) to make three categories for each question. You can accomplish this by creating a new format, as shown in Program 5.4:

Program 5.4: Regrouping Values Using a Format

```
proc format;
   value $Three 1,2    = 'Disagreement'
                3      = 'No opinion'
                4,5    = 'Agreement';
run;
```

You can then apply this to the Question variables in a procedure, as follows:

Program 5.5: Applying the New Format to Several Variables with PROC FREQ

```
title "Question Frequencies Using the Three Format";
proc freq data=Learn.Survey;
   tables Ques1-Ques5;
   format Ques1-Ques5 $Three.;
run;
```

PROC FREQ, as you saw in Chapter 2, is used to count frequencies for the variables listed in the TABLES statement (Ques1–Ques5 in this case). Because of the FORMAT statement in this procedure, the tables have only three categories rather than the original five. Here is a partial listing of the output:

Figure 5.4: Partial Listing from Program 5.5

Question Frequencies Using the Three Format

Ques1	Frequency	Percent	Cumulative Frequency	Cumulative Percent
Disagreement	3	42.86	3	42.86
No opinion	1	14.29	4	57.14
Agreement	3	42.86	7	100.00

Ques2	Frequency	Percent	Cumulative Frequency	Cumulative Percent
Disagreement	2	28.57	2	28.57
No opinion	4	57.14	6	85.71
Agreement	1	14.29	7	100.00

If you look back at Figure 5.2, you can see that the two values of 1 and one value of 2 for Quest1 were combined to give you a frequency of 3 for the category of Disagreement—the two values of 4 and one value of 5 were combined to give you a frequency of 3 for the category of Agreement.

5.4 More on Format Ranges

When you define a format, you can specify individual values or ranges to the left of the equal sign in your VALUE statement. As an example of how flexible this approach is, consider that you have a variable called Grade with values of **A**, **B**, **C**, **D**, **F**, **I**, and **W**. The following VALUE statement creates a format that places these grades into six categories:

```
value $Gradefmt 'A' - 'C' = 'Passing'
                'D'       = 'Borderline'
                'F'       = 'Failing'
                'I','W'   = 'Incomplete or withdrew'
                ' '       = 'Not recorded'
                other     = 'Miscoded';
```

Here you see that grades **A**, **B**, or **C** will be formatted as **Passing**, **D** as **Borderline**, **F** as **Failing**, **I** or **W** as **Incomplete or withdrew**, missing values as **Not recorded**, and any other value as **Miscoded**. You may leave the quotes off the character ranges and the labels if you want. However, as a matter of style, we recommend that you use single or double quotes here.

In Program 5.2, the ranges for the AGE format were defined like this:

```
value Age low-29  = 'Less than 30'
          30-50   = '30 to 50'
          51-high = '51+';
```

This is fine if this format is used with integer values. However, suppose you used this format with a variable that could take on values such as **29.5**? This value falls between the two ranges **low-29** and **30-50**. You can make sure there are no gaps in your ranges like this:

```
value Age low-<30  = 'Less than 30'
          30-<51   = '30 to less than 51'
          51-high  = '51+';
```

The first range includes all values **Less than 30** (which would include **29.5**). The second range includes values from **30 to less than 51** and the last range includes values of **51** and higher.

You can also use a less than (<) sign on the left side of a range.

For the example below, the range 30<-51 does not include 30.

```
value Age low-30   = 'Less than or equal to 30'
          30<-51   = 'Greater than 30 to 51'
          51<-high = 'Greater than 51';
```

Note: A good way to remember how this works is to exclude the first value, then put the < sign after the value—if you want to exclude the last value, then put the < sign before that value.

If you know that your format may be used with non-integer values, be sure that there are no gaps in your ranges.

5.5 Storing Your Formats in a Format Library

As we mentioned earlier, if you place LABEL and FORMAT statements in the DATA step, the labels and formats become permanently associated with their respective variables. If you have user-defined formats with permanent SAS data sets, it is important to make your formats permanent also. Here are the steps to do this:

1. Create a library reference (libref) to indicate where you want to store your SAS formats. This can be the same library where you store your data sets.
2. Use the option LIBRARY=*libref* when you run PROC FORMAT. (Remember, you only have to run this procedure once.)

As an example, suppose you want to make the formats created in Program 5.2, permanent and save them in the **C:\books\learning** folder.

Program 5.6 creates a permanent format library for you.

Program 5.6: Creating a Permanent Format Library

```
libname Myfmts 'C:\books\learning';

proc format library=Myfmts;
   value $Gender 'M'  = 'Male'
                 'F'  = 'Female'
                 ' '  = 'Not entered'
               other  = 'Miscoded';

   value Age low-29   = 'Less than 30'
             30-50    = '30 to 50'
             51-high  = '51+';

   value $Likert '1'  = 'Strongly disagree'
                 '2'  = 'Disagree'
                 '3'  = 'No opinion'
                 '4'  = 'Agree'
                 '5'  = 'Strongly agree';
run;
```

If you run this program on a Windows system, a file called **formats.sas7bcat** will be created in the folder specified by the libref.

5.6 Permanent Data Set Attributes

If you add your LABEL and FORMAT statements in the DATA step, the labels and formats become permanently associated with their respective variables. This makes for a very convenient way to document a data set. Another user could use PROC CONTENTS or the SAS Explorer to list the labels and formats used with each variable.

Anytime you want to use a SAS data set with associated user-defined formats, you need to tell SAS where to look for these formats. By default, SAS will only look for its own formats, formats in a Work library (i.e., temporary formats), or formats in a library with the special name Library.

If you want SAS to also look in one of your own libraries, you need to issue a FMTSEARCH= system option. You can list one or more libraries for SAS to search using this option. For example, if you want to use the formats you placed in the Myfmts library, you would need to submit the following code:

```
options fmtsearch=(Myfmts);
```

If you do this, SAS first looks in the Work library, then the library called Library, and then the Myfmts library. If you want SAS to look in the Myfmts library before it looks in either of the other two libraries, you can name them on the FMTSEARCH statement like this:

```
options fmtsearch=(myfmts Work Library);
```

Now, SAS searches the Myfmts library first and then the Work and Library libraries.

Program 5.7 demonstrates how to make a permanent SAS data set with user-defined formats. (For this example, assume you have already created a permanent SAS library in your **`C:\books\learning`** folder.)

Program 5.7: Adding LABEL and FORMAT Statements in the DATA Step

```
libname Learn 'C:\books\learning';
libname Myfmts 'C:\books\learning';
options fmtsearch=(Myfmts);

data Learn.Survey;
   infile 'C:\books\learning\Survey.txt';
   input ID : $3.
         Gender : $1.
         Age
         Salary
         (Ques1-Ques5)(1.);

   format Gender      $Gender.
          Age         Age.
          Ques1-Ques5 $Likert.
          Salary      Dollar10.0;

   label ID     = 'Subject ID'
         Gender = 'Gender'
         Age    = 'Age as of 1/1/2006'
         Salary = 'Yearly Salary'
         Ques1  = 'The governor is doing a good job?'
         Ques2  = 'The property tax should be lowered'
         Ques3  = 'Guns should be banned'
         Ques4  = 'Expand the Green Acre program'
         Ques5  = 'The school needs to be expanded';
run;
```

Now, run PROC CONTENTS on this data set, as follows:

Program 5.8: Running PROC CONTENTS on a Data Set with Labels and Formats

```
title "Data set Survey";
proc contents data=Learn.Survey varnum;
run;
```

You obtain a listing that helps document the data set, like this (partial listing):

Figure 5.5: Output from Program 5.8

Variables in Creation Order					
#	Variable	Type	Len	Format	Label
1	ID	Char	3		Subject ID
2	Gender	Char	1	$GENDER.	Gender
3	Age	Num	8	AGE.	Age as of 1/1/2006
4	Salary	Num	8	DOLLAR10.	Yearly Salary
5	Ques1	Char	1	$LIKERT.	The governor is doing a good job?
6	Ques2	Char	1	$LIKERT.	The property tax should be lowered
7	Ques3	Char	1	$LIKERT.	Guns should be banned
8	Ques4	Char	1	$LIKERT.	Expand the Green Acre program
9	Ques5	Char	1	$LIKERT.	The school needs to be expanded

You now see the formats and labels associated with each variable.

5.7 Accessing a Permanent SAS Data Set with User-Defined Formats

If you want to use a permanent SAS data set that has user-defined formats, the only requirement is to remember to tell SAS where to find the formats. If you forget the FMTSEARCH= system option, you will get an error message telling you that SAS cannot find the formats.

Note: If you give a copy of a SAS data set with user-defined formats to another user, be sure to also give a copy of the format library to them as well. (On a PC platform, you need to give them a copy of the file **formats.sas7bcat**.)

Here is an example of a program to compute frequencies on the variables Ques1–Ques5 in the permanent SAS data set Survey:

Program 5.9: Using a User-defined Format

```
libname Learn 'C:\books\learning';
libname Myfmts 'C:\books\learning';
options fmtsearch=(Myfmts);

title "Using User-defined Formats";
proc freq data=Learn.Survey;
   tables Ques1-Ques5;
run;
```

Once you submit the FMTSEARCH= option, you can use your own formats just as if they were built-in SAS formats.

5.8 Displaying Your Format Definitions

A useful PROC FORMAT option is FMTLIB. This option creates a listing of each format in the specified library with the ranges and labels. As an example, if you want to display the definitions of all the formats in your Myfmts library, you would submit the following code:

Program 5.10: Displaying Format Definitions in a User-created Library

```
title "Format Definitions in the Myfmts Library";
proc format library=Myfmts fmtlib;
run;
```

You obtain a table like this:

Figure 5.6: Output from Program 5.10

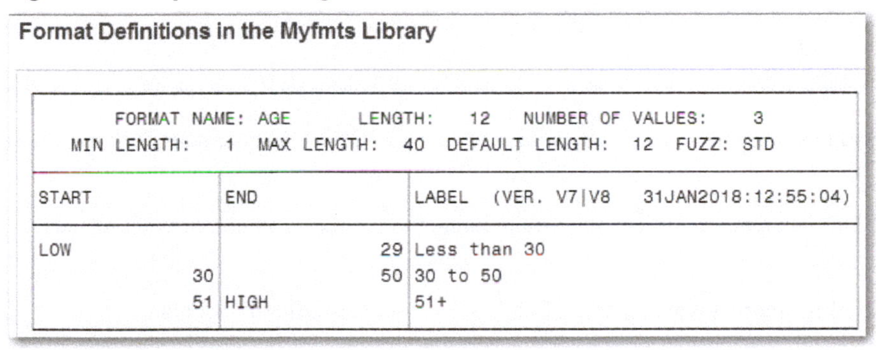

Format Definitions in the Myfmts Library

FORMAT NAME: AGE LENGTH: 12 NUMBER OF VALUES: 3 MIN LENGTH: 1 MAX LENGTH: 40 DEFAULT LENGTH: 12 FUZZ: STD		
START	END	LABEL (VER. V7|V8 31JAN2018:12:55:04)
LOW	29	Less than 30
30	50	30 to 50
51	HIGH	51+

FORMAT NAME: $GENDER LENGTH: 11 NUMBER OF VALUES: 4 MIN LENGTH: 1 MAX LENGTH: 40 DEFAULT LENGTH: 11 FUZZ: 0		
START	END	LABEL (VER. V7|V8 31JAN2018:12:55:04)
		Not entered
F	F	Female
M	M	Male
OTHER	**OTHER**	Miscoded

```
        FORMAT NAME: $LIKERT   LENGTH:    17    NUMBER OF VALUES:     5
   MIN LENGTH:    1  MAX LENGTH:   40  DEFAULT LENGTH:   17  FUZZ:        0
```

START	END	LABEL (VER. V7|V8 31JAN2018:12:55:04)
1	1	Strongly disagree
2	2	Disagree
3	3	No opinion
4	4	Agree
5	5	Strongly agree

If you only want to see specific formats in a format library, you can add a SELECT statement to your PROC FORMAT. You list the formats you want displayed following the keyword SELECT. When you use a SELECT statement, you do not also have to include the FMTLIB option. For example, to display only the AGE and $LIKERT formats, you could use the following program:

Program 5.11: Demonstrating a SELECT Statement with PROC FORMAT

```
proc format library=Myfmts;
   select Age $Likert;
run;
```

Note: This program assumes that you have previously submitted a LIBNAME statement defining the Myfmts library.

There is also an EXCLUDE statement that enables you to name the formats you do not want to see displayed.

Please refer to Chapter 22 for more advanced uses of both SAS and user-defined formats.

5.9 Problems

Solutions to odd-numbered problems are located at the back of this book. Solutions to all problems are available to professors. If you are a professor, visit the book's companion website at support.sas.com/cody for information about how to obtain the solutions to all problems.

1. Run the program here to create a temporary SAS data set called Voter:

```
data Voter;
   input Age Party : $1. (Ques1-Ques4)($1. + 1);
datalines;
23 D 1 1 2 2
45 R 5 5 4 1
67 D 2 4 3 3
39 R 4 4 4 4
19 D 2 1 2 1
75 D 3 3 2 3
57 R 4 3 4 4
;
```

Add formats for Age (0–30, 31–50, 51–70, 71+), Party (D = Democrat, R = Republican), and Ques1–Ques4 (1=Strongly Disagree, 2=Disagree, 3=No Opinion, 4=Agree, 5=Strongly Agree). In addition, label Ques1–Ques4 as follows:

Variable	Label
Ques1	The president is doing a good job
Ques2	Congress is doing a good job
Ques3	Taxes are too high
Ques4	Government should cut spending

Note: Use PROC PRINT to list the observations in this data set and PROC FREQ to list frequencies for the four questions. (The default action of PROC PRINT is to head each column with a variable name, not the label. To use labels as column headings, use the LABEL option with PROC PRINT.)

2. You want to see frequencies for Questions 1 to 4 from the previous question. However, you want only three categories: **`Generally Disagree`** (combine **`Strongly Disagree`** and **`Disagree`**), **`No Opinion`**, and **`Generally Agree`** (combine **`Agree`** and **`Strongly Agree`**). Accomplish this using a new format for Ques1–Ques4.
3. Run the following program to create a SAS data set called Colors (see Chapter 21 for a discussion of the double at signs [@@] in the INPUT statement):

```
data Colors;
   input Color : $1. @@;
datalines;
R R B G Y Y . . B G R B G Y P O O V V B
;
```

Use a format to group the colors as follows:

R, B, G = Group 1
Y, O = Group 2
Missing = Not Given
All others = Group 3

Use PROC FREQ to list the frequencies of the color groups.

4. Make a permanent SAS data set from data set Voter in Problem 1. Place this data set in a folder of your choice. Make the labels and formats permanent attributes in this data set and make your formats permanent as well (place them in the same library as the data set). Use the FMTLIB option with PROC FORMAT when you run this procedure.
5. Write the necessary statements to make three permanent formats in a library of your choice. Use the FMTLIB option to list each of these formats. The formats are defined as follows:

```
YesNo       1 = Yes, 0 = No
$YesNo      Y = Yes, N = No
$Gender     M = Male, F = Female
Age20yr     low-20 = 1, 21-40 = 2, 41-60 = 3, 61-80 = 4,
            81-high = 5
```

Chapter 6: Reading and Writing Data from an Excel Spreadsheet

6.1 Introduction

It is quite common to be given a Microsoft Office Excel spreadsheet as your data source. Luckily, SAS has several methods to easily convert a spreadsheet into a SAS data set. One way is to convert the spreadsheet into a comma-separated values (CSV) file and to read the file using INFILE statements (using the DSD option—see Chapter 3 for more information) and INPUT statements. However, if you have licensed SAS/ACCESS Interface to PC Files, or you are using SAS Enterprise Guide or SAS Studio (perhaps with the University Edition), you can have SAS do the conversion automatically.

6.2 Using the Import Wizard to Convert a Spreadsheet to a SAS Data Set

The workbook used in the next example was created using Microsoft Excel. The Wage column (E) was computed by multiplying the Hours Worked column (C) by the Rate Per Hour column (D). Notice that the column headings are not valid SAS variable names. In addition, this workbook has two worksheets: one named Temporary (for temporary workers) and the other named Permanent (for permanent employees).

Figure 6.1: Excel Workbook Wages

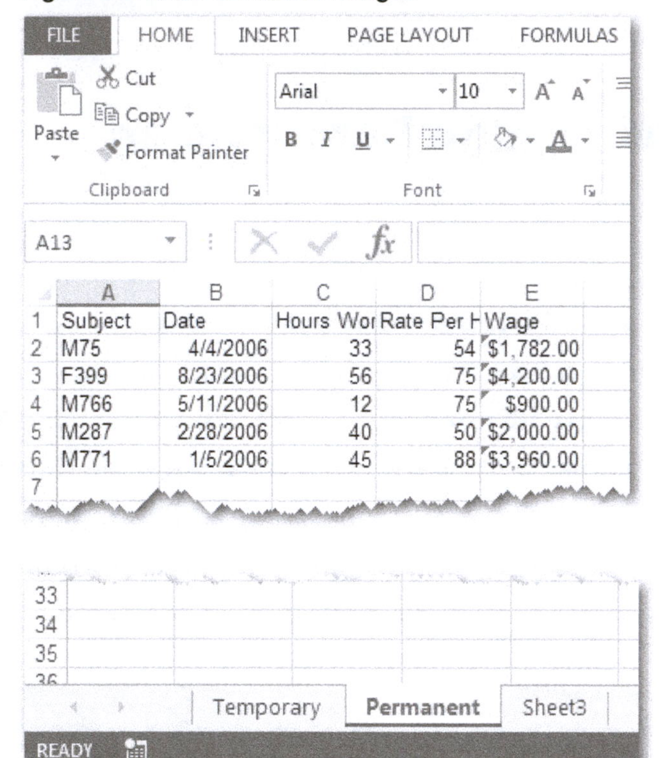

The first step is to select **Import Data** from the SAS **File** menu, as shown here:

Figure 6.2: Select Import Data

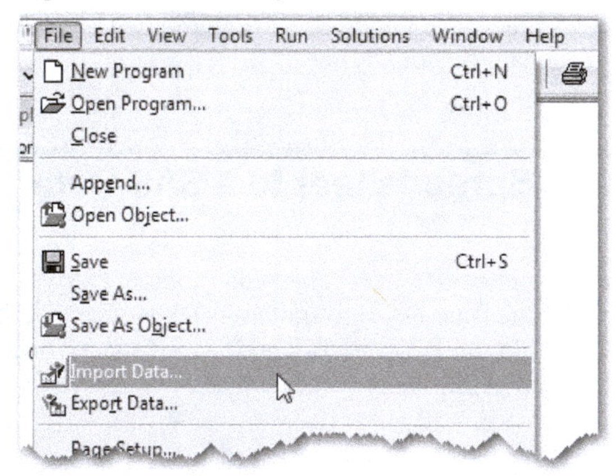

This brings up a screen where you can select from a variety of formats (Excel, Access, dBase, Lotus, and several others). Select **Microsoft Excel** from the pull-down menu.

Figure 6.3: Select Excel

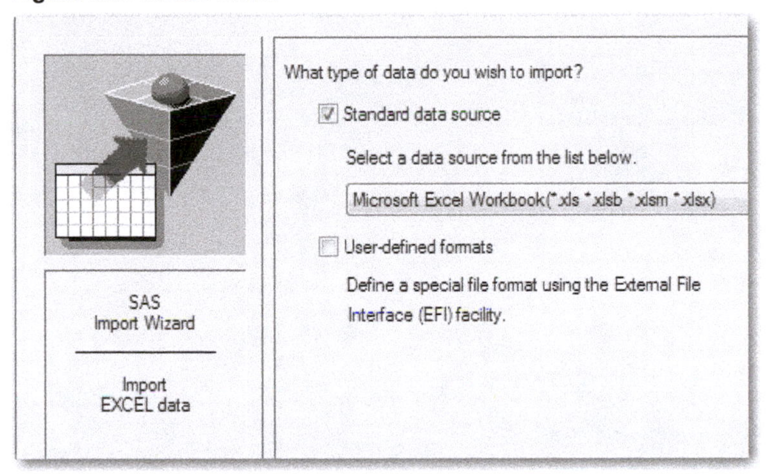

Click **Next** to bring up the next screen. Here you can either type in the name for the Excel file you want to read, or select **Browse** to obtain a list of the Excel spreadsheets on your system.

Figure 6.4: Select Next and Browse for Your File

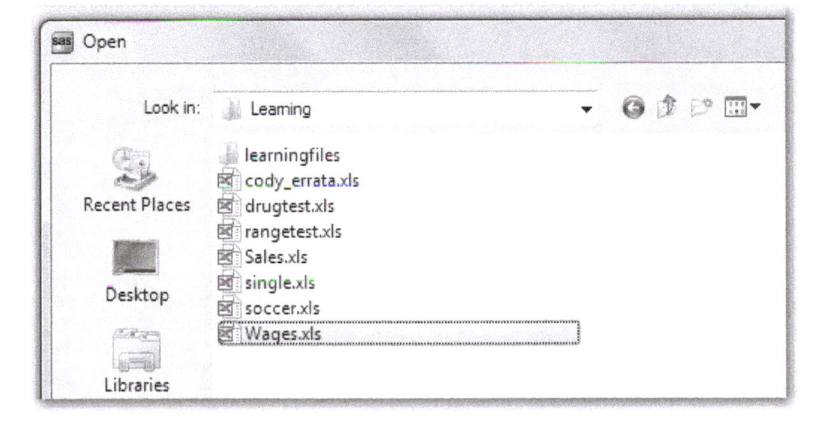

If you have multiple worksheets, you can select the one you want to import. The default worksheet name is Sheet1$ (SAS places a dollar sign after the worksheet name). Any named ranges that have been created in the worksheet are also listed. Worksheet names end in a $; the names of named ranges don't.

Figure 6.5: Select a Worksheet if More than One

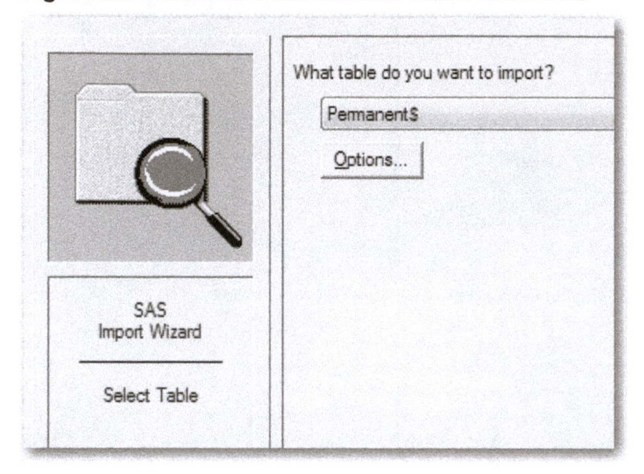

Once you have selected the worksheet you want to import, click **Next**. You are presented with a screen where you can enter a library (Work, if you want a temporary SAS data set, or a libref created with a LIBNAME statement) and the name of the SAS data set (labeled as Member in the destination screen). See the following:

Figure 6.6: Choose a Library and File Name

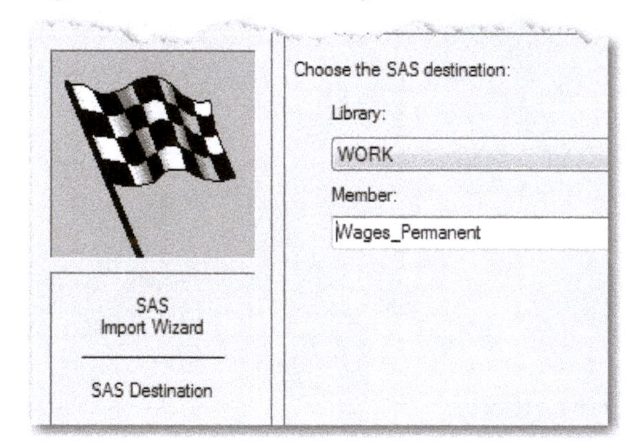

If you then select **Finish**, SAS imports your spreadsheet. If you select **Next**, you can choose to save the conversion program (PROC IMPORT). If you save this program (you will be prompted for the name of a file in which to save the program), you can run it later to perform your conversion.

You are done. That's all there is to it. In this example, you chose the Work library and called your SAS data set Wages_Permanent. It is a good idea to list the contents of this data set using the SAS viewer or PROC PRINT before attempting to use it for reports or further analysis. If the data set is large, you can list the first few observations in the data set by using the OBS=*n* data set option to limit the number of observations you want to process. For example, to see the first four observations in data set Wages_Permanent, you could run the following program:

Program 6.1: Using PROC PRINT to List the First Four Observations in a Data Set

```
title "The First Four Observations of Wages_Permanent";
proc print data=Wages_Permanent(obs=4);
run;
```

Here is the output:

Figure 6.7: Output from Program 6.1

The First Four Observations of Wages_Permanent

Obs	Subject	Date	Hours_Worked	Rate_Per_Hour	Wage
1	M75	04APR2006	33	54	$1,782.00
2	F399	23AUG2006	56	75	$4,200.00
3	M766	11MAY2006	12	75	$900.00
4	M287	28FEB2006	40	50	$2,000.00

Notice that SAS has created valid variable names from the Excel column headings. It replaces any invalid characters in column headings (in this case, blanks) with an underscore (_). It is also a good idea to use the SAS Explorer or PROC CONTENTS to view the type (character or numeric) and length of each variable in this data set. You may need to use PROC DATASETS to change the format of one or more variables or to write a short DATA step to perform a character-to-numeric conversion for variables that you want to be numeric but, for one reason or another, wound up as character.

Using the OBS=*n* data set option is a very useful way to check the data values of very large data sets. By the way, while we are on the topic, you can combine the OBS=*n* data set option with the FIRSTOBS=*m* option. The value of FIRSTOBS= defines the **first** observation you want to process; the value of the OBS= option is the **last** observation you want to process. It is a good idea to think of OBS= as *LASTOBS*. Suppose you want to list observations 100 through 110 in a very large SAS data set called Verybig in a library with a libref of Project. You would combine the FIRSTOBS= and OBS= options like this:

Program 6.2: Using the FIRSTOBS= and OBS= Options Together

```
title "Observations 100 through 110 in VERYBIG";
proc print data=Project.Verybig(firstobs=100 obs=110);
run;
```

Program 6.2 results in a listing containing 11 observations (observation 100 through observation 110).

Importing Excel Worksheets using either Enterprise Guide or SAS Studio is even easier. For example, when using SAS Studio, you don't even have to select an input file format—it looks at the file extension (for example, XLS or XLSX) and automatically selects the correct import method.

6.3 Creating an Excel Spreadsheet from a SAS Data Set

Before we leave this chapter, let's see how you can use the Export Wizard to convert a SAS data set into an Excel spreadsheet. From the **File** menu, select **Export Data** and then select **Microsoft Excel**.

Suppose you want to convert your Sales data set (located in the `c:\books\learning` folder) to an Excel spreadsheet. First, you need to be sure you have a library reference to the folder. For example, issue the following statement:

```
libname Learn 'C:\books\learning';
```

Then, follow these steps:

From the **File** menu, select **Export** Data.

Figure 6.8: First Step in Exporting an Excel Spreadsheet (using the SAS Windowing System)

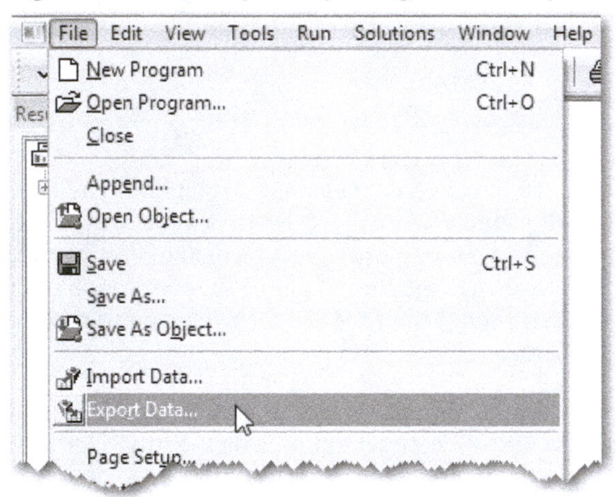

On the next screen, enter the LIBNAME and the name of the SAS data set you want to export (Sales in this example) and click **Next**.

Figure 6.9: Name the Library and Data Set Name

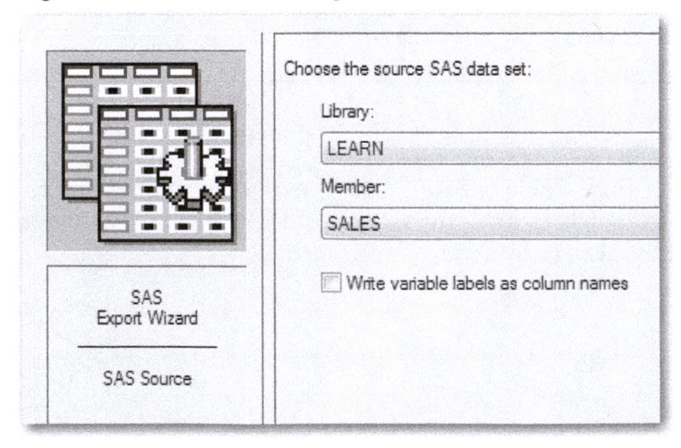

Next, select **Microsoft Excel** from the pull-down menu.

Figure 6.10: Select Microsoft Excel

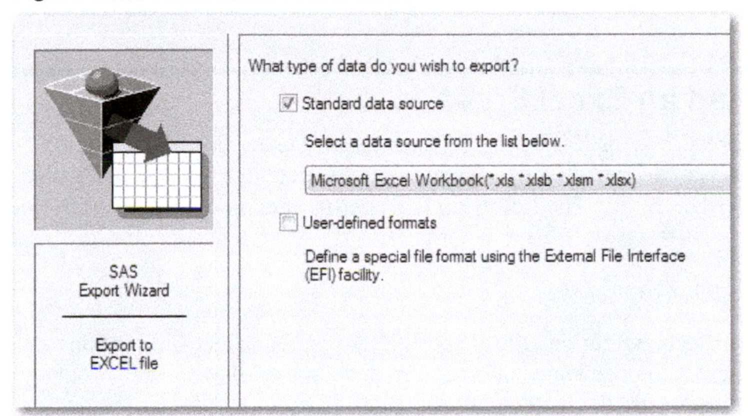

You can browse or enter the name of the Excel spreadsheet.

Figure 6.11: Name the Excel Workbook

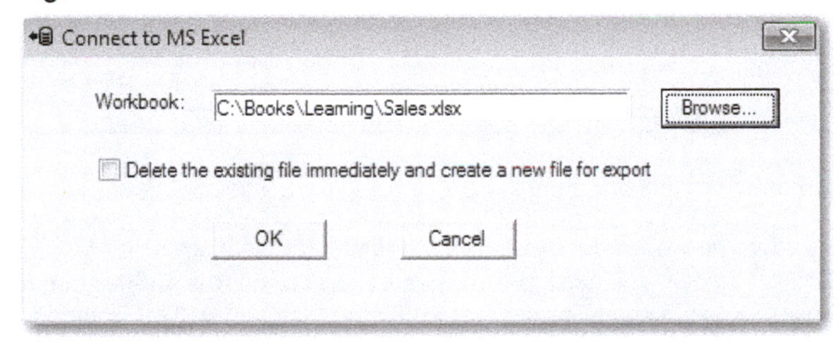

You can also name the specific table.

Figure 6.12: Supply a Table Name

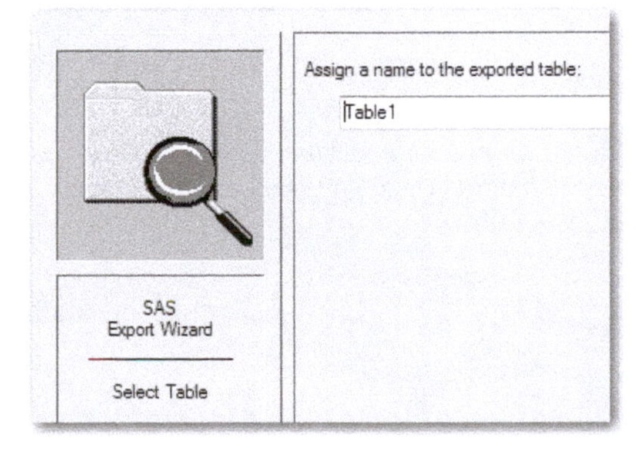

Select **Finish** and SAS writes the data to an Excel spreadsheet (or select **Next** to write the PROC EXPORT statements to a file). Here, the file is called **Sales.xlsx** (in the **C:\books\learning** folder) and the table name is Table1.

6.4 Using an Engine to Read an Excel Spreadsheet

You can have SAS treat an Excel spreadsheet as if it were a SAS data set by using the XLS or XLSX engine. As an example, suppose you want to run a SAS procedure with the data in a spreadsheet called **Wages.xls** in your **C:\books\learning** folder. The following LIBNAME statement enables you to access this spreadsheet directly:

```
libname Read 'C:\books\learning\Wages.xls';
```

You can now access any of the worksheets within this file. This particular spreadsheet file contains two worksheets, Temporary and Permanent. Suppose you want to compute the mean of Wage and Hours Worked in the Permanent worksheet. Here are the SAS statements to do that:

Program 6.3: Reading a Spreadsheet Using an XLSX Engine

```
title "Statistics from Sales Spreadsheet";
libname Read 'C:\books\learning\Wages.xls';

proc means data=Read.'Permanent$'n mean;
   var Wage Hours_Worked;
run;
```

There are several important points to notice in this program.

First, because SAS requires you to follow the worksheet name with a dollar sign and because dollar signs are not normally allowed in SAS data set names, you need to use a name literal to do this. In Program 6.3, you place the worksheet name (**Permanent$**) in single quotes and follow this with an **n**. This notation allows you to use invalid characters as part of SAS names.

Note: In order for this named literal to work, you must have the option VALIDVARNAME set to ANY. To do this, submit the following statement:

```
options validvarname=any;
```

Next, remember that the column heading Hours Worked is not a valid SAS variable name and the rule is that SAS will substitute an underscore for any invalid variable names. Thus, you need to refer to this column as **Hours_Worked**. Here is the output from PROC MEANS:

Figure 6.13: Output from Program 6.3

Statistics from Sales Spreadsheet

Variable	Label	Mean
Wage	Wage	2568.40
Hours_Worked	Hours Worked	37.2000000

Once you have created the appropriate LIBNAME statement, you can treat your spreadsheets as if they were SAS data sets.

6.5 Using the SAS Output Delivery System to Convert a SAS Data Set to an Excel Spreadsheet

You can use the Output Delivery System (ODS) to create CSV files that Excel can open directly. For more information on ODS, refer to Chapter 19.

As an example, suppose you want to send the contents of the permanent SAS data set Survey to Excel. Program 6.4 creates a CSV file from the SAS data set:

Program 6.4: Using ODS to Convert a SAS Data Set into a CSV File (to Be Read by Excel)

```
libname Learn 'C:\books\learning';

ods csv file='C:\books\learning\ODS_Example.csv';

proc print data=Learn.Survey noobs;
run;

ods csv close;
```

The ODS CSV statement opens the CSV file as an output destination. Notice that the NOOBS option of PROC PRINT is used to remove the OBS column from the output. It is important to close the file with an ODS CLOSE statement following the PROC PRINT. Here is a listing of the CSV file **C:\books\learning\ODS_Example.csv**:

```
"ID","Gender","Age","Salary","Ques1","Ques2","Ques3","Ques4","Ques5"
 001,"M",23,28000,1,2,1,2,3
 002,"F",55,76123,4,5,2,1,1
 003,"M",38,36500,2,2,2,2,1
 004,"F",67,128000,5,3,2,2,4
 005,"M",22,23060,3,3,3,4,2
 006,"M",63,90000,2,3,5,4,3
 007,"F",45,76100,5,3,4,3,3
```

You can open this file directly in Excel with the following result:

Figure 6.14: Opening the CSV File in Excel

ID	Gender	Age	Salary	Ques1	Ques2	Ques3	Ques4	Ques5
1	M	23	$28,000	1	2	1	2	3
2	F	55	$76,123	4	5	2	1	1
3	M	38	$36,500	2	2	2	2	1
4	F	67	$128,000	5	3	2	2	4
5	M	22	$23,060	3	3	3	4	2
6	M	63	$90,000	2	3	5	4	3
7	F	45	$76,100	5	3	4	3	3

SAS can read and write Excel data very easily. Be sure to check the resulting files following a transfer to ensure that data values, especially dates, were processed properly. In addition, if you are creating a SAS data set, be sure to run PROC CONTENTS or use the SAS Explorer to verify variable types and lengths.

6.6 A Quick Look at the Import Utility in SAS Studio

As mentioned earlier in this chapter, both SAS Enterprise Guide and SAS Studio can be used to import and export Microsoft Excel files. This section shows some screen shots of importing a spreadsheet into SAS using SAS Studio.

You start by choosing **Tasks and Utilities** in the Navigation Pane.

Figure 6.15: Choose Tasks and Utilities in the Navigation Pane

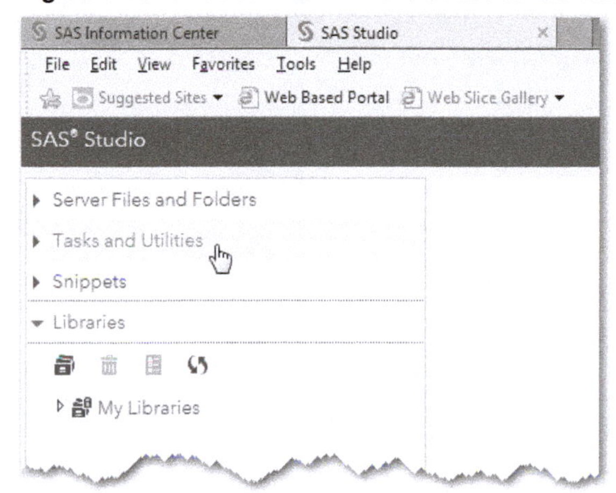

This gives you access to submenus, including **Import Data**.

Figure 6.16: Select Import Data

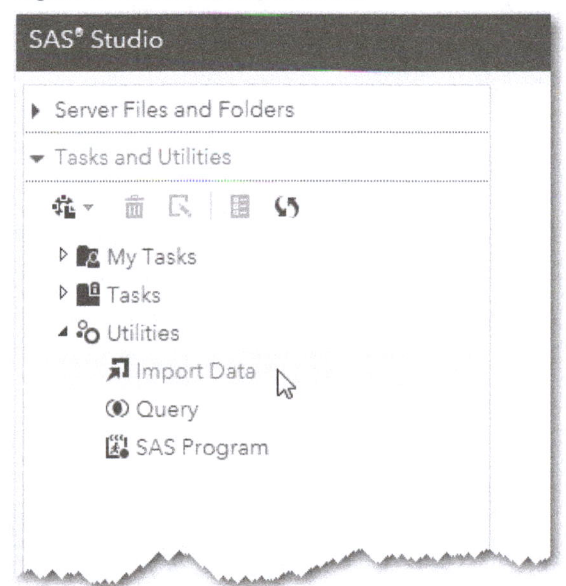

You can drag a file from a list or click **Select File** to choose which file to import.

Figure 6.17: Drag a File to the Drag-and-Drop Area or Click Select File

Select the file you want to import. Remember, this file must either be in the default shared folder (`SASUniversityEdition\Myfolders`) or another shared folder that you have created on your virtual machine.

Figure 6.18: Selecting a File

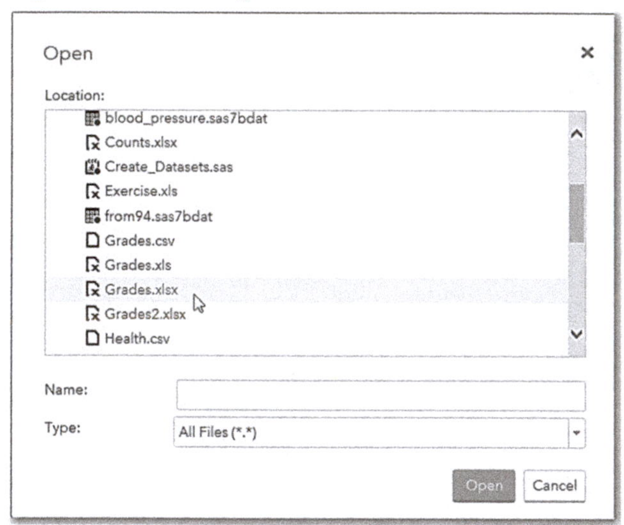

SAS Studio will automatically use the correct import method, based on the file extension (XLS or XLSX, for example).

Figure 6.19: Click the Run Icon

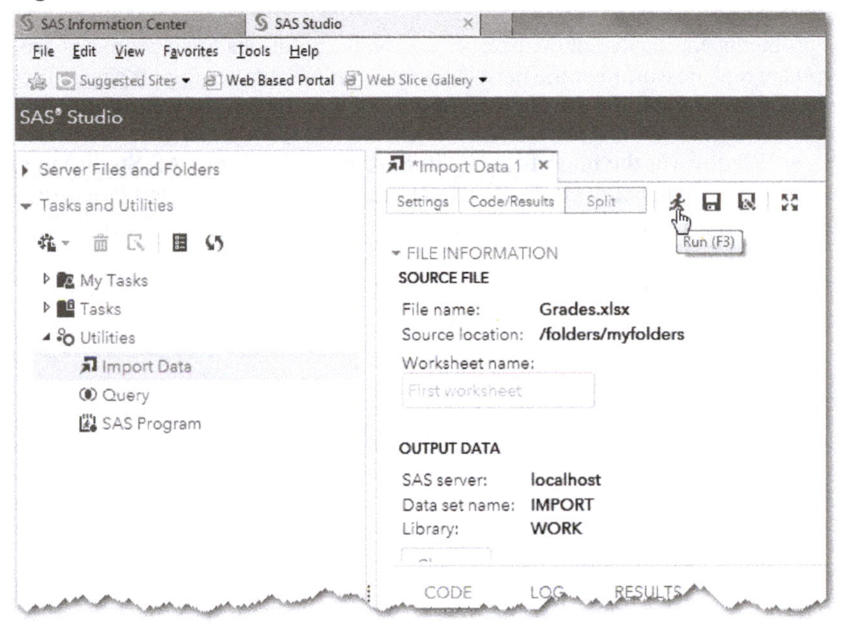

You will be given a choice of naming the SAS data set and the library. Once this is done, click the Run icon.

SAS Studio will automatically show you output from PROC CONTENTS so that you can see the variable names, variable types, and storage lengths of all the variables.

Figure 6.20: Portion of PROC CONTENTS

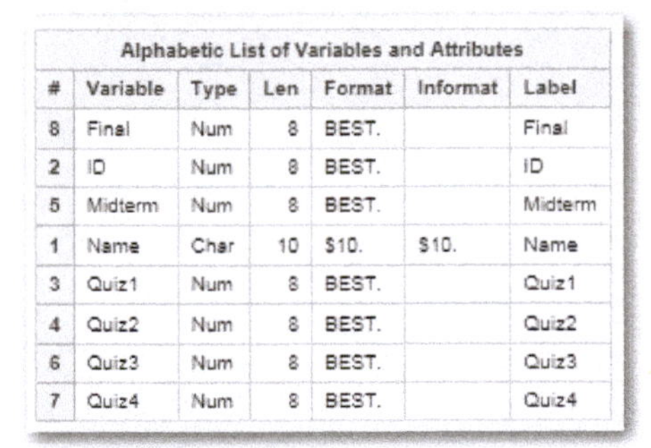

Alphabetic List of Variables and Attributes						
#	Variable	Type	Len	Format	Informat	Label
8	Final	Num	8	BEST.		Final
2	ID	Num	8	BEST.		ID
5	Midterm	Num	8	BEST.		Midterm
1	Name	Char	10	$10.	$10.	Name
3	Quiz1	Num	8	BEST.		Quiz1
4	Quiz2	Num	8	BEST.		Quiz2
6	Quiz3	Num	8	BEST.		Quiz3
7	Quiz4	Num	8	BEST.		Quiz4

Hopefully, this short section will convince you that SAS Studio (with or without SAS University Edition) is easy to use.

6.7 Problems

Solutions to odd-numbered problems are located at the back of this book. Solutions to all problems are available to professors. If you are a professor, visit the book's companion website at support.sas.com/cody for information about how to obtain the solutions to all problems.

1. Use the SAS Explorer Window (or the import facility in Enterprise Guide or SAS Studio) to read the spreadsheet **DrugTest.xls** and convert this to a temporary SAS data set called Drugtest. Use PROC PRINT to list the observations in this data set.

2. Run the following program to create a CSV file. Substitute a folder of your choice for the one specified in the program:

```
data Soccer;
   input Team : $20. Wins Losses;
datalines;
Readington 20 3
Raritan 10 10
Branchburg 3 18
Somerville 5 18
;
options nodate nonumber;
title;
ods listing close;
ods csv file='C:\books\learning\Soccer.csv';
proc print data=Soccer noobs;
run;

ods csv close;
ods listing;
```

Open Excel on your computer and open the CSV file (you will have to change the file type to .csv). It should look like this:

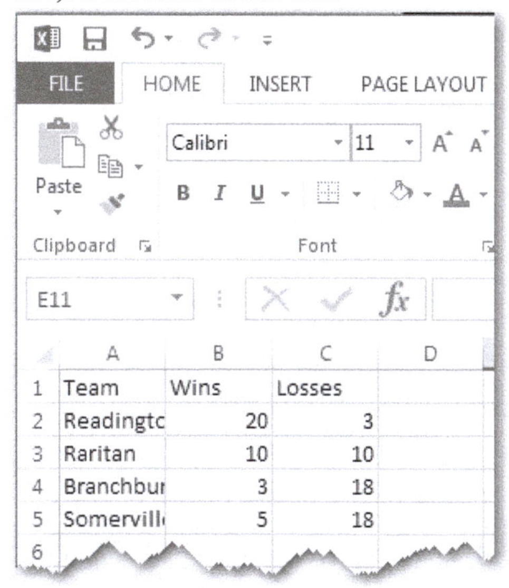

Save this as a spreadsheet using the **File▶Save As** pull down menu and naming the file `Soccer.xls`.

Now, use the SAS IMPORT wizard to convert this spreadsheet into a permanent SAS data set called Soccer in a folder of your choice.

3. Read the file `soccer.xls` created in Problem 2 using an XLS engine. The table name is SOCCER. (You will need to use a name constant 'SOCCER$'n to read this file.)

Chapter 7: Performing Conditional Processing

7.1 Introduction

This chapter describes the tools that allow programs to "make decisions" based on data values. For example, you may want to read a value of Age and create a variable that represents age groups. You may want to determine if values for a particular variable are within predefined limits. Programs that perform any of these operations require conditional processing—the ability to make logical decisions based on data values.

7.2 The IF and ELSE IF Statements

Two of the basic tools for conditional processing are the IF and ELSE IF statements. To understand how these statements work, suppose you have collected the following data on a group of students:

- Age (in years)
- Gender (recorded as M or F)
- Midterm (grade on the midterm exam)
- Quiz (quiz grade from F to A+)
- FinalExam (grade on the final exam)

Your first task is to create a new variable that represents age groups. Here is a first attempt.

Note: This program is not correct.

Program 7.1: First Attempt to Group Ages into Age Groups (Incorrect)

```
data Conditional;
   length Gender $ 1
          Quiz   $ 2;
   input Age Gender Midterm Quiz FinalExam;
   if Age lt 20 then AgeGroup = 1;
   if Age ge 20 and Age lt 40 then AgeGroup = 2;
   if Age ge 40 and Age lt 60 then AgeGroup = 3;
   if Age ge 60 then AgeGroup = 4;
datalines;
21 M 80 B- 82
.  F 90 A  93
35 M 87 B+ 85
48 F  .  .  76
59 F 95 A+ 97
15 M 88 .  93
67 F 97 A  91
.  M 62 F  67
35 F 77 C- 77
49 M 59 C  81
;
title "Listing of Conditional";
proc print data=Conditional noobs;
run;
```

A complete list of the logical comparison operators is displayed below.

Logical Comparison	Mnemonic	Symbol
Equal to	EQ	=
Not equal to	NE	^= or ~= or ¬= *
Less than	LT	<
Less than or equal to	LE	<=
Greater than	GT	>
Greater than or equal to	GE	>=
Equal to one in a list	IN	

* The symbol you use depends on what symbols are available on your keyboard. You can use the mnemonics on any system.

Let's follow the logic of Program 7.1. The first IF statement asks if Age is less than 20. When the logical expression following the keyword IF is true, the statement following the word THEN is executed; if the expression is not true, the program continues to process the next statements in the DATA step.

There is one serious problem with this program's logic and it relates to how SAS treats missing numeric values. Missing numeric values are treated logically as the most negative number you can reference on your computer. Therefore, the first IF statement will be *true for missing values* as well as for all ages less than 20.

Note: This is a very important point. It is a good example of a program that has no syntax errors, runs without any warning or error messages in the log, and produces incorrect results.

There are several ways to prevent your missing values from being included in AgeGroup 1. Here are several options:

```
if Age lt 20 and Age ne . then AgeGroup = 1;

if Age ge 0 and Age lt 20 then AgeGroup = 1;

if 0 le Age lt 20 then AgeGroup = 1;

if Age lt 20 and not missing(Age) then AgeGroup = 1;
```

All of these statements result in the correct value for AgeGroup. Subjects with a missing value for Age will also have a missing value for AgeGroup.

The first IF statement uses the fact that you refer to a numeric missing value in a DATA step by a period. The last IF statement uses the MISSING function. This function returns a value of TRUE if the argument (the variable in parentheses) is missing, and FALSE if the argument is not missing. Note: the MISSING function also works with character data.

This program can be improved further. If a person is less than 20 years of age, the first IF statement is true and AgeGroup is set to `1`. All the remaining IF statements are still executed (although they will all be false and AgeGroup remains `1`). A better way to write this program is to change all the IF statements after the first one to ELSE IF statements.

Here is what the corrected program looks like:

Program 7.2: Corrected Program to Group Ages into Age Groups

```
data Conditional;
   length Gender $ 1
          Quiz   $ 2;
   input Age Gender Midterm Quiz FinalExam;
   if Age lt 20 and not missing(age) then AgeGroup = 1;
   else if Age ge 20 and Age lt 40 then AgeGroup = 2;
   else if Age ge 40 and Age lt 60 then AgeGroup = 3;
   else if Age ge 60 then AgeGroup = 4;
datalines;
```

Using this logic, when any of the IF statements is true, all the following ELSE statements are not evaluated. This saves on processing time.

An alternative way to write this program is to test for a missing value in the first IF statement and use the ELSE statements to advantage, as follows:

Program 7.3: An Alternative to Program 7.2

```
data Conditional;
   length Gender $ 1
          Quiz   $ 2;
   input Age Gender Midterm Quiz FinalExam;
   if missing(Age) then AgeGroup = .;
      else if Age lt 20 then AgeGroup = 1;
      else if Age lt 40 then AgeGroup = 2;
      else if Age lt 60 then AgeGroup = 3;
      else if Age ge 60 then AgeGroup = 4;
datalines;
```

When you write a program like this, you need to "play computer" and make sure your logic is correct. For example, what if a person is 25 years old? The first IF statement is false because Age is not missing. The next ELSE IF statement is evaluated and found to be false as well. Finally, the third IF statement is evaluated and AgeGroup is set equal to **2**. Because this IF statement is true, all the remaining ELSE IF statements are skipped.

If you are working with very large data sets and want to squeeze every last drop out of the efficiency tank, you should place the IF statements in order, from the ones most likely to have a true condition to the ones least likely to have a true condition. This increases efficiency because SAS skips testing all the ELSE conditions when a previous IF condition is true. The disadvantage of using this technique is that it makes the program harder to read and more likely to have an error. With the speed of modern computers, you should consider these extreme efforts at efficiency only for production programs that operate on very large data sets.

7.3 The Subsetting IF Statement

If you want to create a subset of data from a raw data file or from an existing SAS data set, you can use a special form of an IF statement called a *subsetting IF*. As an example, let's use the raw data from Program 7.1, but restrict the resulting data set to females only. Here is the program:

Program 7.4: Demonstrating a Subsetting IF statement

```
data Females;
   length Gender $ 1
          Quiz   $ 2;
   input Age Gender Midterm Quiz FinalExam;
   if Gender eq 'F';
datalines;
21 M 80 B- 82
.  F 90 A  93
35 M 87 B+ 85
48 F  .  .  76
59 F 95 A+ 97
15 M 88 .  93
67 F 97 A  91
.  M 62 F  67
35 F 77 C- 77
49 M 59 C  81
;
title "Listing of Females";
proc print data=Females noobs;
run;
```

Notice that there is no THEN following the IF in this program. If the condition is true, the program continues to the next statement; if the condition is false, control returns to the top of the DATA step. In this case, the only statement following the subsetting IF statement is a DATALINES statement. If the value of Gender is **F**, the end of the DATA step is reached and an automatic output occurs. If the value of Gender is not equal to **F**, control returns to the top of the DATA step and the automatic output does not occur.

Here is the output:

Figure 7.1: Output from Program 7.4

Listing of Females

Gender	Quiz	Age	Midterm	FinalExam
F	A	.	90	93
F		48	.	76
F	A+	59	95	97
F	A	67	97	91
F	C-	35	77	77

A more efficient way to write Program 7.4 would be to read a value of Gender first. Then if it is an **F**, continue reading the rest of the data values; if not, return to the top of the DATA step and do not output an observation. To see how this can be accomplished, look at Section 11 in Chapter 21.

7.4 The IN Operator

If you want to test if a value is one of several possible choices, you can use multiple OR statements, like this:

```
if Quiz = 'A+' or Quiz = 'A' or Quiz = 'B+' or Quiz = 'B'
   then QuizRange = 1;
else if Quiz = 'B-' or Quiz = 'C+' or Quiz = 'C'
   then QuizRange = 2;
else if not missing(Quiz) then QuizRange = 3;
```

These statements can be simplified by using the IN operator, like this:

```
if Quiz in ('A+','A','B+','B') then QuizRange = 1;
else if Quiz in ('B-','C+','C') then QuizRange = 2;
else if not missing(Quiz) then QuizRange = 3;
```

The list of values in parentheses following the IN operator can be separated by commas or blanks. The first line could also be written like this:

```
if Quiz in ('A+' 'A' 'B+' 'B') then QuizRange = 1;
```

You can also use the IN operator with numeric variables. For example, if you had a numeric variable called Subject and you wanted to list observations for Subject numbers **10**, **22**, **25**, and **33**, the following WHERE statement could be used:

```
where Subject in (10,22,25,33);
```

Note: WHERE statements are similar to IF statements but they can be used only when reading observations from a SAS data set, not raw data – see section 7.8 in this chapter.

As with the example using character values, you may separate the values with commas or spaces. You can also specify a range of numeric values, using a colon to separate values in the list. For example, to list observations where Subject is a `10`, `22-25`, or `30`, you can write:

```
where Subject in (10,22:25,30);
```

Remember that you should use a colon and not a dash for this feature to work—a dash would be interpreted as a minus sign in a list of numbers separated by spaces.

> **Note:** The colon notation works only for integer values.

7.5 Using a SELECT Statement for Logical Tests

A SELECT statement provides an alternative to a series of IF and ELSE IF statements.

Here is one way to use a SELECT statement:

```
select (AgeGroup);
   when (1) Limit = 110;
   when (2) Limit = 120;
   when (3) Limit = 130;
   otherwise;
end;
```

The expression following the SELECT statement is referred to as a *select-expression*; the expression following a WHEN statement is referred to as a *when-expression*. In this example, the *select-expression* (AgeGroup) is compared to each of the *when-expressions*. If the comparison is true, the statement following the *when-expression* is executed and control skips to the end of the SELECT group. If the comparison is false, the next *when-expression* is compared to the *select-expression*. If none of the comparisons is evaluated to be true, the expression following the OTHERWISE statement is executed. As you can see in this example, the *otherwise-expression* can be a null statement. It is necessary to include an OTHERWISE statement because the program will terminate if you omit it and none of the preceding comparisons is true.

You can place more than one value in the when-*expression*, like this:

```
select (AgeGroup);
   when (1) Limit = 110;
   when (2) Limit = 120;
   when (3,5) Limit = 130;
   otherwise;
end;
```

In this example, AgeGroup values of `3` or `5` will set Limit equal to `130`.

To help clarify this concept, let's follow some scenarios:

If AgeGroup is equal to `1`, Limit will be `110`. If Agegroup is equal to `3`, Limit will be equal to `130`. If AgeGroup is equal to `4`, Limit will be a missing value (because it is set to a missing value in the PDV at each iteration of the DATA step and it is never assigned a value).

If you do not supply a *select-expression*, each WHEN statement is evaluated to determine if the *when-*expression is true or false. As an example, here is Program 7.5, rewritten using a SELECT statement:

Program 7.5: Demonstrating a SELECT Statement When a Select-Expression is Missing

```
data Conditional;
   length Gender $ 1
          Quiz   $ 2;
   input Age Gender Midterm Quiz FinalExam;
   select;
      when (missing(Age)) AgeGroup = .;
      when (Age lt 20) AgeGroup = 1;
      when (Age lt 40) AgeGroup = 2;
      when (Age lt 60) AgeGroup = 3;
      when (Age ge 60) Agegroup = 4;
      otherwise;
   end;
datalines;
```

Notice that there is no *select-expression* in this SELECT statement. Each *when-expression* is evaluated and, if true, the statement following the expression is executed.

7.6 Using Boolean Logic (AND, OR, and NOT Operators)

You can combine various logical operators (also known as *Boolean operators*) to form fairly complex statements. As an example, the data set Medical contains information on clinic, diagnosis (DX), and weight. A program to list all patients who were seen at the HMC clinic and had either a diagnosis 7 or 9 or weighted over 180 pounds demonstrates how to combine various Boolean operators:

Program 7.6: Combining Various Boolean Operators

```
title "Example of Boolan Expressions";
proc print data=Learn.Medical;
   where Clinic eq 'HMC' and
         (DX in ('7','9') or
         Weight gt 180);
   id Patno;
   var Patno Clinic DX Weight VisitDate;
run;
```

Notice the parentheses around the two statements separated by OR. The AND operator has precedence over (i.e., is performed before) the OR operator. That is, a statement such as the following:

```
if X and Y or Z;
```

is the same as this one:

```
if (X and Y) or Z;
```

If you want to perform the OR operation before the AND operation, use parentheses, like this:

```
if X and (Y or Z);
```

In Program 7.6, you want the clinic to be HMC and you want either the diagnosis to be one of two values or the weight to be over 180 pounds. You need the parentheses to first decide if either the diagnosis or weight condition is true before performing the AND operation with the Clinic variable. Even in cases where parentheses are not needed, it is fine to include them so that the logical statements are easier to read and understand.

The NOT operator has the highest precedence, so it is performed before AND. For example, the statement:

```
if X and not y or z;
```

is equivalent to:

```
if (X and (not y)) or z;
```

Here is the output:

Figure 7.2: Output from Program 7.6

Example of Boolan Expressions

Patno	Patno	Clinic	DX	Weight	VisitDate
004	004	HMC	9	288	11/11/2006
050	050	HMC	123	199	07/06/2006

7.7 A Caution When Using Multiple OR Operators

Look at the short program here:

Program 7.7: A Caution on the Use of Multiple OR Operators

```
data Believe_it_or_Not;
   input X;
   if X = 3 or 4 then Match = 'Yes';
   else Match = 'No';
datalines;
3
7
.
;
title "Listing of Believe_it_or_Not";
proc print data=Believe_it_or_Not noobs;
run;
```

The programmer probably wanted to say:

```
if X = 3 or X = 4 then Match = 'Yes';
```

What happens when you run this program? Many folks would expect to see an error message in the log and would expect the program to terminate. Not so. Here is the output from this program:

Figure 7.3: Output from Program 7.7

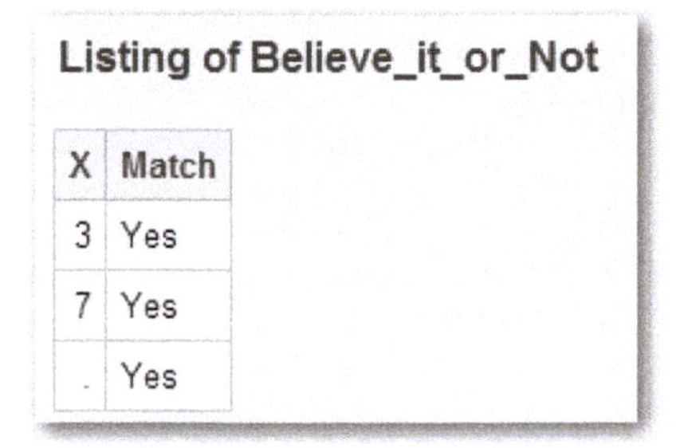

Listing of Believe_it_or_Not

X	Match
3	Yes
7	Yes
.	Yes

In Program 7.7, there is one condition on either side of the OR operator—one is `X = 3`, the other is `4`. In SAS, any value other than `0` or missing is true. Therefore, `4` is evaluated as true and the statement `X = 4` OR `4` is always true.

Note: This bears repeating: Any numeric value in SAS that is not zero or missing is considered TRUE.

7.8 The WHERE Statement

If you are reading data from a SAS data set, you can use a WHERE statement to subset your data. For example, if you started with the SAS data set Conditional (Program 7.3), you could create a data set of all females with the following program:

Program 7.8: Using a WHERE Statement to Subset a SAS Data Set

```
data Females;
   set Conditional;
   where Gender eq 'F';
run;
```

In this example you could use either a WHERE or a subsetting IF statement. There are sometimes advantages to using a WHERE statement instead of a subsetting IF statement. You have a larger choice of operators that can be used with a WHERE statement (to be discussed next) and, if the input data set is indexed, the WHERE statement might be more efficient.

You may also use a WHERE statement in a SAS procedure to subset the data being processed. You cannot use a subsetting IF statement with a SAS procedure.

Note: IF statements are not allowed inside SAS procedures.

7.9 Some Useful WHERE Operators

The table here lists some of the useful operators that you can use with a WHERE statement.

Operator	Description	Example
IS MISSING	Matches a missing value	where Subj is missing
IS NULL	Equivalent to IS MISSING	where Subj is null
BETWEEN AND	An inclusive range	where age between 20 and 40
CONTAINS	Matches a substring	where Name contains Mac
LIKE	Matching with wildcards	where Name like R_n%
=*	Phonetic matching	where Name =* Nick

Note: When using the LIKE operator, the underscore character (_) takes the place of a single character, while the percent sign (%) can be substituted for a string of any length (including a null string).

Here are some examples.

Expression	Matches
where Gender is null	A missing character value
where Age is null	A missing numeric value
where Age is missing	A missing numeric value
where Age between 20 and 40	All values between 20 and 40, including 20 and 40
where Name contains 'mac'	macon immaculate
where Name like 'R_n%'	Ron Ronald Run Running
where Name =* 'Nick'	Nick Nack Nikki

Notes:

- The IS NULL or IS MISSING expression matches a character or a numeric missing value.
- The BETWEEN AND expression matches all the values greater than or equal to the first value and less than or equal to the second value. This works with character as well as numeric variables.
- The CONTAINS expression matches any character value containing the given string.
- The LIKE expression uses two wildcard operators. The underscore (_) is a place holder; enter as many underscores as you need to stand for the same number of characters. The percent (%) matches nothing or a string of any length.

7.10 Problems

Solutions to odd-numbered problems are located at the back of this book. Solutions to all problems are available to professors. If you are a professor, visit the book's companion website at support.sas.com/cody for information about how to obtain the solutions to all problems.

1. Run the program here to create a temporary SAS data set called School:

```
data School;
   input Age Quiz : $1. Midterm Final;
   /* Add you statements here */
datalines;
12 A 92 95
12 B 88 88
13 C 78 75
13 A 92 93
12 F 55 62
13 B 88 82
;
```

   Using IF and ELSE IF statements, create two new variables as follows: Grade (numeric), with a value of `6` if Age is `12` and a value of `8` if Age is `13`.

   The quiz grades have numerical equivalents as follows: `A = 95`, `B = 85`, `C = 75`, `D = 70`, and `F = 65`. Using this information, compute a course grade (Course) as a weighted average of the Quiz (20%), Midterm (30%) and Final (50%).

2. Using the SAS data set Hosp, use PROC PRINT to list observations for Subject values of `5`, `100`, `150`, and `200`. Do this twice, once using OR operators and once using the IN operator.
   Note: Subject is a numeric variable.
3. Using the Sales data set, list the observations for employee numbers (EmpID) `9888` and `0177`. Do this two ways, one using OR operators and the other using the IN operator. **Note**: EmpID is a character variable.
4. Using the Sales data set, create a new, temporary SAS data set containing Region and TotalSales plus a new variable called Weight with values of `1.5` for the North Region, `1.7` for the South Region, and `2.0` for the West and East Regions. Use a SELECT statement to do this.
5. Starting with the Blood data set, create a new, temporary SAS data set containing all the variables in Blood plus a new variable called CholGroup. Define this new variable as follows:

CholGroup	Chol
Low	Low – 110
Medium	111 – 140
High	141 – High

   Use a SELECT statement to do this.
6. Using the Sales data set, list all the observations where Region is `North` and Quantity is less than `60`. Include in this list any observations where the customer name (Customer) is `Pet's are Us`.
7. Using the Bicycles data set, list all the observations for Road Bikes that cost more than $2,500 or Hybrids that cost more than $660. The variable Model contains the type of bike and UnitCost contains the cost.

Chapter 8: Performing Iterative Processing: Looping

8.1 Introduction

Many programming tasks require that blocks of code be run more than once. SAS provides several ways to accomplish this. This chapter covers DO groups, DO loops, DO WHILE statements, and DO UNTIL statements.

8.2 DO Groups

To demonstrate a DO group, we start with a data file containing some information on students: their age, gender, midterm grade, quiz grade, and final exam grade. A listing of the data file **C:\books\learning\Grades.txt** follows:

```
21 M 80 B- 82
.  F 90 A  93
35 M 87 B+ 85
48 F  .  .  76
59 F 95 A+ 97
15 M 88 .  93
67 F 97 A  91
.  M 62 F  67
35 F 77 C- 77
49 M 59 C  81
```

You want to read values from this file and compute two new variables—age group (AgeGroup) and a value (Grade) computed from the midterm and final exam grades. If the age is less than or equal to 39, you want to set AgeGroup equal to **`Younger group`** and the grade to be computed as a weighted average of the midterm grade (40%) and the final exam grade (60%). If the age is greater than 39, you want to set AgeGroup equal to **`Older group`** and compute the grade as a simple average of the midterm and final exam grades.

Take a look at Program 8.1 that solves this problem:

Program 8.1: Example of a Program That Does Not Use a DO Group

```
data Grades;
   length Gender $ 1
          Quiz   $ 2
          AgeGroup $ 13;
   infile 'C:\books\learning\Grades.txt' missover;
   input Age Gender Midterm Quiz FinalExam;
   if missing(Age) then delete;
   if Age le 39 then AgeGroup = 'Younger group';
   if Age le 39 then Grade    = .4*Midterm + .6*FinalExam;
   if Age gt 39 then AgeGroup = 'Older group';
   if Age gt 39 then Grade    = (Midterm + FinalExam)/2;
run;

title "Listing of Grades";
proc print data=Grades noobs;
run;
```

Notice that the first two IF statements and the last two IF statements test the same condition. You would like to be able to test a condition and then perform several operations. By using DO and END statements, you can do this. Program 8.2 works identically to Program 8.1 except it uses DO and END statements to make the code more efficient and easier to read:

Program 8.2: Demonstrating a DO Group

```
data Grades;
   length Gender $ 1
          Quiz   $ 2
          AgeGroup $ 13;
   infile 'C:\books\learning\Grades.txt' missover;
   input Age Gender Midterm Quiz FinalExam;
   if missing(Age) then delete;
   if Age le 39 then do;
      AgeGroup = 'Younger group';
      Grade = .4*Midterm + .6*FinalExam;
   end;
   else if Age gt 39 then do;
      AgeGroup = 'Older group';
      Grade = (Midterm + FinalExam)/2;
   end;
run;

title "Listing of Grades";
proc print data=Grades noobs;
run;
```

All the statements between DO and END form a DO group. When the IF condition is true, all the statements in the DO group execute. A good way to think of this structure is "If the condition is true, do the following statements until you reach the end." It is standard practice to indent all the statements in the DO group as shown here.

The DO group coding is not only more efficient than multiple IF statements, it is also easier to read. Here is a listing of data set Grades:

Figure 8.1: Output from Program 8.2

Listing of Grades

Gender	Quiz	AgeGrp	Age	Midterm	FinalExam	Grade
M	B-	Younger group	21	80	82	81.2
M	B+	Younger group	35	87	85	85.8
F		Older group	48	.	76	.
F	A+	Older group	59	95	97	96.0
M		Younger group	15	88	93	91.0
F	A	Older group	67	97	91	94.0
F	C-	Younger group	35	77	77	77.0
M	C	Older group	49	59	81	70.0

You should take note that both of these programs work properly if there is a missing value for Age. The MISSING function returns a true value if its argument is a missing character or numeric value. The DELETE statement does two things: first, it prevents the current observation from being added to the data set, and second, it forces a return to the top of the DATA step.

8.3 The Sum Statement

The programs in the next few sections all make use of the SUM statement. This seemingly simple statement is extremely useful—you may find you use it in a majority of your SAS programs—yet many programmers who use it don't fully appreciate how it works.

There are two primary uses for a SUM statement: one is to accumulate totals such as a month-to-date total, and the other is to create a counter—a variable that is incremented by a fixed amount on each iteration of the DATA step.

Suppose you have a data set with one observation for each day of the week, and you want a program that will read in these values and compute a cumulative sum. The following program creates a test data set (Revenue) and attempts to create the cumulative sum.

Program 8.3: Attempt to Create a Cumulative Total (First Attempt)

```
data Revenue;
   input Day : $3.
         Revenue : dollar6.;
   Total = Total + Revenue; /* Note: this does not work */
   format Revenue Total dollar8.;
datalines;
Mon $1,000
Tue $1,500
Wed  .
Thu $2,000
Fri $3,000
;

title "Listing of Revenue";
proc print data=Revenue noobs;
run;
```

Here is the output:

Figure 8.2: Output from Program 8.3

Listing of Data Set Revenue

Day	Revenue	Total
Mon	$1,000	.
Tue	$1,500	.
Wed	.	.
Thu	$2,000	.
Fri	$3,000	.

Remember that variables read from raw data or created by assignment statements are initialized to a missing value for each iteration of the DATA step. On the first iteration of the DATA step, you are adding a missing value (Total) to a Revenue value ($1,000) and the result is a missing value. Using this same logic, you see that Total is missing for every observation.

You can use a RETAIN statement to tell SAS not to do this. A RETAIN statement also enables you to set an initial value for a variable. Here is attempt number two (a bit better but still not there):

Program 8.4: Creating a Cumulative Total with the RETAIN Statement (Second Attempt)

```
data Revenue;
   retain Total 0;
   input Day : $3.
         Revenue : dollar6.;
   Total = Total + Revenue; /* Note: this does not work */
   format Revenue Total dollar8.;
datalines;
Mon $1,000
Tue $1,500
Wed  .
Thu $2,000
Fri $3,000
;

title "Listing of Revenue";
proc print data=Revenue noobs;
run;
```

In this program, Total is retained and initialized at 0. Here is the output:

Figure 8.3: Output from Program 8.4

Listing of Data Set Revenue

Total	Day	Revenue
$1,000	Mon	$1,000
$2,500	Tue	$1,500
.	Wed	.
.	Thu	$2,000
.	Fri	$3,000

Everything works fine until the program encounters a missing value for Revenue. This sets Total to a missing value, where it remains for the rest of the DATA step. You can fix this by adding a statement to test the value of Revenue before you attempt to add it to the cumulative Total, like this:

Program 8.5: Creating a Cumulative Total with RETAIN and IF Statements (Third Attempt)

```
data Revenue;
   retain Total 0;
   input Day : $3.
         Revenue : dollar6.;
   if not missing(Revenue) then Total = Total + Revenue;
   format Revenue Total dollar8.;
datalines;
Mon $1,000
Tue $1,500
Wed  .
Thu $2,000
Fri $3,000
;

title "Listing of Revenue";
proc print data=Revenue noobs;
run;
```

Here is the output:

Figure 8.4: Output from Program 8.5

Listing of Data Set Revenue

Total	Day	Revenue
$1,000	Mon	$1,000
$2,500	Tue	$1,500
$2,500	Wed	.
$4,500	Thu	$2,000
$7,500	Fri	$3,000

This time the program worked correctly. But, there is an easier way: use a SUM statement. A SUM statement takes the following form:

```
variable + expression;
```

Notice there is no equal sign in this statement. That's what identifies this as a SUM statement to SAS. This statement does the following:

- *Variable* is retained
- *Variable* is initialized at 0
- Missing values (of *increment*) are ignored

So, rewriting Program 8.5, you have the following

Program 8.6: Using a SUM Statement to Create a Cumulative Total

```
data Revenue;
   input Day : $3.
         Revenue : dollar6.;
   Total + Revenue;
   format Revenue Total dollar8.;
datalines;
Mon $1,000
Tue $1,500
Wed  .
Thu $2,000
Fri $3,000
;
```

The output from this program is identical to the output from Program 8.5.

Another very common use of a SUM statement is to create counters, for example:

Program 8.7: Using a SUM Statement to Create a Counter

```
data Test;
   input x;
   if missing(x) then MissCounter + 1;
datalines;
2
.
7
.
;
```

Here is the output:

Figure 8.5: Listing of Data Set Test

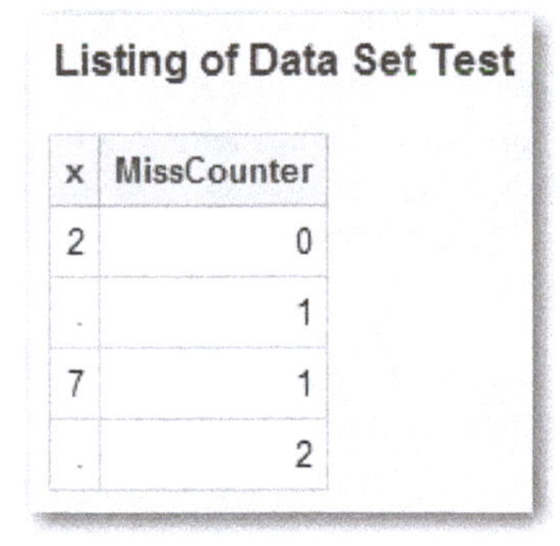

Listing of Data Set Test

x	MissCounter
2	0
.	1
7	1
.	2

MissCounter is counting the number of missing values for *x*.

8.4 The Iterative DO Loop

Although it is useful to have the capability of executing a group of code when a condition is true, there are times when you would like to execute a group of SAS statements multiple times. The program here is written without any iterative loops.

Program 8.8: Program Without Iterative Loops

```
data Compound;
   Interest = .0375;
   Total = 100;

   Year + 1;
   Total + Interest*Total;
   output;

   Year + 1;
   Total + Interest*Total;
   output;

   Year + 1;
   Total + Interest*Total;
   output;

   format Total dollar10.2;
run;

title "Listing of Compound";
proc print data=compound noobs;
run;
```

The purpose of this program should be obvious: you want to compute the total amount of money you will have if you start with $100 and invest it at a 3.75% interest rate for 3 years. (Yes, there are formulas for compound interest as well as SAS functions, but this makes for a good example.)

As we just discussed, the statement **`Year + 1`**; is a SUM statement. It increments the value of Year by 1 each time it executes. The value of Total is also computed using a SUM statement. Notice here that the increment can be an expression (**`Interest * Total`**). Remember that SAS will perform the multiplication before the addition because multiplication has a higher precedence than addition. Feel free to include parentheses if it makes the program easier for you to understand.

The OUTPUT statement is an instruction for SAS to write out an observation to the output data set. An output usually occurs automatically at the bottom of the DATA step. But here, you want to output an observation each time you compute a new Total.

> **Note:** When you include an OUTPUT statement anywhere in a DATA step, SAS does not execute an automatic output at the bottom of the DATA step.

Next, notice that the group of three statements, starting with this SUM statement, is repeated three times. This is a clue that there is probably a better way to write this program. That better way is to use an iterative DO loop, like this:

Program 8.9: Demonstrating an Iterative DO Loop

```
data Compound;
   Interest = .0375;
   Total = 100;
   do Year = 1 to 3;
      Total + Interest*Total;
      output;
   end;
   format Total dollar10.2;
run;

title "Listing of Data Set Compound";
proc print data=Compound noobs;
run;
```

When this program executes, Year is first set to `1`, the lower limit in the iterative DO range. All the statements up to the END statement are executed and Year is automatically incremented by 1 (the default increment value). SAS then tests if the new value of Year is between the lower and the upper limit (the value after the keyword TO). If it is, the statements in the DO group execute again; if not, the program continues at the first line following the END statement.

This is a good time to mention that the FORMAT statement in this program is executed at compile time and you could have placed this statement anywhere between the DATA statement and the RUN statement that ends the DATA step.

Here is the output:

Figure 8.6: Output from Program 8.9

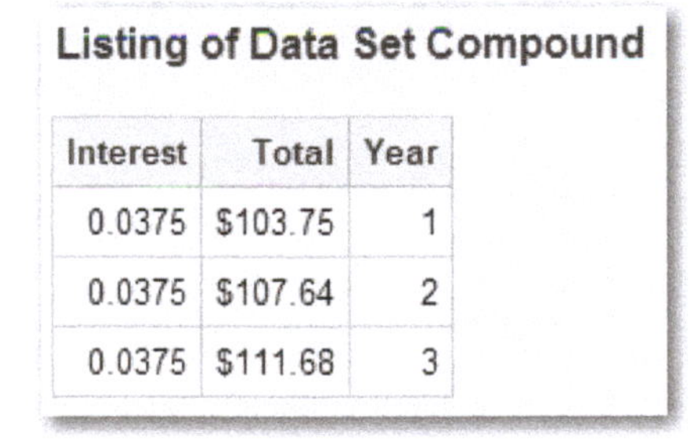

Listing of Data Set Compound

Interest	Total	Year
0.0375	$103.75	1
0.0375	$107.64	2
0.0375	$111.68	3

Each of the three executions of the OUTPUT statement writes out an observation to data set Compound and the value of Interest times Total is added to the Total for each Year.

One form of an iterative DO statement follows:

```
do index-variable = start to stop by increment;
```

If you leave off the *increment*, it defaults to `1`.

Suppose you want to generate a table of the integers from 1 to 10, along with their squares and square roots. An iterative DO loop makes simple work of this, as follows:

Program 8.10: Using an Iterative DO Loop to Make a Table of Squares and Square Roots

```
data Table;
   do n = 1 to 10;
      Square = n*n;
      SquareRoot = sqrt(n);
      output;
   end;
run;

title "Table of Squares and Square Roots";
proc print data=table noobs;
run;
```

Notice that this program does not have any input data. It generates the value of *n* in the DO loop, computes the squares and square roots (SQRT is a square root function—it returns the square root of its argument), and outputs an observation to the data set. This continues for all the values from `1` to `10`. Here is the output:

Figure 8.7: Output from Program 8.10

Table of Squares and Square Roots

n	Square	SquareRoot
1	1	1.00000
2	4	1.41421
3	9	1.73205
4	16	2.00000
5	25	2.23607
6	36	2.44949
7	49	2.64575
8	64	2.82843
9	81	3.00000
10	100	3.16228

What if you want a table where *n* has values of `0`, `10`, `20`, `30`, and so forth up to `100`. The following DO statement does the trick:

```
do n = 0 to 100 by 10;
```

DO loops can also count backward. For example, the following statement produces values of `10`, `8`, `6`, `4`, and `2` for the index variable:

```
do index = 10 to 1 by -2;
```

When you use a negative increment value, the index value is decremented by this value for each iteration of the DO loop. When the value of the index variable is less than the stop value, the loop stops.

Here is another example. You have an equation relating X and Y and want a graph of Y versus X for values of X from `-10` to `+10`. Again, an iterative DO loop makes this very easy:

Program 8.11: Using an Iterative DO Loop to Graph an Equation

```
data Equation;
   do X = -10 to 10 by .01;
      Y = 2*X**3 - 5*X**2 + 15*X -8;
      output;
   end;
run;

title "Plot of Y = 2*X**3 - 5*X**2 + 15*X -8";
proc sgplot;
   series x=X y=Y;
run;
```

The DO loop starts X at `-10` and increments the value by .01 until X reaches `+10`. The OUTPUT statement inside the loop writes an observation out to the data set for each iteration of the loop.

PROC SGPLOT is used to plot the line. You can obtain information on PROC SGPLOT in Chapter 20. Here is the graph:

Figure 8.8: Output from Program 8.11

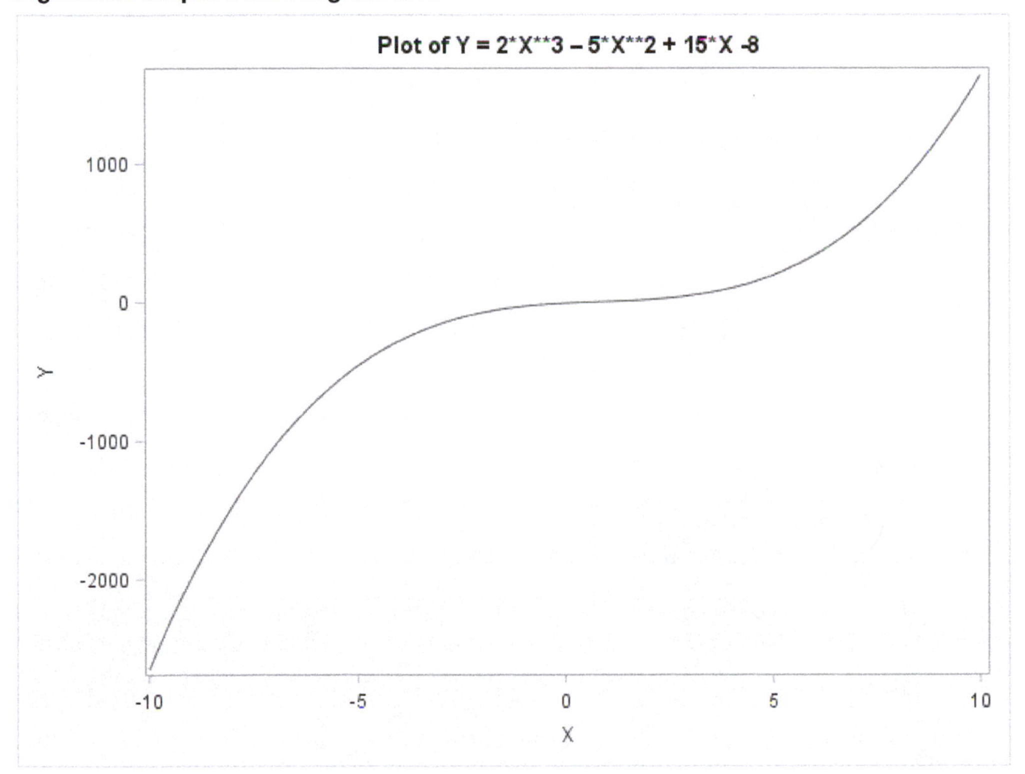

8.5 Other Forms of an Iterative DO Loop

SAS provides several other methods of specifying how a DO loop operates. You can provide a list of numeric or character values following the index variable. Here are some examples:

```
do x = 1,2,5,10;
(values of x are: 1, 2, 5, and 10)

do Month = 'Jan','Feb','Mar';
(values of Month are: 'Jan', 'Feb', and 'Mar')


do n = 1,3, 5 to 9 by 2, 100 to 200 by 50;
(values of n are: 1, 3, 5, 7, 9, 100, 150, and 200)
```

If you use character values in the DO statement, the length of the first character value determines the storage length of the index variable. Therefore, you may need to use a LENGTH statement to set the storage length for the index variable.

Using character values for DO loop indices can be especially useful, as illustrated by this example.

You have five scores for patients on a placebo drug and five scores for patients on an active drug. The raw data values are arranged like this:

```
250 222 230 210 199
166 183 123 129 234
```

An easy way to read these values is to use a DO loop with character values, like this:

Program 8.12: Using Character Values for DO Loop Index Values

```
data Easyway;
   do Group = 'Placebo','Active';
      do Subj = 1 to 5;
         input Score @;
         output;
      end;
   end;
datalines;
250 222 230 210 199
166 183 123 129 234
;

title "Listing of Data Set Easyway";
proc print data=Easyway noobs;
run;
```

Before we discuss the DO loops in this program, let's take a moment to tell you about the at (@) sign at the end of the INPUT statement. (This is referred to as a single trailing @ sign and you can find a detailed explanation in Chapter 21, Section 11.) To keep this program compact, it was convenient to place several scores on a single line. Without the @ sign, each time SAS executes an INPUT statement, it goes to a new line of data. The single trailing @ sign is an instruction to "hold the line" for another INPUT statement in the DATA step. Try running this program without the trailing @ sign to see what happens.

This program demonstrates two things: first, you can use character values in a DO loop, and second, you can nest one DO loop inside another. Let's "play computer" and follow the execution of this program.

The outer loop first sets Group equal to `Placebo`. Next, the inner loop iterates five times, reading score values and outputting an observation each time (remember that Group is equal to `Placebo` for each of these five observations). When the inner loop is finished, control returns to the top of the outer loop where Group is set to `Active`. Five more values of Score are read and five observations are written to the SAS data set. Here is a listing of data set Easyway:

Figure 8.9: Output from Program 8.12

Listing of Easyway

Group	Subj	Score
Placebo	1	250
Placebo	2	222
Placebo	3	230
Placebo	4	210
Placebo	5	199
Active	1	166
Active	2	183
Active	3	123
Active	4	129
Active	5	234

8.6 DO WHILE and DO UNTIL Statements

Instead of choosing a stopping value for an iterative DO loop, you can stop a loop when a condition is met or while a condition is true. Enter the DO UNTIL and DO WHILE statements.

Let's revisit the compound interest problem from earlier in this chapter. Instead of asking how much money you have after *x* years, you want to know how many years you need to keep your $100 in the bank at 3.75% interest to double your money. The two programs here solve this problem:

Program 8.13: Demonstrating a DO UNTIL Loop

```
data Double;
   Interest = .0375;
   Total = 100;
   do until (Total ge 200);
      Year + 1;
      Total = Total + Interest*Total;
      output;
   end;
   format Total dollar10.2;
run;

title "Listing of Double";
proc print data=Double noobs;
run;
```

The condition is placed in parentheses following the keyword UNTIL. In this example, the loop continues to repeat until the value of Total is greater than or equal to **200**. Here is the output:

Figure 8.10: Output from Program 8.13

Listing of Double

Interest	Total	Year
0.0375	$103.75	1
0.0375	$107.64	2
0.0375	$111.68	3
0.0375	$115.87	4
0.0375	$120.21	5
0.0375	$124.72	6
0.0375	$129.39	7
0.0375	$134.25	8
0.0375	$139.28	9
0.0375	$144.50	10
0.0375	$149.92	11
0.0375	$155.55	12
0.0375	$161.38	13
0.0375	$167.43	14
0.0375	$173.71	15
0.0375	$180.22	16
0.0375	$186.98	17
0.0375	$193.99	18
0.0375	$201.27	19

Note: An important point to remember about DO UNTIL is that the condition, placed in parentheses after the keyword UNTIL, is tested at the **bottom** of the loop. Therefore, a **DO UNTIL loop always executes at least once.**

To make this clear, suppose you started with $300. What happens when you run the program?

Program 8.14: Demonstrating That a DO UNTIL Loop Always Executes at Least Once

```
data Double;
   Interest = .0375;
   Total = 300;
   do until (Total gt 200);
      Year + 1;
      Total = Total + Interest*Total;
      output;
   end;
   format Total dollar10.2;
run;

proc print data=double noobs;
   title "Listing of Double";
run;
```

The condition is true even before the loop starts, but because the condition is tested at the bottom of the loop, this program outputs one observation (as shown here):

Figure 8.11: Output from Program 8.14

Listing of Double

Interest	Total	Year
0.0375	$311.25	1

An alternative to DO UNTIL is DO WHILE. As you might expect, a DO WHILE loop iterates as long as the condition following WHILE is true. There is another difference between DO WHILE and DO UNTIL—the **WHILE condition is tested at the top of the loop** rather than at the bottom.

Unlike a DO UNTIL block that always iterates at least once, a DO WHILE block does not execute even once if the condition is false. You can rewrite Program 8.13 using a DO WHILE statement, like this:

Program 8.15: Demonstrating a DO WHILE Statement

```
data Double;
   Interest = .0375;
   Total = 100;
   do while (Total le 200);
      Year + 1;
      Total = Total + Interest*Total;
      output;
   end;
   format Total dollar10.2;
run;

proc print data=double noobs;
   title "Listing of Double";
run;
```

The block of code between the DO WHILE and END statements executes as long as Total is less than or equal to `200`. Output from this program is identical to the output from Program 8.13.

To reinforce the idea that DO WHILE conditions are tested at the top of the loop, look at this program:

Program 8.16: Demonstrating That DO WHILE Loops Are Evaluated at The Top

```
data Double;
   Interest = .0375;
   Total = 300;
   do while (Total lt 200);
      Year + 1;
      Total = Total + Interest*Total;
      output;
   end;
   format Total dollar10.2;
run;
```

Because the WHILE condition is never true, the statements inside the DO WHILE block never execute and the data set Double has no observations.

8.7 A Caution When Using DO UNTIL Statements

It is very important that the condition you place on a DO UNTIL statement becomes true at some point. For example, if you change the DO UNTIL statement in Program 8.14 to read as follows, the condition is never true and you have what is called an ***infinite loop***:

```
do until (Total eq 200);
```

Depending on whether or not you are paying for your computer time, this could be a bad (expensive) thing. On a PC platform, you can usually interrupt the program by selecting the icon to stop a SAS program (the exclamation point on the taskbar) or simultaneously clicking the CTRL and C keys. The lesson here is to be very careful when using a DO UNTIL statement: make sure the condition you specify eventually returns a true value.

One way to prevent infinite loops is to combine a regular DO loop with an UNTIL condition. You could rewrite the program like this:

Program 8.17: Combining a DO UNTIL and Iterative DO Loop

```
data Double;
   Interest = .0375;
   Total = 100;
   do Year = 1 to 100 until (Total gt 200);
      Total = Total + Interest*Total;
      output;
   end;
   format Total dollar10.2;
run;
```

There are two advantages to this structure: first, even if the UNTIL condition never becomes true, the loop ends when Year reaches `100`, and second, you don't have to assign a value to Year inside the loop as you did in Program 8.13.

8.8 LEAVE and CONTINUE Statements

The LEAVE statement inside a DO loop shifts control to the statement following the END statement at the bottom of the loop. The CONTINUE statement halts further statements within the DO loop from executing and continues iterations of the loop.

Note: You can also use a LEAVE statement inside a SELECT group.

Program 8.18 demonstrates how the LEAVE statement works.

Program 8.18: Demonstrating the LEAVE Statement

```
data Leave_it;
   Interest = .0375;
   Total = 100;
   do Year = 1 to 100;
      Total = Total + Interest*Total;
      output;
      if Total ge 200 then leave;
   end;
   format Total dollar10.2;
run;
```

In this program, the loop continues until Total is greater than or equal to **200**. At this point, the LEAVE statement terminates the loop and you are at the bottom of the RUN statement at the bottom of the DATA step.

To demonstrate a CONTINUE statement, take a look at the following program:

Program 8.19: Demonstrating a CONTINUE Statement

```
data Continue_on;
   Interest = .0375;
   Total = 100;
   do Year = 1 to 100 until (Total ge 200);
      Total = Total + Interest*Total;
      if Total le 150 then continue;
      output;
   end;
   format Total dollar10.2;
run;

title "Listing of Data Set Continue_on";
proc print data=Continue_on noobs;
run;
```

As long as Total is less than or equal to `150`, the CONTINUE statement causes execution to drop to the bottom of the loop (skipping the OUTPUT statement) and the loop continues. When Total is greater than

`150`, output occurs and the outer loop continues until Total is greater than `200`. Thus, this program prints values of Total greater than `150` until Total reaches or exceeds `200`. Here is the output:

Figure 8.12: Output from Program 8.19

Listing of Data Set Continue_on

Interest	Total	Year
0.0375	$155.55	12
0.0375	$161.38	13
0.0375	$167.43	14
0.0375	$173.71	15
0.0375	$180.22	16
0.0375	$186.98	17
0.0375	$193.99	18
0.0375	$201.27	19

As you saw in this chapter, iterative statements in SAS can make your programs shorter and easier to understand. They also allow you to write DATA steps that generate data for creating tables or plotting functions.

8.9 Problems

Solutions to odd-numbered problems are located at the back of this book. Solutions to all problems are available to professors. If you are a professor, visit the book's companion website at support.sas.com/cody for information about how to obtain the solutions to all problems.

1. Run the program here to create a temporary SAS data set called Vitals:

```
data Vitals;
   input ID    : $3.
         Age
         Pulse
         SBP
         DBP;
   label SBP = "Systolic Blood Pressure"
         DBP = "Diastolic Blood Pressure";
datalines;
001 23 68 120 80
002 55 72 188 96
003 78 82 200 100
004 18 58 110 70
005 43 52 120 82
006 37 74 150 98
007  . 82 140 100
;
```

Using this data set, create a new data set (NewVitals) with the following new variables:

For subjects less than 50 years of age:

> If Pulse is less than 70, set PulseGroup equal to Low;
> otherwise, set PulseGroup equal to High.
> If SBP is less than 130, set SBPGroup equal to Low;
> otherwise, set SBPGroup equal to High.

For subjects greater than or equal to 50 years of age:

> If Pulse is less than 74, set PulseGroup equal to Low;
> otherwise, set PulseGroup equal to High.
> If SBP is less than 140, set SBPGroup equal to Low;
> otherwise, set SBPGroup equal to High.

You may assume there are no missing values for Pulse or SBP.

2. Run the program here to create a temporary SAS data set (MonthSales):

```
data MonthSales;
   input month sales @@;
   /* add your line(s) here */
datalines;
1 4000 2 5000 3 . 4 5500 5 5000 6 6000 7 6500 8 4500
9 5100 10 5700 11 6500 12 7500
;
```

   Modify this program so that a new variable, SumSales, representing Sales to date, is added to the data set. Be sure that the missing value for Sales in month 3 does not result in a missing value for SumSales.

3. Modify the program below so that each observation contains a subject number (Subj), starting with 1:

```
data Test;
   input Score1-Score3;
   /* add your line(s) here */
datalines;
90 88 92
75 76 88
88 82 91
72 68 70
;
```

4. Count the number of missing values for the variables A, B, and C in the Missing data set. Add the cumulative number of missing values to each observation (use variable names MissA, MissB, and MissC). Use the MISSING function to test for the missing values.
5. Create and print a data set with variables N and LogN, where LogN is the natural log of N (the function is LOG). Use a DO loop to create a table showing values of N and LogN for values of N going from `1` to `20`.
6. Repeat Problem 5, except have the range of N go from `5` to `100` by `5`.

7. Use an iterative DO loop to plot the following equation:

```
y = 3*x² - 5*x + 10
```

Use values of x from 0 to 10, with an increment of .10. Use the code below to display the resulting equation.

```
title "Problem 7";
proc sgplot data=plotit;
   series x=x y=y;
run;
```

8. Use an iterative DO loop to plot the following equation:

```
Logit = log(p / (1 - p))
```

Use values of p from `0` to `1` (with a point at every .05). Using the following PROC SGPLOT statements will produce a very nice plot.

```
title "Logit Plot";
proc sgplot data = Logitplot;
   series x=p y=Logit;
run;
```

9. You have the following seven values for temperatures for each day of the week, starting with Monday: 70, 72, 74, 76, 77, 78, and 85. Create a temporary SAS data set (Temperatures) with a variable (Day) equal to **`Mon`**, **`Tue`**, **`Wed`**, **`Thu`**, **`Fri`**, **`Sat`**, and **`Sun`** and a variable called **`Temp`** equal to the listed temperature values. Use a DO loop to create the Day variable.

10. You are testing three speed-reading methods (A, B, and C) by randomly assigning 10 subjects to each of the three methods. You are given the results as three lines of reading speeds, each line representing the results from each of the three methods, respectively. Here are the results:

```
250 255 256 300 244 268 301 322 256 333
267 275 256 320 250 340 345 290 280 300
350 350 340 290 377 401 380 310 299 399
```

Create a temporary SAS data set from these three lines of data. Each observation should contain Method (A, B, or C), and Score. There should be 30 observations in this data set. Use a DO loop to create the Method variable and remember to use a single trailing @ in your INPUT statement. Provide a listing of this data set using PROC PRINT.

11. You have daily temperatures for each hour of the day for two cities (Dallas and Houston). The 48 temperature values are strung out in several lines like this:

```
80 81 82 83 84 84 87 88 89 89
91 93 93 95 96 97 99 95 92 90 88
86 84 80 78 76 77 78
80 81 82 82 86
88 90 92 92 93 96 94 92 90
88 84 82 78 76 74
```

The first 24 values represent temperatures from Hour 1 to Hour 24 for Dallas and the next 24 values represent temperatures for Hour 1 to Hour 24 for Austin. Using the appropriate DO loops, create a data set (Temperature) with 48 observations, each observation containing the variables City, Hour, and Temp.

Note: For this problem, you will need to use a single trailing @ on your INPUT statement (see Chapter 21, Section 21.11 for an explanation).

12. You place money in a fund that returns a compound interest of 4.25% annually. You deposit $1,000 every year. How many years will it take to reach $30,000? Do not use compound interest formulas. Rather, use "brute force" methods with DO WHILE or DO UNTIL statements to solve this problem.

13. You invest $1,000 a year at 4.25% interest, compounded *quarterly*. How many years will it take to reach $30,000? Do not use compound interest formulas. Rather, use "brute force" methods with DO WHILE or DO UNTIL statements to solve this problem.

14. Generate a table of integers and squares starting at `1` and ending when the square value is greater than `100`. Use either a DO UNTIL or DO WHILE statement to accomplish this.

Chapter 9: Working with Dates

9.1 Introduction

Most data sets contain date information, such as a date of birth or a transaction date. SAS can read dates in almost any commonly used format. It can calculate intervals between dates and much, much more. Read on.

9.2 How SAS Stores Dates

SAS can read dates in almost any form, such as the following:

Date	Description
10/21/1950	Month – Day – Year
21/10/1950	Day – Month – Year
21Oct1950	Day – Month Abbreviation – Year
50294	Julian Date

However, SAS does not normally store dates in any of these forms—it converts all of these dates into a single number—the number of days from January 1, 1960. Dates after January 1, 1960 are positive integers; dates before January 1, 1960 are negative integers. For example, the following table shows some dates and the internal values stored by SAS:

Date	SAS Internal Value
January 1, 1960	0
January 2, 1960	1
December 31, 1959	-1
December 21, 2017	21,174
October 21, 1950	-3,359

9.3 Reading Date Values from Text Data

As you saw in Chapter 3, formatted input (or list input with an appropriate informat) allows SAS to read nonstandard numeric data. SAS has many built-in date informats that allow you to read dates in any of the forms in the table in Section 9.2 (and more) and automatically convert these date values into SAS dates. Let's look at an example.

You have a raw data file **C:\books\learning\Dates.txt**, as shown here:

```
         1         2         3
1234567890123456789012345678901234567 Columns
001 10/21/1950 05122003 08/10/65 23Dec2005
002 01/01/1960 11122009 09/13/02 02Jan1960
```

The first three dates (starting in columns 5, 16, and 25) are in the month-day-year form; the last date (starting in column 34) starts with the day of the month, a three-letter month abbreviation, and a four-digit year. Notice that some of the dates include separators between the values, while others do not. Also, the third date value has only a two-digit year. Here is a program to read these dates:

Program 9.1: Program to Read Dates from Text Data

```
data Four_Dates;
   infile 'C:\books\learning\Dates.txt' truncover;
   input @1  Subject    $3.
         @5 DOB         mmddyy10.
         @16 VisitDate mmddyy8.
         @26 TwoDigit   mmddyy8.
         @34 LastDate   date9.;
run;
```

The TRUNCOVER option was added to the INFILE statement. You typically want to use either the TRUNCOVER or PAD option when reading raw data files with data in fixed columns (see Chapter 21). Both of these options prevent errors when you have a short record.

Each of the four dates is read with an appropriate date informat. The first three dates are all in month-day-year form so they all use the MMDDYY*w*. informat. Because the date of birth (DOB) takes up 10 columns, MMDDYY10. is the proper informat to use. The second date (VisitDate) uses a format similar to the DOB, except that there are no separators between the month, day, and year values. Therefore, the number of columns occupied by these dates is 8. The third date uses separators between the values but has only two-digit year values (this is illustration purposes only—always use four-digit years). Therefore, these dates also take up 8 columns. Finally, the last date uses the day of the month, a three-character month abbreviation, and a four-digit year. The number of columns used for these dates is 9, and the informat name is DATE.

Each of these dates is read with an appropriate SAS informat. As these dates are read, SAS converts all of the dates to their corresponding numerical value. To see this, here is a listing, produced by PROC PRINT:

Figure 9.1: Listing of Data Set Four_Dates

Listing of Data Set Four_Dates

Subject	DOB	VisitDate	TwoDigit	LastDate
001	-3359	15837	2048	16793
002	0	18213	15596	1

The numbers in this listing represent the internally stored values for the four dates that SAS read from the raw data and converted. These values are stored the same way SAS stores all numeric values.

You can choose any SAS date format to display these dates—they do not have to match the informat that was used to read the dates. To demonstrate this, the next program adds a FORMAT statement in the DATA step, as follows:

Program 9.2: Adding a FORMAT Statement to Format Each of the Date Values

```
data Four_Dates;
   infile 'C:\books\learning\Dates.txt' truncover;
   input @1  Subject   $3.
         @5 DOB        mmddyy10.
         @16 VisitDate mmddyy8.
         @26 TwoDigit  mmddyy8.
         @34 LastDate  date9.;
   format DOB VisitDate date9.
          TwoDigit LastDate mmddyy10.;
run;
```

Two very popular date formats are DATE9. and MMDDYY10. Again, it is important to separate the way the dates are read with the way you choose to display them. Here is the listing, with formatted date values:

Figure 9.2: Listing of Four_Dates with Formatted Dates

Listing of Data Set Four_Dates

Subject	DOB	VisitDate	TwoDigit	LastDate
001	21OCT1950	12MAY2003	08/10/1965	12/23/2005
002	01JAN1960	12NOV2009	09/13/2002	01/02/1960

There are several things to notice about how SAS reads these four date values. First, two of the dates (VisitDate and TwoDate) both used the same informat (MMDDYY8.) because both of these date values occupied 8 columns. SAS was clever enough to realize that VisitDate did not use delimiters, so the year must have been four digits. The values for TwoDate did include delimiters; therefore, the year values must have been two digits. All of these dates would also have been read correctly if the day of the month or the month values did not include leading 0s.

Next, notice the values for TwoDigit in this listing. The first observation shows the year as 1965, while the second observation shows the year as 2002. How did SAS figure out whether to make the first two digits 19 or 20?

Whenever you use a two-digit year, it is impossible to know if the date is from 1900 to 1999 or from 2000 to 2999 (or some other century). There is a system option called YEARCUTOFF that enables SAS to compute these values. The default value for this option was 1920 in SAS 8 and is 1926 in SAS 9.4. This value determines the start of a 100-year interval that SAS uses when it encounters a two-digit year. With a YEARCUTOFF value of 1926, all two-digit years are in the interval from 1926 to 2025. That is why the first date (8/10/65) is given the value 8/10/1965 and the second date (9/13/02) is given the value 9/13/2002.

Note: If you use four-digit years in your data, the YEARCUTOFF option has no effect. Always use four-digit years.

9.4 Computing the Number of Years between Two Dates

Suppose you want to compute a person's age, given his or her date of birth. The following program uses the Four_Dates data set as input and computes the subject's Age as of the visit date (VisitDate):

Program 9.3: Compute a Person's Age in Years

```
data Ages;
   set Four_dates;
   Age = yrdif(DOB,VisitDate);
run;

title "Listing of Ages";
proc print data=Ages noobs;
run;
```

Program 9.3 uses a SAS function, YRDIF, to compute the difference in years between the date of birth and the visit date.

Here is the listing:

Figure 9.3: Output from Program 9.3

Listing of Ages

DOB	VisitDate	Age
21OCT1950	12MAY2003	52.5562
01JAN1960	12NOV2009	49.8630

If you want the Age as of the person's last birthday (dropping any fractional part of a year), you can use the INT (integer) function:

```
Age = int(yrdif(DOB,VisitDate));
```

If you want to round the Age to the nearest year, you can use the ROUND function:

```
Age = round(yrdif(DOB,VisitDate));
```

You may see the following expression in some older programs:

```
Age = (VisitDate - DOB) / 365.25;
```

This expression gives an approximate value for the number of years between two dates (the .25 accounts for a leap year every four years). Because the YRDIF function has the ability to return an exact value, you should use it rather than this expression.

9.5 Demonstrating a Date Constant

How do you compute a person's age as of a certain date, January 1, 2017, for example? You need to be able to enter this date as the second argument of the YRDIF function. SAS allows you to enter dates in a DATA step by using a date constant. The date January 1, 2017, is written as follows:

```
'01Jan2017'd
```

The general form of a date constant is a one- or two-digit day of the month, a three- character month abbreviation, and a two- or four-digit year (always use four-digit years) in single or double quotation marks, followed by an upper- or lowercase d. This is the only form allowed as a date constant. You cannot use '01/01/2017'd, for example. You can use date constants in any expression involving dates.

Rewriting Program 9.3 to compute Age as of January 1, 2017, you have the following program:

Program 9.4: Demonstrating a Date Constant

```
data Ages;
   set Four_Dates;
   Age = yrdif(DOB,'01Jan2017'd);
run;

title "Listing of Ages";
proc print data=Ages;
   format Age 5.1;
run;
```

In this program, a date constant represents the date January 1, 2017. This program also formats Age so that it prints the value to the nearest 10th. Here is the listing:

Figure 9.4: Output from Program 9.4

Listing of Ages

Obs	Subject	DOB	VisitDate	TwoDigit	LastDate	Age
1	001	21OCT1950	12MAY2003	08/10/1965	12/23/2005	66.2
2	002	01JAN1960	12NOV2009	09/13/2002	01/02/1960	57.0

9.6 Computing the Current Date

Suppose you want to compute a quantity based on the current date. The TODAY function returns the value of the current date. Here is an example where the current date is substituted for January 1, 2017, in Program 9.4:

Program 9.5: Using the TODAY Function to Return the Current Date

```
data Ages;
   set Four_Dates;
   Age = yrdif(DOB,today());
run;
```

Even though you do not pass any arguments to the TODAY function, the parentheses following the function name are required (so that SAS can tell you want to use the function rather than a variable named Today). The DATE function performs an identical task as the TODAY function. Feel free to use either name.

9.7 Extracting the Day of the Week, Day of the Month, Month, and Year from a SAS Date

There are SAS functions to extract the day of the week, day of the month, month, and year from a SAS date. Program 9.6 demonstrates these functions:

Program 9.6: Extracting the Day of the Week, Day of the Month, Month, and Year from a SAS Date

```
data Extract;
   set Four_Dates;
   Day = weekday(DOB);
   DayOfMonth = day(DOB);
   Month = Month(DOB);
   Year = year(DOB);
run;

title "Listing of Extract";
proc print data=Extract noobs;
   var DOB Day -- Year;
run;
```

The WEEKDAY function returns the day of the week, with `1 = Sunday, 2 = Monday`, and so on, and the DAY function returns the day of the month (a number from 1 to 31). Be careful not to get these two functions mixed up. The MONTH function returns a number from 1 to 12 and the YEAR function returns a four-digit year value.

If you haven't seen it before, the variable list used in the VAR statement uses the double dash (--) form of a variable list. This VAR statement includes all of the variables from Day through Year in the order they are stored in the SAS data set. Be careful when using the double dash form of a variable list because the variable order may not be what you expect. It is best to generate a list of the variables in a SAS data set by running PROC CONTENTS (use the VARNUM option) before using this notation.

A listing of data set Extract is shown next:

Figure 9.5: Output from Program 9.6

Listing of Extract

DOB	Day	DayOfMonth	Month	Year
21OCT1950	7	21	10	1950
01JAN1960	6	1	1	1960

9.8 Creating a SAS Date from Month, Day, and Year Values

A very useful function, MDY (month day year), allows you to create a SAS date value by supplying month, day, and year values. This function is especially useful if you have a SAS data set that contains these values but does not contain the corresponding SAS date value or if you have values for month, day, and year in a raw data file that do not conform to any of the SAS date informats. Data set Month_Day_Year contains the variables Month, Day, and Year. You can create a SAS date like this:

Program 9.7: Using the MDY Function to Create a SAS Date from Month, Day, and Year

```
data MDY_Example;
   set Learn.Month_Day_Year;
   Date = mdy(Month, Day, Year);
   format Date mmddyy10.;
run;
```

In this program, the three arguments in the MDY function are MONTH, DAY, and YEAR values from the input data set.

Here is a listing of data set MDY_Example:

Figure 9.6: Listing of Data Set MDY_Example

Listing of Data Set MDY_Example

Obs	Month	Day	Year	Date
1	10	21	1950	10/21/1950
2	3	.	2005	.
3	5	7	2000	05/07/2000

In Observation 2, where the day of the month is missing, the value for Date is also missing.

9.9 Substituting the 15th of the Month when the Day Value Is Missing

There are occasions where you have a missing value for the day of the month but still want to compute an approximate date. Many people use the 15th of the month to substitute for a missing Day value. You can use the Month data set from the previous section to demonstrate how this is done. Here is the program:

Program 9.8: Substituting the 15th of the Month When a Day Value is Missing

```
data Substitute;
   set Learn.Month_Day_Year;
   if missing(Day) then Date = mdy(Month,15,Year);
   else Date = mdy(Month,Day,Year);
   format Date mmddyy10.;
run;
```

Here the MISSING function tests if there is a missing value for the variable Day. If so, the number `15` is used as the second argument to the MDY function. The resulting listing shows the 15th of the month for the date in the second observation:

Figure 9.7: Listing of Data Set Substitute

Listing of Data Set Substitute

Obs	Month	Day	Year	Date
1	10	21	1950	10/21/1950
2	3	.	2005	03/15/2005
3	5	7	2000	05/07/2000

9.10 Using Date Interval Functions

Two functions, INTCK and INTNX, deal with date intervals (such as months, quarters, years). The INTCK function computes the number of interval boundaries that are crossed going from one date to another date; the INTNX function computes a date after a given number of intervals.

First of all, nobody knows how to pronounce either of these two functions—you're on your own. Next, these functions can be very complicated. For a more complete treatment of these two functions, please see *SAS Functions by Example*, 2nd edition (Cody – published by SAS Press 2010) or SAS Help Center at http://go.documentation.sas.com.

To understand even the most basic use of these two functions, you must understand that they both deal with interval boundaries. For example, an interval boundary for the interval YEAR is January 1, the boundary for WEEK is Sunday. A few examples make this clear. Look at the following table:

Expression	Value Returned
`INTCK('year','01Jan2005'd,'31Dec2005)`	0
`INTCK('year','31Dec2005'd,'01Jan2006)`	1
`INTCK('month','01Jan2005'd,'31Jan2005'd)`	0
`INTCK('month','31Jan2005'd,'01Feb2005'd)`	1
`INTCK('qtr','25Mar2005'd,'15Apr2005'd)`	1

The INTCK function takes three arguments. The first is the desired interval (see Figure 9.8 for a list of interval values). The next two arguments represent a starting date and an ending date. The function returns the number of interval boundaries that have been crossed by going from the first date to the second date. The first entry in the table starts at January 1, 2005, and ends at December 31, 2005. The value returned is `0` because a year boundary (January 1) was not crossed in going from the first date to the second. The next entry, from December 31, 2005, to January 1, 2006, returns a `1` because a year boundary was crossed (even though this represents only one day).

The same logic holds for the two INTCK functions with MONTH as the interval. A month boundary is the first of any month. The last table entry asks how many quarters are crossed going from March 25, 2005 to April 15, 2005. Because the boundaries for quarters are January 1, April 1, July 1, and October 1, this expression returns a `1` (the boundary of April 1 was crossed).

Besides YEAR, MONTH, and QUARTER, the two date functions INTCK and INTNX support several other intervals. A partial list of intervals supported by both the INTCK and INTNX functions is displayed below

Figure 9.8: A Partial List of Intervals Supported by the INTCK and INTNX Functions

Interval	Description
Week	Weekly intervals with Sunday as the first day of the week
Weekday	Weekdays (default Sat. and Sun. not included)
Month	Months (with the first of each month as the boundary)
Qtr	Quarters (Jan. 1, April 1, July 1, and Oct. 1)
Semi-year	Semi-annual intervals
Year	Yearly intervals

Here is an example. Data set Hosp contains 1,000 observations. Variables in this data set include a date of birth (DOB) and an admission date (AdmitDate). You want to see a graph showing the number of admissions for each quarter starting with January 1, 2003, and ending with June 30, 2006. Without the INTCK function, this task would require a lot of programming; with the INTCK function, this becomes a more straightforward problem. Here is one possible solution:

Program 9.9: Demonstrating the INTCK Function

```
data Frequency;
   set Learn.Hosp(keep=AdmitDate
                  where=(AdmitDate between '01Jan2003'd and
                        '30Jun2006'd));
   Quarter = intck('qtr','31Dec2002'd,AdmitDate);
run;

title "Admissions from January 1, 2003 to June 30, 2006";
proc sgplot data=Frequency;
   vbar Quarter;
run;
```

The key to this program is using the INTCK function to count the number of quarters from the starting date of December 31, 2002. (This date was used so that any admission in the first quarter of 2003 would be quarter 1.) PROC SGPLOT can produce a vertical bar chart (hence the keyword VBAR) showing how many visits occurred in each quarter. The resulting chart is shown next:

Figure 9.9: Output from Program 9.9

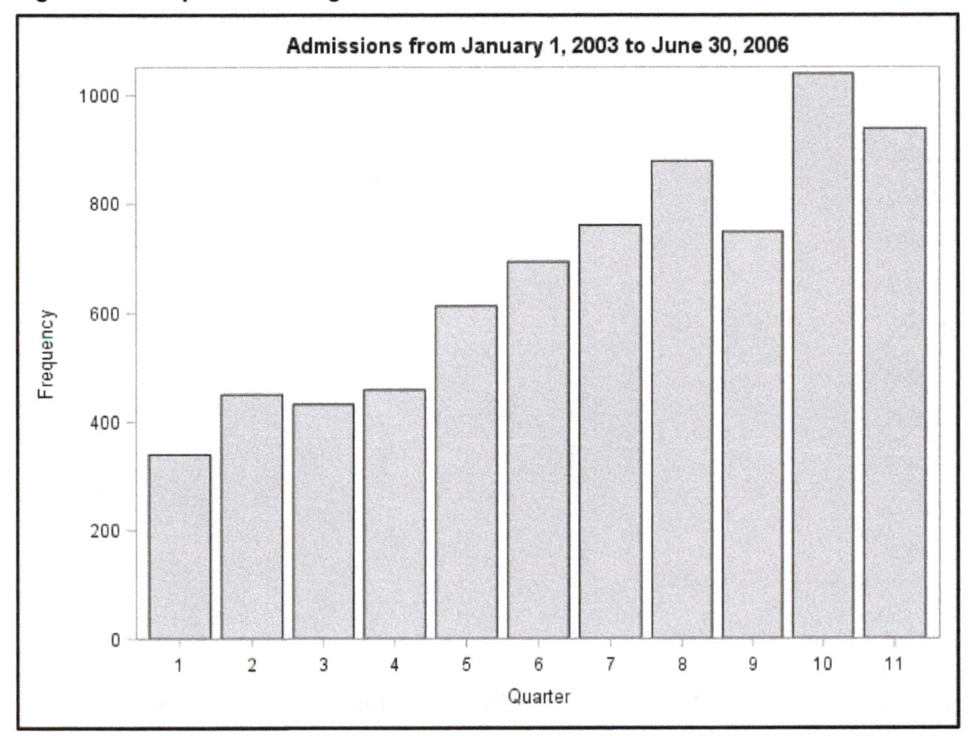

The converse of the INTCK function is INTNX. This function takes a starting date and returns the date after which a given number of interval boundaries have been crossed. Here is an example.

You have a data set (Hosp_Discharge) containing patient numbers (Patient) and discharge dates (Discharge). You want to notify each patient on the first day of the month 6 months after discharge. First, here is a listing of the Hosp_Discharge data set:

Figure 9.10: Listing of Data Set Hosp_Discharge

Listing of Data Set Hosp_Discharge

Patient	Discharge
879	29NOV2003
880	30NOV2003
883	04SEP2003
884	28AUG2003
885	04SEP2003
886	26AUG2003
887	31AUG2003
888	25AUG2003
913	16NOV2003
914	15NOV2003

Program 9.10 computes notification dates:

Program 9.10: Using the INTNX Function to Compute Dates 6 Months After Discharge

```
data Followup;
   set Learn.Hosp_Discharge;
   FollowDate = intnx('month',Discharge,6);
   format FollowDate date9.;
run;
```

Here the INTNX function uses MONTH as the interval and returns a date 6 months after the discharge date. A listing follows:

Figure 9.11: Listing of Data Set Follow-Up

Listing of Data Set Follow-Up

Discharge	Patient	FollowDate
29NOV2003	879	01MAY2004
30NOV2003	880	01MAY2004
04SEP2003	883	01MAR2004
28AUG2003	884	01FEB2004
04SEP2003	885	01MAR2004
26AUG2003	886	01FEB2004
31AUG2003	887	01FEB2004
25AUG2003	888	01FEB2004
16NOV2003	913	01MAY2004
15NOV2003	914	01MAY2004

Notice that each Follow-Up date falls on the first of the month (a boundary for a MONTH interval). If you want the patient to return on the same day of the month after an interval of 6 months has passed, you can add SAMEDAY, an optional fourth argument to the INTNX function. This fourth argument is referred to in SAS documentation as an *alignment parameter*. Choices are BEGINNING, MIDDLE, END, and SAMEDAY. SAMEDAY is a particularly useful option since you can find dates on the same day of the week or the same day of the month as the starting date. To see how this can help you recall the patients on the same day of the month, 6 months hence, modify Program 9.10 like this:

Program 9.11: Demonstrating the SAMEDAY Alignment with the INTNX Function

```
data Followup;
   set Learn.Hosp_Discharge;
   FollowDate = intnx('month',Discharge,6,'sameday');
   format FollowDate date9.;
run;
```

The resulting output is:

Figure 9.12: Listing of Data Set Follow-Up

Listing of Data Set Follow-Up

Discharge	Patient	FollowDate
29NOV2003	879	29MAY2004
30NOV2003	880	30MAY2004
04SEP2003	883	04MAR2004
28AUG2003	884	28FEB2004
04SEP2003	885	04MAR2004
26AUG2003	886	26FEB2004
31AUG2003	887	29FEB2004
25AUG2003	888	25FEB2004
16NOV2003	913	16MAY2004
15NOV2003	914	15MAY2004

The follow-up date is now on the same day of the month as the discharge date.

These two functions can be very confusing. And, we haven't even mentioned multiple and shifted intervals! A parting word: be very careful when you use these two functions and try running your programs with test data.

9.11 Problems

Solutions to odd-numbered problems are located at the back of this book. Solutions to all problems are available to professors. If you are a professor, visit the book's companion website at support.sas.com/cody for information about how to obtain the solutions to all problems.

1. You have several lines of data, consisting of a subject number and two dates (date of birth and visit date). The subject number starts in column 1 (and is 3 bytes long), the date of birth starts in column 4 and is in the form month-day-year, and the visit date starts in column 14 and is in the form of a two-digit day, a three-character month abbreviation, followed by a four-digit year (see sample lines below). Read the following lines of data to create a temporary SAS data set called Dates. Format both dates using the DATE9. format. Include the subject's age at the time of the visit in this data set.

```
0011021195011Nov2006
0020102195525May2005
0031225200525Dec2006
```

2. Using the following lines of data, create a temporary SAS data set called ThreeDates. Each line of data contains three dates, the first two in the form mm/dd/yyyy and the last in the form of a two-

digit day, a three-character month abbreviation, followed by a four-digit year. Name the three date variables Date1, Date2, and Date3. Format all three using the MMDDYY10. format. Include in your data set the number of years from Date1 to Date2 (call it Year12) and the number of years from Date2 to Date3 (call it Year23). Round these values to the nearest year. Here are the lines of data (note that the columns do not line up):

```
01/03/1950 01/03/1960 03Jan1970
05/15/2000    05/15/2002    15May2003
10/10/1998 11/12/2000    25Dec2005
```

3. You have several dates that range from 1910 to 2006 in a raw data file. Unfortunately, all of the dates only have two-digit years. Read these dates and be sure that the resulting data set (call it Dates1910_2006) are in this range. **Hint:** Remember the option YEARCUTOFF.

   Here are the values (all starting in column 1).

   ```
   01/01/11
   02/23/05
   03/15/15
   05/09/06
   ```

4. Using the Hosp data set, compute the subject's ages two ways: as of January 1, 2006 (call it AgeJan1), and as of today's date (call it AgeToday). The variable DOB represents the date of birth. Take the integer portion of both ages. List the first 10 observations.
5. Using the Hosp data set, compute the frequencies for the days of the week, months of the year, and year, corresponding to the admission dates (variable AdmitDate). Supply a format for the days of the week and months of the year. Use PROC FREQ to list these frequencies.
6. Using the Medical data set, compute frequencies for the days of the week for the date of the visit (VisitDate). Supply a format for the days of the week and months of the year.
7. Using the Hosp data set, list all the observations with admission dates (AdmitDate) before July 15, 2002. Write your statement so that if there were any missing values for AdmitDate, they are not included in this list.
8. Using the values for Day, Month, and Year in the raw data below, create a temporary SAS data set containing a SAS date based on these values (call it Date) and format this value using the MMDDYY10. format. Here are the Day, Month, and Year values:

   ```
   25 12 2005
   1    1    1960
   21    10    1946
   ```

9. Repeat Problem 8, except use the following data. If there is a missing value for the day, substitute the 15th of the month.

   ```
   25 12 2005
   .  5  2002
   12 8      2006
   ```

10. Using the Hosp data set, compute the number of months from the admission date (AdmitDate) and December 31, 2007 (call it MonthsDec). Also, compute the number of months from the admission date to today's date (call it MonthsToday). Use a date interval function to solve this problem. List the first 20 observations for your solution.

11. The data set Medical contains a variable called VisitDate. Create a temporary SAS data set (Interval) with the variables in Medical plus a new variable (Quarter), representing the number of quarters from January 1, 2006.
12. You want to see each patient in the Medical data set on the same day of the week 5 weeks after they visited the clinic (the variable name is VisitDate). Provide a listing of the patient number (Patno), the visit date, and the date for the return visit.
13. You want to see each patient in the Medical data set on the same day of the month 6 month after they visited the clinic (the variable name is VisitDate). Provide a listing of the patient number (Patno), the visit date, and the date of the return visit. Remember, the SAMEDAY alignment, when used with the MONTH interval, results in a date on the same day of the month.

Chapter 10: Subsetting and Combining SAS Data Sets

10.1 Introduction

This chapter describes how to subset a SAS data set and how to combine data from several data sets into a single data set. You will learn how to concatenate (add observations or rows) to SAS data sets, how to add variables from one data set to another, based on an ID or other identifier, and how to update values in a SAS data set based on transaction data in another SAS data set.

10.2 Subsetting a SAS Data Set

Subsetting a SAS data set involves selecting observations from one data set by defining selection criteria, usually in a WHERE or subsetting IF statement. As an example, suppose you want to select all observations from the permanent SAS data set Survey where the value of Gender is **F**. One way to do this is with a WHERE statement, as follows:

Program 10.1: Subsetting a SAS Data Set Using a WHERE Statement

```
data Females;
   set Learn.Survey;
   where Gender = 'F';
run;
```

Remember that the variables used in a WHERE statement must all come from a SAS data set. Variables that are created by reading raw data or from an assignment statement may not be used in this fashion. In Program 10.1, the data set Females contains all the observations in the data set Survey where Gender has a value of **F**. Here is a listing of this data set:

Figure 10.1: Listing of Data Set Females

Listing of Data Set Females

ID	Gender	Age	Salary	Ques1	Ques2	Ques3	Ques4	Ques5
002	Female	51+	$76,123	Agree	.	Strongly agree	.	Disagree
004	Female	51+	$128,000	Strongly agree	.	No opinion	.	Disagree
007	Female	30 to 50	$76,100	Strongly agree	.	No opinion	.	Agree

Notice that the data set Females contains all the variables found in the data set Survey. If you do not need all the variables from the input data set (the data set on the SET statement), you can use a KEEP= or DROP= data set option. For example, to keep ID, Gender, Age, and Ques1-Ques5, use the following method:

Program 10.2: Demonstrating a KEEP= Data Set Option

```
data Females;
   set Learn.Survey(keep=ID Gender Age Ques1-Ques5);
   where Gender = 'F';
run;
```

There is an important difference between using a KEEP= data set option on the input data set and placing a KEEP statement somewhere in the DATA step. In this example, the variable Salary is not present in the Program Data Vector (PDV) because it was not included in the list of variables to keep. An alternative to listing eight variables on the KEEP= data set option would be to use DROP=Salary. While this would save some typing, using a KEEP= data set option instead of a DROP= data set option allows someone reading your program to know what variables you are bringing in from the Survey data set. If you use DROP=Salary, you do not know what variables are being brought in from the Survey data set unless you already have a list of the variables in that data set. See Chapter 2, "How SAS Works (a Look Inside the "Black Box")."

Note: If you use a KEEP statement listing the eight variables instead of the KEEP= data set option, all of the variables, including Salary would be in the PDV but not written out to the Females data set. If the input data set contains a large number of variables and you want only a few of these variables in the new data set, using the KEEP= data set option is more efficient than a KEEP statement.

You must remember that when you use the KEEP= data set option, the variables not in the keep list are also not available in the DATA step.

10.3 Creating More Than One Subset Data Set in One DATA Step

You can create multiple SAS data sets from one input data set (something that SQL cannot do). Following the previous example, you can create a data set containing only data on females and one containing only data on males, in one step like this:

Program 10.3: Creating Two Data sets in One DATA Step

```
data Males Females;
   set Learn.Survey;
   if Gender = 'F' then output Females;
   else if Gender = 'M' then output Males;
run;
```

Notice that you must name the data set following the OUTPUT statement. If you do not, SAS outputs the observation to all the data sets listed in the DATA statement.

10.4 Adding Observations to a SAS Data Set

Suppose you want to create a single data set from several similar data sets. For example, your company may collect data each month into separate data sets and you want to analyze a year's worth of data. You can list as many data sets as you want on a SET statement and SAS will add all the observations together to form a single data set. For example, look at the listings of data sets One and Two here:

Figure 10.2: Listing of Data Sets One and Two

Data Set One

Obs	ID	Name	Weight
1	7	Adams	210
2	1	Smith	190
3	2	Schneider	110
4	4	Gregory	90

Data Set Two

Obs	ID	Name	Weight
1	9	Shea	120
2	3	O'Brien	180
3	5	Bessler	207

Each of the data sets contains the same variables. Also, the storage length for Name is the same in each of the two data sets. One way to combine these data sets is like this:

Program 10.4: Using a SET Statement to Combine Observations from Two Data Sets

```
data One_Two;
   set One Two;
run;
```

Here is the output:

Figure 10.3: Listing of Data Set One_Two

Listing of Data Set One_Two

Obs	ID	Name	Weight
1	7	Adams	210
2	1	Smith	190
3	2	Schneider	110
4	4	Gregory	90
5	9	Shea	120
6	3	O'Brien	180
7	5	Bessler	207

All the observations from data set One are followed by all the observations from data set Two. SAS refers to this process as *concatenating data sets*. This seems simple enough. But what happens if you use the SET statement on two data sets that don't contain all the same variables? To see what happens, here is data set Three:

Figure 10.4: Listing of Data Set Three

Listing of Data Set Three

Obs	ID	Gender	Name
1	10	M	Horvath
2	15	F	Stevens
3	20	M	Brown

Data set Three contains a new variable, Gender, and does not contain the variable Weight. You can't tell from the listing, but the variable Name in both data sets is the same length. Let's use the SET statement on data sets One and Three and see what happens.

Program 10.5: Using a SET Statement on Two Data Sets Containing Different Variables

```
data One_Three;
   set One Three;
run;
```

Here is the output:

Figure 10.5: Listing of Data Set One_Three

Listing of Data Set One_Three

Obs	ID	Name	Weight	Gender
1	7	Adams	210	
2	1	Smith	190	
3	2	Schneider	110	
4	4	Gregory	90	
5	10	Horvath	.	M
6	15	Stevens	.	F
7	20	Brown	.	M

Looking at this output helps you understand what is going on. At compile time, SAS looks at every data set listed in the SET statement. First comes data set One. This brings ID, Name, and Weight into the PDV (and all their attributes). Next SAS looks at data set Three. Are there any variables in data set Three that are not already in the PDV? Yes, Gender. Therefore, Gender is added to the PDV.

Let's follow what goes on as Program 10.5 executes. First, all the variables in the PDV are set to missing. Then, each observation from data set One is read and then written to the output data set. The PDV is not set back to missing as each observation in data set One is processed, but this really doesn't matter because each new observation replaces the values in the PDV. (This would be true even if one of the values in the current observation was missing.)

After all the observations in data set One are read and written out to data set One_Three and SAS prepares to read observations from data set Three, all the variables in the PDV are again set to missing. This is important because data set Three does not contain the variable Weight. If this initialization (setting values in the PDV to missing) did not take place, every person coming from data set Three would wind up at 90 pounds (the last Weight value in data set One). SAS continues reading observations from data set Three until it reaches the end of that data set.

If the length of a variable is different in any of the input data sets, the length of the variable in the output data set is equal to the length that variable has in the first data set encountered in the DATA step. It is a good idea to check lengths of character variables when you are combining several data sets to be sure you will not truncate any values. If necessary, place a LENGTH statement **before** the SET statement to be sure the resulting length is adequate to hold all of your values. (Remember that the length of a character variable is determined as soon as that variable enters the PDV; it cannot be changed after that.) Finally, if you have a variable in two data sets, one character and the other numeric, SAS prints an error message in the log and the program terminates.

10.5 Interleaving Data Sets

There is another way to add observations from several data sets. If each of the data sets to be combined is already sorted, you can take advantage of that fact and wind up with a sorted output data set. All you need to do is to follow the SET statement with a BY statement. SAS selects observations from each of the input data sets in order, with the resulting data set already sorted. The advantage of this method is that you don't have to sort the resulting data set. There are times when the resulting data set would be too large to sort conveniently or at all.

To demonstrate the *interleaving* of data sets, let's combine data sets One and Two, with the resulting data set in ID order. Here is the code:

Program 10.6: Interleaving Data Sets

```
proc sort data=One;
   by ID;
run;

proc sort data=Two;
   by ID;
run;

data Interleave;
   set One Two;
   by ID;
run;
```

Remember, each of the data sets in the SET statement must be in order of the BY variable(s). Here is the output:

Figure 10.6: Listing of Data Set Interleave

Listing of Data Set Interleave

Obs	ID	Name	Weight
1	1	Smith	190
2	2	Schneider	110
3	3	O'Brien	180
4	4	Gregory	90
5	5	Bessler	207
6	7	Adams	210
7	9	Shea	120

Notice that this output data set is in ID order. There is a minor difference between obtaining a data set by interleaving and concatenating the data sets and then sorting. When you use PROC SORT to sort a SAS data set, a sort flag is set (you can see this on the first page of output from PROC CONTENTS) and SAS does not re-sort this data set if you attempt to sort it again by the same BY variables. When you interleave data sets, this sort flag is not set (which should not cause you any problems). Interleaving is especially useful when the concatenated data set is so large that it may be difficult or impossible to sort. By sorting each of the individual data sets first, you still obtain a concatenated data set in sorted order.

10.6 Combining Detail and Summary Data

Suppose you have a SAS data set and want to express values of a variable as a percentage of the mean for all observations. For example, you want to express each value of cholesterol in the Blood data set as a percentage of the mean for all subjects. Problems like this involve combining detail data (data on individuals) with summary data (the mean of all subjects).

One solution is to perform a SET operation conditionally. Here is an example:

Program 10.7: Combining Detail and Summary Data: Using a Conditional SET Statement

```
proc means data=Learn.Blood noprint;
   var Chol;
   output out = Means(keep=Chol_Mean)
          mean = / autoname;
run;

data Percent;
   set Learn.Blood(keep=Subject Chol);
   if _n_ = 1 then set Means;
   PercentChol = Chol / Chol_Mean;
   format PercentChol percent8.;
run;
```

The PROC MEANS step creates a SAS data set (Means) with one observation and one variable. Chol_Mean is the mean cholesterol value for all observations in the Blood data set. (You can read more about creating summary data sets in Chapter 16.)

To save you the trouble of flipping through pages to Chapter 16, the OUTPUT option AUTONAME names the variables in the output data set by using the variable names in the VAR statement and adding an underscore and the name of the statistic. In this example, because you are asking for a mean, the name of the variable holding the mean in the output data set is Chol_Mean.

A listing of the data set Means follows:

Figure 10.7: Listing of Data Set Means

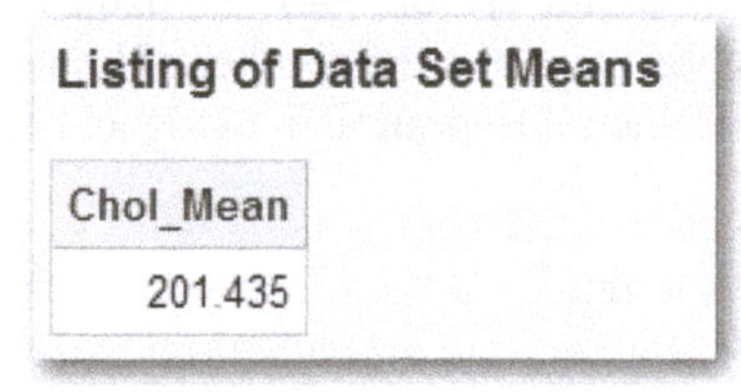

Listing of Data Set Means

Chol_Mean
201.435

To combine this single value with every observation in the Blood data set, you execute a SET statement **conditionally**. Here's how it works.

The first SET statement brings in an observation from the Blood data set. The automatic variable *_n_* counts iterations of the DATA step. You can use this variable to conditionally perform the SET operation on the MEANS data set. During the first iteration of the DATA step, *_n_* is equal to `1` and the first (and only) observation from the Means data set is brought into the PDV. The first observation in the data set Percent contains all the variables in the Blood data set plus the variable Chol_Mean.

For the second iteration of the DATA step, the next observation from the Blood data set is brought into the PDV. Because *_n_* is now equal to `2`, the conditional SET statement does not execute. (Without the conditional SET statement, the DATA step would end when SAS tried to read a second observation from the Means data set—see Section 10.11 for details on when a DATA step ends.) However, because variables that come from SAS data sets are automatically retained (that is, the value in the PDV is not set to a missing value), the Chol_Mean value is still in the PDV and it is added to the second observation. This process continues until the Blood data set reaches the end of file.

You may wonder why the computation of PercentChol doesn't multiply the result by 100. The answer is that the PERCENT format not only adds a percent sign to the value, it also multiplies the value by 100.

To help make this clear, here are listings of the first few observations in the Percent set:

Figure 10.8: Listing of the First Few Observations in the Data Set Percent

Listing of Data Set Percent

Subject	Chol	Chol_Mean	PercentChol
1	258	201.435	128%
2	.	201.435	.
3	184	201.435	91%
4	.	201.435	.
5	187	201.435	93%
6	142	201.435	70%
7	290	201.435	144%

10.7 Merging Two Data Sets

SAS uses the term *merge* to describe the process of combining variables (columns) from two or more data sets. For example, you could have an employee data set (Employee) containing ID numbers and names. If you had another data set (Hours) containing ID numbers, along with a job class and the number of hours worked, you might want to add the name from the Employee data set to each observation in the Hours data set.

To demonstrate this process, take a look at a listing of the two data sets here:

Figure 10.9: Listing of Data Sets Employee and Hours

Listing of Data Set Employee

ID	Name
7	Adams
1	Smith
2	Schneider
4	Gregory
5	Washington

Listing of Data Set Hours

ID	JobClass	Hours
1	A	39
4	B	44
9	B	57
5	A	35

You want to merge the Employee and Hours data sets as follows:

Program 10.8: Merging Two SAS Data Sets

```
proc sort data=Employee;
   by ID;
run;

proc sort data=Hours;
   by ID;
run;

data Combine;
   merge Employee Hours;
   by ID;
run;
```

You first sort each data set by the variable or variables that link the two data sets. Next, you name each of the data sets in a MERGE statement. Be sure to follow the MERGE statement with a BY statement, naming the variable or variables that tell SAS which observations to place in the same observation.

Here is the data set Combine:

Figure 10.10: Listing of Data Set Combine

Listing of Data Set Combine

ID	Name	JobClass	Hours
1	Smith	A	39
2	Schneider		.
4	Gregory	B	44
5	Washington	A	35
7	Adams		.
9		B	57

It's pretty clear what is happening. When the BY variable is present in both data sets, the merged observation contains the corresponding values from both data sets. When the BY variable is missing from the Employee data set, Name will have missing values; when the BY variable is missing from the Hours data set, JobClass and Hours will have missing values.

For those readers who are familiar with SQL, the MERGE statement, by default, performs a full join.

10.8 Omitting the BY Statement in a Merge

This is a good time to discuss what happens when you omit a BY statement when performing a merge.

Note: Without the BY statement in Program 10.8, the first three observations in data set Employee would be matched with the first three observations in data set Hours. The result would be completely wrong.

To be sure this is clear, here is a listing of the merge when you omit the BY statement:

Figure 10.11: Data Set Created by Program 10.8 When the BY Statement is Omitted

Listing of Data Set Combine
Where the BY Statement is Omitted

ID	Name	JobClass	Hours
1	Smith	A	39
2	Schneider		.
4	Gregory	B	44
5	Washington	A	35
7	Adams		.
9		B	57

What you see here is the first observation in data set Employee combined with the first observation in data set Hours—the second observation in data set Employee combined with the second observation in data set Hours, and so forth.

Performing a merge without a BY statement is almost always a mistake and almost always a bad idea. You can set a system option called MERGENOBY to values of NOWARN, WARN, or ERROR to control what happens when you attempt to perform a merge without a BY statement. The default value is NOWARN—that is, SAS performs the merge and does not issue a warning message. If you set this option to WARN, SAS still performs the merge, but a warning message is printed in the SAS log. Finally, if you set this option to ERROR, the program terminates if you attempt a merge that is not followed with a BY statement. At the very minimum, you should set this option to WARN, like this:

```
options mergenoby = warn;
```

To be even safer, set it to ERROR, like this:

```
options mergenoby = error;
```

Note: This option is usually set at the beginning of your SAS session. If you have a file called AUTOEXEC.SAS in the directory where SAS starts, this would be a good place to put this statement—all statements in the special file AUTOEXEC.SAS are executed when you start a new SAS session.

10.9 Controlling Observations in a Merged Data Set

You may want the merged data set to contain only those employees who are in the Hours data set or you may want to see if any employees in the Hours data set are not listed in the Employee data set.

SAS provides a method of controlling which observations you want in the merged data set. To demonstrate how this works, let's merge the two data sets as before, but add a data set option called IN= and examine the result:

Program 10.9: Demonstrating the IN= Data Set Option

```
data New;
   merge Employee(in=In_Employee)
         Hours    (in=In_Hours);
   by ID;
   file print;
   put ID= In_Employee= In_Hours= Name= JobClass= Hours=;
run;
```

An IN= data set option follows each data set name. Following the keyword IN= is a variable name that you make up. In this example, the names In_Employee and In_Hours were chosen. These two variables are temporary variables (SAS will not add them to the output data set). They have a value of `1` (true) if the data set they refer to is making a contribution to the current observation and `0` (false) otherwise. Because these variables are not in the output data set, Program 10.9 uses a PUT statement to list the values of these variables, along with other variables in the data set. Here is the output:

Figure 10.12: Inspecting the IN= Variables

Listing of Variables from Data Set New
Including the IN= Variables

```
ID=1 In_Employee=1 In_Hours=1 Name=Smith JobClass=A Hours=39
ID=2 In_Employee=1 In_Hours=0 Name=Schneider JobClass=  Hours=.
ID=4 In_Employee=1 In_Hours=1 Name=Gregory JobClass=B Hours=44
ID=5 In_Employee=1 In_Hours=1 Name=Washington JobClass=A Hours=35
ID=7 In_Employee=1 In_Hours=0 Name=Adams JobClass=  Hours=.
ID=9 In_Employee=0 In_Hours=1 Name=  JobClass=B Hours=57
```

Because Employee 1 is in both data sets, In_Employee and In_Hours are both equal to `1` in the first observation. Employee 2 is in the Employee data set but not in the Hours data set, so in the second observation In_Employee is equal to `1` and In_Hours is equal to `0`. Finally, in the last observation (`ID = 9`), there is no corresponding ID in the Employee data set, so In_Employee is equal to `0`.

You can use IN= variables to control which observations are written to the output data set. For example, if you want only those employees who are in both data sets, you can add a subsetting IF statement to the DATA step, like this:

Program 10.10: Using IN= Variables to Select IDs That Are In Both Data Sets

```
data Combine;
   merge Employee(in=In_Employee)
         Hours(in=In_Hours);
   by ID;
   if In_Employee and In_Hours;
run;
```

You can, alternatively, write the subsetting IF statement like this:

```
if In_Employee = 1 and In_Hours = 1;
```

The subsetting IF statement in this program (see Chapter 7 for more details) allows observations in the Combine data set only where both In_Employee and In_Hours are true. When either of these IN= variables is false, no observation is written to the output data set.

The resulting data set contains only Employees 1, 4, and 5. Here is a listing of the data set Combine:

Figure 10.13: Listing of Data Set Combine

Listing of Data Set Combine

ID	Name	JobClass	Hours
1	Smith	A	39
4	Gregory	B	44
5	Washington	A	35

For those readers who are familiar with SQL, this would be called an inner join.

10.10 More Uses for IN= Variables

You can use IN= variables to check if an ID is missing from a particular data set as well. The program here accomplishes this:

Program 10.11: More Examples of Using IN= Variables

```
data In_Both
   Missing_Name(drop = Name);
   merge Employee(in=In_Employee)
         Hours(in=In_Hours);
    by ID;
    if In_Employee and In_Hours then output In_Both;
    else if In_Hours and not In_Employee then
       output Missing_Name;
run;
```

In this program, you are creating two SAS data sets in one DATA step. Data set In_Both contains all observations where the ID variable was found in both data sets. Data set Missing_Name contains all the employees who were in the Hours data set but were not in the Employee data set. You can also write the ELSE IF statement like this:

```
else if In_Hours = 1 and In_Employee = 0 then
   output Missing_Name;
```

Either way, you are asking for all IDs that are in the Hours data set and not in the Employee data set. Use whatever form of this statement makes most sense to you.

To be sure this is clear, here are the listings of both data sets:

Figure 10.14: Listing of Data Set In_Both

List of Employees Who are in Both Files

ID	Name	JobClass	Hours
1	Smith	A	39
4	Gregory	B	44
5	Washington	A	35

Figure 10.15: Listing of Data Set Missing_Hours

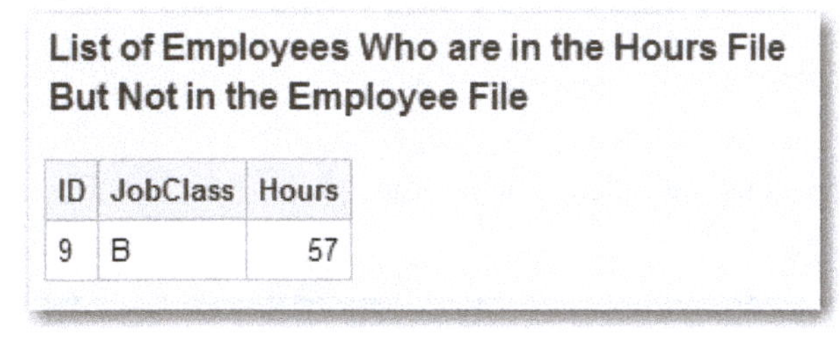

List of Employees Who are in the Hours File But Not in the Employee File

ID	JobClass	Hours
9	B	57

10.11 When Does a DATA Step End?

When any data set reaches an end of file, it signals the end of the DATA step. Take a look at the short program that follows:

Program 10.12: Demonstrating When a DATA Step Ends

```
data Short;
   input x;
datalines;
1
2
;
data Long;
   input x;
datalines;
3
4
5
6
;
data New;
   set Short;
   output;
   set Long;
   output;
run;
```

Data set Short has two observations and data set Long has four. How many observations are in data set New? Each SET statement u a pointer to keep track of which observation it is reading. In this program, an observation is first read from the Short data set, an observation is written out to the New data set, an observation is read from the Long data set, and another observation is written to the New data set. You might expect that this would continue until all the observations from both data sets were read. However, when the end of file on data set Short is encountered, it signals an end to the DATA step, with the result that data set New has only four observations, with values of x equal to 1, 2, 3, and 4. In more complicated programs, you need to be sure that your DATA step doesn't end prematurely when an input data set reaches an end of file.

10.12 Merging Two Data Sets with Different BY Variable Names

In a perfect world, multiple data sets would all use the same name for the variable needed to put them together. However, you will sometimes find yourself trying to merge two data sets where the name of the variable you want to use to join them has a different name in each data set. For example, one data set may call a variable ID and the other EmpID. Luckily, this is an easy problem to solve; you can use a RENAME= data set option to rename the variable in one data set to be consistent with the name in the other. Also be sure that if you have character variables, the lengths of these variables is the same in each of the data sets.

Here is an example:

Data set Bert has two variables: ID (a character variable) and X (a numeric variable). Data set Ernie also has two variables: EmpNo (a character variable) and Y (a numeric variable). You want to merge these two data sets on the ID (or EmpNo) variable. Listings of these two data sets are shown here:

Figure 10.16: Listing of Data Sets Bert and Ernie

Listing of Data Set Bert

ID	X
123	90
222	95
333	100

Listing of Data Set Ernie

EmpNo	Y
123	200
222	205
333	317

You can use the RENAME= data set option to either rename ID to EmpNo or vice versa. Here's one solution:

Program 10.13: Merging Two Data Sets by Renaming a Variable in One Data Set

```
data Sesame;
   merge Bert
         Ernie(rename=(EmpNo = ID));
   by ID;
run;
```

Here, the name of the variable EmpNo from data set Ernie is changed to ID as it is read into the PDV so that the data sets can be merged on a common variable. (Names of variables in data set Ernie are not affected.) Data set Sesame contains three observations, as shown in the following listing:

Figure 10.17: Listing of Data Set Sesame

Listing of Data Set Sesame

ID	X	Y
123	90	200
222	95	205
333	100	317

Notice that the name of the ID variable in the merged data set is ID.

10.13 Merging Two Data Sets with Different BY Variable Data Types

A slightly more complicated situation occurs when the variable you want to use as the BY variable in a merge is a different data type in the two data sets. SAS performs a merge only when the BY variables in both data sets are the same type—either character or numeric.

As an example, two divisions of your company store Social Security numbers differently (they do this just to make your life interesting). Division 1 stores the Social Security number as a numeric variable and Division 2 stores the Social Security number as a character variable. This is not an unusual situation. Here are listings of two sample data sets to demonstrate this problem:

Figure 10.18: Listings of Data Sets Division1 and Division2

Listing of Data Set Division1

SS	DOB	Gender
111223333	11/14/1956	M
123456789	05/17/1946	F
987654321	04/01/1977	F

Listing of Data Set Division2

SS	JobCode	Salary
111-22-3333	A10	45123
123-45-6789	B5	35400
987-65-4321	A20	87900

In this example, not only are the data types different, but one of the data sets (Division2) includes dashes in the SS values. Luckily, SAS can handle this easily.

You can create a character variable based on the numeric value of SS in the Division1 data set or you can create a numeric variable based on the character variable in the Division2 data set. The following solution uses the first approach:

Program 10.14: Merging Two Data Sets When the BY Variables Are Different Data Types

```
data Division1C;
   set Division1(rename=(SS = NumSS));
   SS = put(NumSS,ssn11.);
   drop NumSS;
run;

data Both_Divisions;
   ***Note: Both data sets already in order
      of BY variable;
   merge Division1C Division2;
   by SS;
run;
```

Because the two data sets use the same name for the BY variable, first you have to rename this variable and then create a new character variable with the variable name of SS. The PUT function takes the variable named in the first argument, applies the format named in the second argument, and returns the formatted value (see Chapter 11 for more details on this function). The SSN11. format is a built-in SAS format that prints leading 0s and adds dashes as required for Social Security numbers. Thus, the variable SS in data set Division1C is identical in form to the SS variable in Division2. You can now proceed with the merge in the standard manner. Here is a listing of the resulting data set:

Figure 10.19: Listing of Data Set Divisions

Listing of Data Set Both_Divisions

DOB	Gender	SS	JobCode	Salary
11/14/1956	M	111-22-3333	A10	45123
05/17/1946	F	123-45-6789	B5	35400
04/01/1977	F	987-65-4321	A20	87900

You may choose to create a numeric variable from the character value of SS in data set Division2 instead. The choice of which method to use depends on which data type you want in the merged data set. Here is the alternative approach—creating a numeric variable for the merge:

Program 10.15: An Alternative to Program 10.14

```
data Division2N;
   set Division2(rename=(SS = CharSS));
   SS = input(compress(CharSS,,'kd'),9.);
   ***Alternative:
   SS = input(CharSS,comma11.);
   drop CharSS;
run;
```

Because the character value of SS contains dashes, you need to first use the COMPRESS function to remove the dashes from the value, and then use the INPUT function to perform the character-to-numeric conversion. (You can read about the COMPRESS function in Chapter 12—all you need to know now is that this function with the 'k' and 'd' modifiers, takes a character string and "**k**eeps the **d**igits", removing all other characters.) The INPUT function takes the first argument (a character variable), and applies the informat specified in the second argument, resulting in a numeric variable.

A very clever alternative (inspired by a good friend, Mike Zdeb) is to use the comma informat directly in the INPUT function. It is not commonly known that this informat not only removes commas and dollar signs when reading a value, it also removes dashes. You can therefore use the COMMA11. informat in the INPUT function directly without first having to remove the dashes. You can now merge data sets Division1 and Division2N as before. We will not bother to show a listing of the result—it contains a numeric variable called SS.

10.14 One-to-One, One-to-Many, and Many-to-Many Merges

In the merge examples shown previously, there was only one observation for each value of the BY variable in both data sets. This is called a *one-to-one merge*. You may have a situation where one data set has only one observation for each value of the BY variable, but the other has more than one observation for each value of the BY variable. This is referred to as a *one-to-many merge*. As an example, suppose you want to merge data set Bert with a data set with multiple values of ID. To demonstrate this, take a look at the following two data sets:

Figure 10.20: Listing of Data Sets Bert and Oscar

Listing of Data Set Bert

ID	X
123	90
222	95
333	100

Listing of Data Set Oscar

ID	Y
123	200
123	250
222	205
333	317
333	400
333	500

Notice that data set Oscar has multiple observations with the same ID. What happens when you perform a merge? The resulting data set is shown next:

Figure 10.21: Result of Merging Bert and Oscar

Listing of Data Set Combine

ID	X	Y
123	90	200
123	90	250
222	95	205
333	100	317
333	100	400
333	100	500

Inspection of the merged data set shows that the correct value of X is added to each observation in Oscar, based on the value of ID. This type of merge also works if you reverse the order of the two data sets in the MERGE statement.

> **Note:** A *many-to-many merge* has the potential to be catastrophic. If you have **exactly** the same number of observations for each BY value in the two data sets, a many-to-many merge works as expected. If you have more than one observation with a given BY value in each of the two data sets and the number of observations differs in the two data sets, you should not attempt to use a MERGE statement to combine the data sets.

To see what happens in this situation, look at the two data sets here:

Figure 10.22: Listing of Data Sets Many_One and Many_Two

Listing of Data Set Many_One

ID	X
123	90
123	80
222	95
333	100
333	150
333	200

Listing of Data Set Many_Two

ID	Y
123	3
123	4
123	5
222	6
333	7
333	8

An attempt to merge these two data sets, with ID as the BY variable, results in the following:

Figure 10.23: Listing of Many_to_Many

Listing of Data Set Many_to_Many

ID	X	Y
123	90	3
123	80	4
123	80	5
222	95	6
333	100	7
333	150	8
333	200	8

Notice that the first two observations in the merged data set seem OK. Because there are only two observations with ID equal to 123 in data set One, SAS uses the last value for X (`80`) for all remaining observations in data set Two with ID equal to 123. This result is unlikely to be of use. Notice also that there are no error or warning messages to indicate that you are producing this strange result.

10.15 Updating a Master File from a Transaction File

If you have two data sets that have some common variables and you perform a data set merge, values in the second data set replace values in the first data set, even if the values in the second data set are missing values. If you use an UPDATE statement instead, missing values in the second data set do not replace values in the first data set. This makes the UPDATE statement perfect for updating values in a master data set from new values in a transaction data set. Here is an example.

You have a data set called Prices containing item codes, descriptions, and prices. Here is a listing of this data set:

Figure 10.24: Listing of Data Set Prices

Listing of Data Set Prices

ItemCode	Description	Price
150	50 foot hose	19.95
175	75 foot hose	29.95
200	greeting card	1.99
204	25 lb. grass seed	18.88
208	40 lb. fertilizer	17.98

You also have a data set called New15Dec2017 with item codes and some new prices as follows:

Figure 10.25: Listing of Data Set New15Dec2017

Listing of Data Set New15Dec2017

ItemCode	Price
204	17.87
175	25.11
208	.

You want to use this data set to update the prices in data set Prices. You can use the UPDATE statement like this:

Program 10.16: Updating a Master File From a Transaction File

```
proc sort data=Prices;
   by ItemCode;
run;

proc sort data=New15Dec2017;
   by ItemCode;
run;

data Prices_15dec2017;
   update Prices New15Dec2017;
   by ItemCode;
run;
```

Here is the result:

Figure 10.26: Listing of Data Set Prices_15Dec2017

Listing of Data Set Prices_15Dec2017

ItemCode	Description	Price
150	50 foot hose	19.95
175	75 foot hose	25.11
200	greeting card	1.99
204	25 lb. grass seed	17.87
208	40 lb. fertilizer	17.98

Only nonmissing values of Price in the transaction data set replaced values in the master file, as shown in this listing.

You may want to take a look at Chapter 26, "PROC SQL," to see other ways to combine data from multiple data sets.

10.16 Problems

Solutions to odd-numbered problems are located at the back of this book. Solutions to all problems are available to professors. If you are a professor, visit the book's companion website at support.sas.com/cody for information about how to obtain the solutions to all problems.

1. Using the SAS data set Blood, create two temporary SAS data sets called Subset_A and Subset_B. Include in both of these data sets a variable called Combined equal to .001 times WBC plus RBC. Subset_A should consist of observations from Blood where Gender is equal to **`Female`** and BloodType is equal to **`AB`**. Subset_B should consist of all observations from Blood where Gender is equal to **`Female`**, BloodType is equal to **`AB`**, and Combined is greater than or equal to **`14`**.
2. Using the SAS data set Hosp, create a temporary SAS data set called Monday2002, consisting of observations from Hosp where the admission date (AdmitDate) falls on a Monday and the year is 2002. Include in this new data set a variable called Age, computed as the person's age as of the admission date, rounded to the nearest year.
3. Using the SAS data set Blood, create two temporary SAS data sets by selecting all subjects with cholesterol levels (Chol) below 100. Place the male subjects in Lowmale and the female subjects in Lowfemale. Do this using a single DATA step.

   Note: Values for Gender are **`Male`** and **`Female`**.

   Careful, some of the cholesterol values are missing. Print the resulting data sets.
4. Using the SAS data set Bicycles, create two temporary SAS data sets as follows: Mountain_USA consists of all observations from Bicycles where Country is **`USA`** and Model is **`Mountain Bike`**. Road_France consists of all observations from Bicycles where Country is **`France`** and Model is **`Road Bike`**. Print these two data sets.
5. Print out the observations in the two data sets Inventory and NewProducts. Next, create a new temporary SAS data set (Updated) containing all the observations in Inventory followed by all the observations in NewProducts. Sort the resulting data set and print out the observations.
6. Repeat Problem 5, except this time sort Inventory and NewProducts first (create two temporary SAS data sets for the sorted observations). Next, create a new, temporary SAS data set (Updated) by interleaving the two temporary, sorted SAS data sets. Print out the result.
7. Using the Gym data set, create a new, temporary SAS data set (Percent) that contains all the variables found in Gym plus a new variable (call it CostPercent) that represents the Cost as a percentage of the average cost for all subjects. Use PROC MEANS to create a SAS data set containing the mean cost (see Program 10.7 to see how to do this). Round this value to the nearest percent.
8. Run the program here to create a SAS data set called Markup:

   ```
   data Markup;
      input Manuf : $10. Markup;
   datalines;
   Cannondale 1.05
   Trek 1.07
   ;
   ```

   Combine this data set with the Bicycles data set so that each observation in the Bicycles data set now has a markup value of 1.05 or 1.07, depending on whether the bicycle is made by Cannondale or Trek. In this new data set (call it Markup_Prices), create a new variable (NewTotal) computed as TotalSales times Markup.

9. Merge the Purchase and Inventory data sets to create a new, temporary SAS data set (Pur_Price) where the Price value found in the Inventory data set is added to each observation in the Purchase data set, based on the Model number (Model). There are some models in the Inventory data set that were not purchased (and, therefore, are not in the Purchase data set). Do not include these models in your new data set. Based on the variable Quantity in the Purchase data set, compute a total cost (TotalCost) equal to Quantity times Price in this data set as well.
10. Using the Purchase and Inventory data sets, provide a list of all Models (and the Price) that were not purchased.
11. Merge the Purchase and Inventory data sets without sorting either one and omitting the BY Model statement. Check the SAS log and list the observations in Purchase, Inventory, and the merged data set. Now, run this same program with the system option MERGENOBY set to WARN. Check the SAS log. Finally, set MERGENOBY to ERROR and run this program a third time. Check the SAS log. **Note:** You might want to give the merged data set a different name for each of the three runs so that it is clear which programs run and which programs do not run.
12. You want to merge two SAS data sets, Demographic_ID and Survey1, based on an identifier. In Demographic_ID, this identifier is called ID; in Survey1, the identifier is called Subj. Both are character variables.
13. You want to merge two SAS data sets, Demographic_ID and Survey2, based on an identifier. In Demographic_ID, this identifier is called ID and it is character; in Survey2, the identifier is also called ID, but it is numeric. **Hint:** If you choose to convert the numeric identifier to a character variable, use a Z3. format so that the leading 0s are present in the character value.
14. Data set Inventory contains two variables: Model (an 8-byte character variable) and Price (a numeric value). The price of Model M567 has changed to 25.95 and the price of Model X999 has changed to 35.99. Create a temporary SAS data set (call it NewPrices) by updating the prices in the Inventory data set.

Chapter 11: Working with Numeric Functions

11.1 Introduction

SAS functions are essential tools in DATA step programming. They perform such tasks as rounding numbers, computing dates from month-day-year values, summing and averaging the values of SAS variables, and hundreds of other tasks. This chapter focuses on functions that primarily operate on numeric values—the next chapter describes some of the functions that work with character data. Date functions were discussed separately in Chapter 9.

There is so much to say about SAS functions that one could write a whole book on the subject. As a matter of fact, someone did. Please check out *SAS Functions by Example, 2nd edition* by this author. Information about SAS functions can be found on the SAS Help Center at http://go.documentation.sas.com. These two chapters touch only the surface of the enormous power of SAS functions.

11.2 Functions That Round and Truncate Numeric Values

Three of the most useful functions in this category are the ROUND, INT, and CEIL functions. As the names suggest, ROUND is used to round numbers, either to the nearest integer or to other values such as 10ths or 100ths. The INT function returns the integer portion of a numeric value. (That is different from

Ents, which are talking trees in Hobbit land.) The CEIL function (stands for ceiling) returns the next highest integer.

As an example, suppose you have raw data values on age (in years) and weight (in pounds). You want a data set with age as of the last birthday (that is, throw away any fractional part of a year) and weight in pounds, rounded to the nearest pound. In addition, you also want to compute weight in kilograms, rounded to the nearest 10th. Here is the program:

Program 11.1: Demonstrating the ROUND and INT Truncation Functions

```
data Truncate;
   input Age Weight Cost;
   Age = int(Age);
   WtKg = round(Weight/2.2, .1);
   Weight = round(Weight);
   Next_Dollar = Ceil(Cost);
datalines;
18.8 100.7 98.25
25.12 122.4 5.99
64.99 188 .0001
;
```

The ROUND function is used twice in this program. When it is used to round the WtKg values, it has two arguments, separated by a comma. The first argument is the value to be rounded—the second argument is the round-off unit. Typical values for round-off units are .1, .01, .001, and so forth. However, you can use other values here as well. For example, a round-off unit of 2 rounds values to the nearest even number—a value of 100 rounds a value to the nearest 100. When ROUND is used with the Weight variable, there is no second argument. Therefore, the default action—rounding to the nearest whole number—occurs. Finally, the CEIL function returns the next highest integer.

Here is a listing of the resulting data set:

Figure 11.1: Listing of Data Set Truncate

Listing of Data Set Truncate

Age	Weight	Cost	WtKg	Next_Dollar
18	101	98.2500	45.8	99
25	122	5.9900	55.6	6
64	188	0.0001	85.5	1

Notice that the fractional part of all the age values has been dropped, the WtKg values are rounded to the nearest 10th and the Weight values are rounded to the nearest whole number. In this program, it was important to compute the value of WtKg before Weight was rounded. Each value of Cost was "rounded up" to the next highest integer (even the value of .0001 returned a value of 1).

11.3 Functions That Work with Missing Values

In SAS, you can refer to a missing numeric value with a period and a missing character value by a single blank (inside single or double quotes). For example, take a look at the short program here that tests to see if several variables contain a missing value:

Program 11.2: Testing for Missing Numeric and Character Values (without the MISSING Function)

```
data Test_Miss;
   set Learn.Blood;
   if Gender = ' ' then MissGender + 1;
   if WBC = . then MissWBC + 1;
   if RBC = . then MissWBC + 1;
   if Chol lt 200 and Chol ne . then Level = 'Low ';
   else if Chol ge 200 then Level = 'High';
run;
```

Instead of using separate values for a missing numeric and character value, you can use the MISSING function instead. This function returns a value of true if its argument is a missing value and false otherwise. The amazing thing about this function is that the argument can be either character or numeric—a real rarity among SAS functions. You can rewrite Program 11.2 using the MISSING function like this:

Program 11.3: Demonstrating the MISSING Function

```
data Test_Miss;
   set Learn.Blood;
   if missing(Gender) then MissGender + 1;
   if missing(WBC) then MissWBC + 1;
   if missing(RBC) then MissWBC + 1;
   if Chol lt 200 and not missing(Chol) then
      Level = 'Low ';
   else if Chol ge 200 then Level = 'High';
run;
```

Besides making programs easier to read, the MISSING function also returns a value of true for the alternate numeric values (`.A, .B, ... .Z,` and `._`). These alternative numeric missing values can be useful when you have different categories of missing information.

For example, you could assign `.A` to `did not answer` and `.B` to `Not applicable`. SAS treats all of these values as missing for numeric calculations but allows you to distinguish among them.

11.4 Setting Character and Numeric Values to Missing

The CALL routine CALL MISSING is a handy way to set one or more character and/or numeric variables to missing. When you use a variable list such as X1–X10 or Char1–Char5, you need to precede the list with the keyword OF. Here are some examples:

Examples of the CALL MISSING Routine

call missing(X,Y,Z,of A1–A10);	X, Y, and Z are numeric and A1–A10 are character. X, Y, and Z are set to a numeric missing value; A1–A10 are set to a character missing value.
call missing(of X1–X10);	X1–X10 are numeric and are set to a numeric missing value.
call missing(of _all_);	All variables defined up to the point of the function call are set to missing.
call missing(of X1–X5,of Y1–Y5);	X1–X5 and Y1–Y5 are all numeric and are set to a numeric missing value.

For those readers interested in the details, the following is a brief discussion on the difference between SAS functions and CALL routines.

A SAS function can only return a single value and, typically, the arguments do not change their value when the function is executed. A CALL routine, unlike a function, cannot be used in an assignment statement. Also, the arguments used in a CALL routine can change their value after the call executes. Therefore, if you want to retrieve more than a single value or, as in the CALL MISSING examples above, setting all of the arguments to a missing value, you need to use a CALL routine instead of a function.

11.5 Descriptive Statistics Functions

One group of SAS functions is called *descriptive statistics functions*. While these functions are capable of computing statistical results such as means and standard deviations, many functions in this category provide extremely useful non-statistical tasks.

Suppose you want to score a psychological test and the scoring instructions state that you should take the average (mean) of the first 10 questions (labeled Q1–Q10). This calculation should be performed only if there are seven or more nonmissing values. In addition, you want to identify the two questions that resulted in the highest and lowest scores, respectively. Here is a program that performs all of these tasks:

Program 11.4: Demonstrating the N, MEAN, MIN, and MAX Functions

```
data Psych;
   input ID $ Q1-Q10;
   if n(of Q1-Q10) ge 7 then Score = mean(of Q1-Q10);
   MaxScore = max(of Q1-Q10);
   MinScore = min(of Q1-Q10);
datalines;
001 4 1 3 9 1 2 3 5 . 3
002 3 5 4 2 . . . 2 4 .
003 9 8 7 6 5 4 3 2 1 5
;
```

What seemed like a difficult problem was solved in just three lines of SAS code, using some of the descriptive statistics functions. The N function returns the number of nonmissing numeric values among its arguments. As with all of the functions in this category, you must precede any list of variables in the form Var1–Var*n* with the keyword OF. Without this, SAS assumes that you want to subtract the two values. If there are seven or more nonmissing values, you use the MEAN function to compute the mean of the nonmissing values. Otherwise, Score will be a missing value. The MEAN function, as well as the other functions discussed in this section, ignores missing values. For example, subject 001 has nine nonmissing values so the mean is computed by adding up the nine values and dividing by 9.

The MAX and MIN functions return the largest and smallest (nonmissing) value of its arguments. Here is a listing of data set Psych:

Figure 11.2: Listing of Data Set Psych

Listing of Data Set Psych

ID	Q1	Q2	Q3	Q4	Q5	Q6	Q7	Q8	Q9	Q10	Score	MaxScore	MinScore
001	4	1	3	9	1	2	3	5	.	3	3.44444	9	1
002	3	5	4	2	.	.	.	2	4	.	.	5	2
003	9	8	7	6	5	4	3	2	1	5	5.00000	9	1

Another function similar to the N function is NMISS. This function returns the number of missing values in the list of variables.

What if you want the second or third largest value in a group of values? The LARGEST function enables you to extract the *n*th largest value, given a list of variables. For example, to find the sum of the three largest scores in the Psych data set, you could use the SUM and LARGEST functions together like this:

Program 11.5: Finding the Sum of the Three Largest Values in a List of Variables

```
data Three_Large;
   set Psych(keep=ID Q1-Q10);
   SumThree = sum(largest(1,of Q1-Q10),
                  largest(2,of Q1-Q10),
                  largest(3,of Q1-Q10));
run;
```

The first argument of the LARGEST function tells SAS which value you want—`1` gives you the largest value, `2` gives you the second largest, and so forth. This function ignores missing values, the same as the other descriptive statistics functions.

> **Note:** LARGEST(1, *variable-list*) is the same as MAX(*variable-list*).

Here is a listing of Three_Large:

Figure 11.3: Listing of Data Set Three_Large

Listing of Data Set Three_Large

ID	Q1	Q2	Q3	Q4	Q5	Q6	Q7	Q8	Q9	Q10	SumThree
001	4	1	3	9	1	2	3	5	.	3	18
002	3	5	4	2	.	.	.	2	4	.	13
003	9	8	7	6	5	4	3	2	1	5	24

Are you surprised to know that SAS also has a SMALLEST function? SMALLEST(1, *variable-list*) gives you the smallest nonmissing value in a list of variables (equal to the result of the MIN function); SMALLEST(2, *variable-list*) gives you the second smallest value, and so on.

11.6 Computing Sums within an Observation

One way to compute the sum of several variables is to write a statement such as this:

```
SumCost = Cost1 + Cost2 + Cost3;
```

What if one of the Cost values is missing? This causes the sum to be missing. If you want to ignore missing values, you can either write some DATA step logic, or simply use the SUM function, like this:

```
SumCost = sum(of Cost1-Cost3);
```

If one or two of the Cost values are missing, SumCost is the sum of the nonmissing values. If all three Cost values are missing, the SUM function returns a missing value. If you want the sum to be 0 if all the arguments are missing, a nice trick is to include a 0 in the list of arguments, like this:

```
SumCost = sum(0, of Cost1-Cost3);
```

You can also see how useful the SUM function is when you have a large number of variables to sum. Here is an example.

You have a SAS data set (EndofYear) containing Pay1–Pay12 and Extra1–Extra12. You want to compute the sum of these 24 values. In addition, you want the sum to be 0 if all 24 values are missing. Here is the program:

Program 11.6: Using the SUM Function to Compute Totals

```
data Sum;
   set Learn.EndofYear;
   Total = sum(0, of Pay1-Pay12, of Extra1-Extra12);
run;
```

11.7 Mathematical Functions

We'll start out with four straightforward functions: ABS (absolute value), SQRT (square root), EXP (exponentiation), and LOG (natural log). The following program demonstrates these four functions—a short explanation follows:

Program 11.7: Demonstrating the ABS, SQRT, EXP, and LOG Functions

```
data Math;
   input x @@;
   Absolute = abs(x);
   Square = sqrt(x);
   Exponent = exp(x);
   Natural = log(x);
datalines;
2 -2 10 100
;
```

To take an absolute value of a number, you throw away the minus sign (if the number is negative). Although you can raise a number to the .5 power to compute a square root, using the SQRT function may make your program easier to read (and to write). The EXP function raises e (the base of natural logarithms) to the value of its argument. The LOG function takes the natural logarithm of its argument. (If you need a base 10 log, use the LOG10 function.)

The @@ signs at the end of the INPUT statement are called a double trailing @. Notice that you are reading four values of **x** on a single line and creating four observations. Without the @@ at the end of the INPUT statement, SAS would go to a new line each time SAS reached the bottom of the DATA step. The @@ at the end of the line is an instruction to "hold the line" and keep reading values on the same line until there are no more values to read.

Note: The double trailing @ is different from the single trailing @ (described in Chapter 21, sections 11 and 12). The single trailing @ holds the line for another INPUT statement in the DATA step but releases the line at the bottom of the DATA step—the double trailing @ does not release the line at the bottom of the DATA step. One of the only reasons for using a double trailing @ is to create more than one observation from a single line of data.

Here is a listing of the Math data set:

Figure 11.4: Listing of Data Set Math

Listing of Data Set Math

x	Absolute	Square	Exponent	Natural
2	2	1.4142	7.39	0.69315
-2	2	.	0.14	.
10	10	3.1623	22026.47	2.30259
100	100	10.0000	2.6881E43	4.60517

Notice that there are missing values for the SQRT and LOG functions when the argument to these functions is a negative value. SAS also prints a note in the log to inform you about this fact.

11.8 Computing Some Useful Constants

The CONSTANT function returns values of commonly used mathematical constants such as pi and e. For a SAS programmer, perhaps the most useful feature of this function is its ability to compute the largest integer that can be stored exactly in less than 8 bytes. The next program demonstrates some of the more common uses of this function:

Program 11.8: Computing Some Useful Constants with the CONSTANT Function

```
data Constants;
   Pi = constant('pi');
   e = constant('e');
   Integer3 = constant('exactint',3);
   Integer4 = constant('exactint',4);
   Integer5 = constant('exactint',5);
   Integer6 = constant('exactint',6);
   Integer7 = constant('exactint',7);
   Integer8 = constant('exactint',8);
run;
```

Pi and e are computed as shown here. To compute the largest integer stored in *n* bytes, you provide a second argument indicating the number of bytes. Output from this program is shown as follows:

Figure 11.5: Listing of Data Set Constants

Listing of Data Set Constants

Pi	e	Integer3	Integer4	Integer5	Integer6	Integer7	Integer8
3.14159	2.71828	8192	2097152	536870912	137438953472	3.5184E13	9.0072E15

To be sure the exact integer feature of this function is clear, if you use a LENGTH statement to set the length of a numeric variable to 3, the largest integer you can represent without losing accuracy is 8,192; with a length of 4, you can represent integers up to 2,097,152. (These values may vary, depending on your operating system.)

11.9 Generating Random Numbers

We will discuss only one of the random number functions here—RAND. This function generates random numbers based on any one of 29 probability distributions. Some of the more commonly used distributions are listed in figure 11.6:

Figure 11.6: Some of the Commonly Used Probability Distributions

Distribution	Example	Description
Uniform	X = rand('uniform');	Uniform random numbers between 0 and 1
Normal	HR = rand('normal',70,10);	Normally distributed numbers with a mean of 70 and a standard deviation of 10
Bernoulli	Heads = rand('Bernoulli',.5);	A Bernoulli distribution with p = .5 (simulates a coin toss)
Binomial	N = rand('Binomial',.3,10)	The number of "successes" in 10 trials with a probability of .3

> **Note:** The RAND function replaces some of the older random number functions such as RANUNI (generates uniform random numbers) and RANNOR (generates normally distributed random numbers).

You may wonder why you would ever need to generate random numbers. Some possible uses are to create data sets for benchmarking, to select random samples, and to assign subjects randomly to two or more groups.

Computers do not really generate true random numbers. However, they are capable of generating series of numbers that are very close to random (called pseudo-random numbers by snobs). In order to get the computer started, you need to provide a *seed* number to be used to generate the first number in the random sequence. If you use the RAND function without first setting a seed with a `call streaminit(n)` command (where *n* is an integer, preferably a large one), SAS uses the computer's clock to supply the seed. If you choose any positive integer in the `call streaminit`, that number is used as a seed. SAS recommends that you choose a seed of at least 7 digits. If you supply the seed, the program generates the same sequence of random numbers every time you run the program. Without the call to `streaminit`, the sequence is different every time you run the program. Hopefully, several examples will make this clear.

As the first example, you want to generate 5 uniform random numbers.

Program 11.9: Program to Generate Five Uniform Random Numbers

```
data Uniform;
   do i = 1 to 5;
      X = rand('uniform');
      output;
   end;
run;
```

First, notice that this DATA step does not read any data. The DO loop causes the RAND function to execute 5 times and output 5 uniform random numbers. It is important that the OUTPUT statement is placed inside the loop, otherwise the program would only output one random number. Let's run this program once and list the output data set:

Figure 11.7: Listing of Data Set Uniform (First Run)

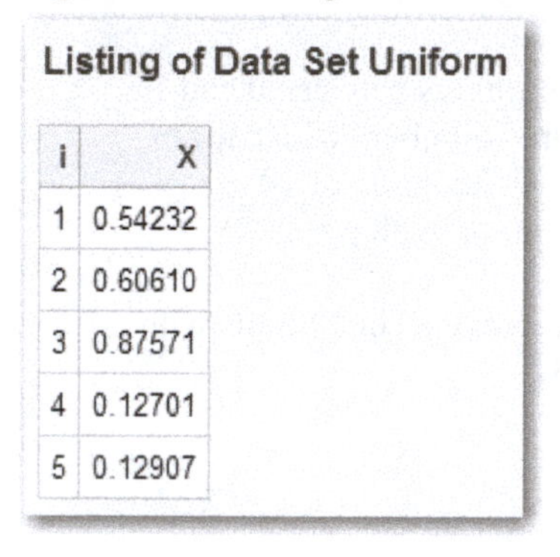

Listing of Data Set Uniform

i	X
1	0.54232
2	0.60610
3	0.87571
4	0.12701
5	0.12907

Notice that all the random numbers are between 0 and 1. Let's run the program again and list the data:

Figure 11.8: Running Program 11.9 Again and Listing the Results

Listing of Data Set Uniform

i	X
1	0.88584
2	0.62149
3	0.92060
4	0.08345
5	0.36384

Notice that the 5 random numbers are different from the ones produced by the first run. The reason is that you did not include the call to `streaminit`. Let's run Program 11.9 twice more, but specify a seed value with a call to `streaminit`:

Program 11.10: Including a Call to Streaminit

```
data Uniform;
   call streaminit(1234567);
   do i = 1 to 5;
      X = rand('uniform');
      output;
   end;
run;
```

We will run Program 11.10 twice and list the output:

Figure 11.9: Listing of Data Set Uniform with Call to Streaminit (First Run)

Listing of Data Set Uniform

i	X
1	0.88584
2	0.62149
3	0.92060
4	0.08345
5	0.36384

Figure 11.10: Listing of Data Set Uniform with Call to Streaminit (Second Run)

Listing of Data Set Uniform

i	X
1	0.88584
2	0.62149
3	0.92060
4	0.08345
5	0.36384

Notice that by setting a seed value, the program generates the same random sequence each time it is run.

You may wonder why would anyone want to do this? One possible reason is to generate random groupings for a clinical trial. When the trial is over, you want to be able to generate the exact same random assignments, so you use the same seed value that you used originally.

Suppose you want to select approximately 10% of the observations from the Blood data set. You can use uniform random numbers like this:

Program 11.11: Using the RAND function to randomly select observations

```
data Subset;
   set Learn.Blood;
   if rand('uniform') le .1;
run;
```

Because values from a uniform distribution are always between 0 and 1, approximately 10% of these numbers will be less than .1 and the observation will be written out to data set Subset. Here is part of the SAS log after this program was run:

```
NOTE: There were 1000 observations read from the data set LEARN.BLOOD.

    NOTE: The data set WORK.SUBSET has 104 observations and 7 variables.
```

Notice that the Subset data set is not exactly 10% of the Blood data set. If you need an exact 10% sample, Use PROC SURVEYSELECT as shown next.

PROC SURVEYSELECT, which is available in SAS/STAT software, is the best (and most efficient) way to create random subsets. This procedure is quite flexible and offers many options for generating these subsets. As an example, to obtain a simple random sample of size 100 from the Blood data set, you could use the following program:

Program 11.12: Using PROC SURVEYSELECT to Obtain a Random Sample

```
proc surveyselect data=Learn.Blood
                  out=Subset
                  method=srs
                  sampsize=100;
run;
```

The procedure options DATA= and OUT= are pretty clear. METHOD= allows you to choose a method for selecting your sample (SRS is a simple random sample, a sample taken without replacement). SAMPSIZE= allows you to choose the size of the sample. One additional option, which is not used here, is SEED=. If you supply a value for this option, this value is used as the seed. Also, if you prefer to select a percentage of the original data set, substitute the option SAMPRATE=*Percent*, (where *Percent* is the proportion of the data set you want to sample) instead of SAMPSIZE=. For more details on this procedure, see the SAS Help Center at http://go.documentation.sas.com.

To generate random **integers** in the range from 1 to 10, you could use:

```
RandomInteger = ceil(rand('uniform')*10);
```

Remember that the CEIL function rounds up to the next highest integer. For example, suppose that your first random number is almost 0 (remember, it can never be 0). Ten times almost 0 is also a small number and the CEIL function returns a 1. If the RAND function returns a number almost equal to 1, ten times almost 1 is almost 10. However, the CEIL function will return a 10 in this case.

11.10 Special Functions

The name *special functions* (a SAS category) makes these functions seem esoteric. That is far from the case. The INPUT and PUT functions, in particular, are extremely useful functions that you will use all the time.

The INPUT function enables you to "read" a character value using a SAS or user-defined informat and assign the resulting value to a SAS numeric variable. One of the most common uses of this function is to perform a character-to-numeric conversion. Another is to convert a date as a character string to a SAS date value. Here is an example.

You are given a SAS data set (Chars) that contains the variables Height, Weight, and Date. All three variables are character variables. You want to create a new data set called Nums with the same variables, except that they are numeric variables.

A listing of data set Chars is shown here:

Figure 11.11: Listing of Data Set Chars

Listing of Data Set Chars

Height	Weight	Date
58	155	10/21/1950
63	200	5/6/2005
45	79	11/12/2004

Although it may not be obvious from the listing, all three variables are character variables.

Program 11.13: Using the INPUT Function to Perform a Character-to-Numeric Conversion

```
data Nums;
   set Learn.Chars (rename=
                   (Height = Char_Height
                    Weight = Char_Weight
                    Date   = Char_Date));
   Height = input(Char_Height,8.);
   Weight = input(Char_Weight,8.);
   Date   = input(Char_Date,mmddyy10.);
   drop Char_Height Char_Weight Char_Date;
run;
```

The technique used in this program is called *swap and drop*. Because the same variable cannot be both character and numeric, this technique allows you to wind up with numeric variables with the same names as the original character variables. You use the data set option RENAME= to rename all three variables. Next, you use the INPUT function to do the character-to-numeric conversions. A good way to understand the INPUT function is to think about what an INPUT statement does. It reads character data from a raw data file using an informat to determine how the value should be read and assigns the result to a SAS variable. Think of the INPUT function as "reading" a value of a character variable according to whatever informat you supply as the second argument of the function. The INPUT function, when used for the Height and Weight variables, uses an 8. informat. It doesn't matter if this value is larger than you need. Because the dates are in the month-day-year form, you use the MMDDYY10. informat to do the conversion. Because you don't want the original character values, you drop them. That's why this technique is called swap and drop.

Before we leave this function, we should point out that you can shorten the DROP statement as follows:

```
drop Char_:;
```

The colon notation is a SAS wildcard. This statement says to drop all variables that begin with `Char_`.

The other special function we discuss here is the PUT function. Just as a PUT statement can send the formatted value of a variable to an external file, a PUT function takes a value (the first argument), formats this value using the format supplied (the second argument), and "writes" the result to a variable. The result of a PUT function is always a character value. One common use of a PUT function is to perform a numeric-to-character conversion.

The following program demonstrates some possible uses of the PUT function:

Program 11.14: Demonstrating the PUT Function

```
proc format;
   value Agefmt low-<20 = 'Group One'
                20-<40  = 'Group Two'
                40-high = 'Group Three';
run;

data Convert;
   set Learn.Numeric;
   Char_Date = put(Date,date9.);
   AgeGroup = put(Age,Agefmt.);
   Char_Cost = put(Cost,dollar10.);
   drop Date Cost;
run;
```

Data set Numeric contains three numeric variables: Date, Age, and Cost. In data set Convert, the three variables Char_Date, AgeGroup, and Char_Cost are all character variables. The second argument of the PUT function specifies the format to apply to the first argument. Therefore, Char_date is a date in the DATE9. format, and Char_Cost is a value written using the DOLLAR10. format. Finally, AgeGroup applies a user-written format to place the age values into one of three groups. Here is a listing of the resulting data set.

Figure 11.12: Listing of Data Set Convert

Listing of Data Set Convert

Age	Char_Date	AgeGroup	Char_Cost
23	15OCT2000	Group Two	$12,345
55	12NOV1923	Group Three	$39,393

11.11 Functions That Return Values from Previous Observations

Because SAS processes data from raw data files and SAS data sets line by line (or observation by observation), it is difficult to compare a value in the present observation with one from a previous observation. Two functions, LAG and DIF, are useful in this regard.

Let's start out with a short program that demonstrates how the LAG function works:

Program 11.15: Demonstrating the LAG and LAG*n* Functions

```
data Look_Back;
   input Time Temperature;
   Prev_Temp = lag(Temperature);
   Two_Back = lag2(Temperature);
datalines;
1 60
2 62
3 65
4 70
;
```

A listing of data set Look_Back follows:

Figure 11.13: Listing of Data Set Look_Back

Listing of Data Set Look_Back

Time	Temperature	Prev_Temp	Two_Back
1	60	.	.
2	62	60	.
3	65	62	60
4	70	65	62

As you can see from this listing, the LAG function returns the temperature from the previous time and the LAG2 function returns the temperature from the time before that. (There is a whole family of LAG functions: Lag, LAG2, LAG3, and so on.) This program might give you the idea that the LAG function returns the value of its argument from the previous observation. This is not always true.

> **Note:** The correct definition of the LAG function is that it returns the value of its argument **the last time the LAG function executed.**

To help clarify this somewhat clunky sounding definition, see if you can predict the values of X and Last_X in the program that follows:

Program 11.16: Demonstrating What Happens When You Execute a LAG Function Conditionally

```
data Laggard;
   input X @@;
   if X ge 5 then Last_X = lag(X);
datalines;
9 8 7 1 2 12
;
```

Here is a listing of data set Laggard:

Figure 11.14: Listing of Data Set Laggard

Listing of Data Set Laggard

Obs	X	Last_X
1	9	.
2	8	9
3	7	8
4	1	.
5	2	.
6	12	7

Are you surprised? The value of Last_X in the first three observations is clear. But, what happened in Observation 6? To understand this, you need to read the definition carefully. The IF statement is not true in Observations 4 and 5; therefore, Last_X, which is set to a missing value at each iteration of the DATA step, remains missing. In Observation 6, the IF statement is true and the LAG function returns the value of X from the last time this function executed, which was back at Observation 3, where X was equal to 7.

Note: The take-home message is this: "Be careful if you execute a LAG function conditionally." In most cases, you want to execute the LAG function for each iteration of the DATA step. When you do, this function returns the value of its argument from the previous observation.

A common use of the LAG function is to compute differences between observations. For example, you can modify Program 11.15 to compute the difference in temperature from one time to the next, as follows:

Program 11.17: Using the LAG Function to Compute Inter-observation Differences

```
data Diff;
   input Time Temperature;
   Diff_Temp = Temperature - lag(Temperature);
datalines;
1 60
2 62
3 65
4 70
;
```

Here is a listing of Diff:

Figure 11.15: Listing of Data Set Diff

Listing of Data Set Diff

Time	Temperature	Diff_Temp
1	60	.
2	62	2
65	4	-58

The LAG function is often used to compute differences between values from one observation to the next in the form:

```
x - lag(x);
```

Therefore, SAS has a set of DIF functions (DIF, DIF2, DIF3, and so on) that is equal to X – LAG(X). You could, therefore, rewrite Program 11.17 like this:

Program 11.18: Demonstrating the DIF Function

```
data Diff;
   input Time Temperature;
   Diff_Temp = dif(Temperature);
datalines;
1 60
2 62
3 65
4 70
;
```

For more examples using the LAG and DIF functions, please see the examples in Chapter 24.

11.12 Sorting Within an Observations—a Game Changer

There are some functions that are useful and can save you some programming time. There are other functions and/or call routines that I would call "game changers." CALL SORTN is in the latter category.

Suppose you give a quiz every week for eight weeks and, if the students take all of the quizzes, you are willing to drop the two lowest quizzes. If they take just 7 quizzes, you will drop only one single lowest quiz grade. You could always use an approach like this:

Program 11.19: Solving the Quiz Problem the Hard Way

```
data Quizzes;
   input ID $ Quiz1 - Quiz8;
   if n(of Quiz1-Quiz8) ge 7 then Quiz_Score = mean(
      largest(1, of Quiz1-Quiz8), largest(2, of Quiz1-Quiz8),
      largest(3, of Quiz1-Quiz8), largest(4, of Quiz1-Quiz8),
      largest(5, of Quiz1-Quiz8), largest(6, of Quiz1-Quiz8));
datalines;
001 80 70 90 100 88 90 90 51
002 80 70 90 100 88 90 . .
003 60 60 70 70 70 70 80 .
;
```

You are using the LARGEST function to select the six highest quiz scores. The first thing you notice about this code is that it violates what I call "Cody's Law of SAS Programming":

> **Note:** If a program is getting very tedious to write, check for an easier way (perhaps an ARRAY, a FUNCTION, a CALL ROUTINE or a MACRO).

In order to simplify the program above, we need to introduce the CALL SORTN routine. Given a list of arguments, this call routine returns the values of each of the arguments in ascending order.

For example if `x1 = 7`, `x2 = .`, `x3=3`, and `x4 = 9`

If you execute the line `call sortn(of x1-x4);` the four x values are now:

`x1 = .`, `x2 =3`, `x3 = 7`, and `x4 = 9`. That is, the x-values are now **sorted within an observation**. Do you see how this reduces the problem of computing the quiz grades above?

Let's rewrite the program above using the CALL SORN routine.

Program 11.20: Repeating Program 11.19 Using the CALL SORTN Routine

```
data Quizzes;
   input ID $ Quiz1 - Quiz8;
   call sortn(of Quiz1 - Quiz8);
   /***Scores are now in ascending order***/
   if n(of Quiz1-Quiz8) ge 7 then Quiz_Score = mean(of Quiz3-Quiz8);
datalines;
001 80 70 90 100 88 90 90 51
002 80 70 90 100 88 90 . .
003 60 60 70 70 70 70 80 .
;
```

This works because the call routine rearranged all the Quiz scores (including missing values) from lowest to highest.

> **Note:** Remember that when you use the CALL SORTN routine, the original values of the arguments are lost. If you need the original values and the values are in sorted order, copy the original values, using different variable names (i.e. Orig_Quiz1-Orig_Quiz8).

To make the program a bit easier to read, you can simulate a descending sort by listing the Quiz scores in reverse order like this:

```
call sortn(of Quiz8 - Quiz1); /***This is a descending sort***/
if n(of Quiz1-Quiz8) ge 7 then Quiz_Score = mean(of Quiz1 - Quiz6);
```

You will find many applications where using CALL SORTN will save you from writing a tedious program.

11.13 Problems

Solutions to odd-numbered problems are located at the back of this book. Solutions to all problems are available to professors. If you are a professor, visit the book's companion website at support.sas.com/cody for information about how to obtain the solutions to all problems.

1. Using the SAS data set Health, compute the body mass index (BMI) defined as the weight in kilograms divided by the height (in meters) squared. Create four other variables based on BMI: 1) BMIRound is the BMI rounded to the nearest integer, 2) BMITenth is the BMI rounded to the nearest tenth, 3) BMIGroup is the BMI rounded to the nearest 5, and 4) BMITrunc is the BMI with a fractional amount truncated. Conversion factors you will need are: 1 Kg equals 2.2 Lbs and 1 inch = .0254 meters.
2. Count the number of missing values for WBC, RBC, and Chol in the Blood data set. Use the MISSING function to detect missing values.
3. Create a new, temporary SAS data set (Miss_Blood) based on the SAS data set Blood. Set Gender, RBC, and Chol to a missing value if WBC is missing. Use the MISSING and CALL MISSING functions in this program.
4. The SAS data set Psych contains an ID variable, 10 question responses (Ques1–Ques10), and 5 scores (Score1–Score5). You want to create a new, temporary SAS data set (Evaluate) containing the following:

   a. A variable called QuesAve computed as the mean of Ques1–Ques10. Perform this computation only if there are seven or more nonmissing question values.
   b. If there are no missing Score values, compute the minimum score (MinScore), the maximum score (MaxScore), and the second highest score (SecondHighest).

5. The SAS data set Psych contains an ID variable, 10 question responses (Ques1–Ques10), and 5 scores (Score1–Score5). You want to create a new, temporary SAS data set (Evaluate) containing the following:

   a. A value (ScoreAve) consisting of the mean of the three highest Score values. If there are fewer than three nonmissing score values, ScoreAve should be missing.
   b. An average of Ques1–Ques10 (call it QuesAve) if there are seven or more nonmissing values.
   c. A composite score (Composit) equal to ScoreAve plus 10 times QuesAve.

6. Write a short DATA _NULL_ step to determine the largest integer you can score on your computer in 3, 4, 5, 6, and 7 bytes.

7. Given values of x, y, and z, compute the following (using a DATA _NULL_ step):
   d. AbsZ = absolute value of z
   e. Expx = e raised to the x power
   f. Circumference = 2 times pi times y

   Use values of x, y, and z equal to 10, 20, and –30, respectively. Round the values for b and c to the nearest .001.
8. Create a temporary SAS data set (Random) consisting of 1,000 observations, each with a random integer from 1 to 5. Make sure that all integers in the range are equally likely. Run PROC FREQ to test this assumption.
9. Using the random functions, create a temporary SAS data set (Fake) with 100 observations. Each observation should contain a subject number (Subj) starting from 1, a random gender (with approximately 40% females and 60% males), and a random age (integers from 10 to 50). Compute the frequencies for Gender and list the first 10 observations in the data set.
10. Data set Char_Num contains character variables Age and Weight and numeric variables SS and Zip. Create a new, temporary SAS data set called Convert with new variables NumAge and NumWeight that are numeric values of Age and Weight, respectively, and CharSS and CharZip that are character variables created from SS and Zip. CharSS should contain leading 0s and dashes in the appropriate places for Social Security numbers and CharZip should contain leading 0s. **Hint:** The Z5. format includes leading 0s for the ZIP code.
11. Repeat Problem 10, except this time use the same variable names for the converted variables. **Hint:** Swap and drop.
12. Using the Stocks data set (containing variables Date and Price), compute daily changes in the prices. Use the statements here to create the plot.

```
title "Plot of Daily Price Differences";
proc sgplot data=Difference;
   series x=Date y=Date;
run;
```

13. Plot the daily stock prices in data set Stocks along with a moving average of the prices using a three-day moving average. Use the PLOT statements here to produce the plots.

```
title "Plot of Price and Moving Average";
proc sgplot data=Moving;
   series x=Date y=Price;
   series x=Date y=Moving;
run;
```

Chapter 12: Working with Character Functions

12.1 Introduction

This chapter covers some of the basic functions that work with character values. These functions enable you to search for character strings, take strings apart and put them back together again, and remove selected characters from a string. With SAS 9, you can search for or remove classes of characters, such as digits, punctuation marks, or space characters. You can even perform fuzzy matches between two character values—useful in matching names that may be misspelled. For more information, see *SAS Functions by Example,* 2nd edition by this author and SAS Help Center at http://go.documentation.sas.com.

12.2 Determining the Length of a Character Value

The LENGTHN function returns the length of a character value, not counting trailing blanks. An older function, LENGTH, performed exactly as the LENGTHN function with one exception—when the argument is a missing value. The older LENGTH function returned 1 in this situation—the newer LENGTHN function returns a 0 (the preferred action).

For those curious readers, the 'N' at the end of the LENGTHN function stands for "null string," that is, a string of zero length. This concept was added with SAS 9. You will see other functions in this chapter that are familiar to you except for the fact that they now end with an 'N'. In most cases, use the 'N' version. Here is an example of the LENGTHN function:

You have a SAS data set (Sales) that includes a variable called Name. You want to see all the names that are longer than 12 characters. Here is the program:

Program 12.1: Determining the Length of a Character Value

```
data Long_Names;
   set Learn.Sales;
   if lengthn(Name) gt 12;
run;
```

The subsetting IF statement is true whenever the number of characters in Name exceeds 12 (not counting trailing blanks but including any blanks within the string).

Another function, LENGTHC, returns the storage length of a string. You may want to test storage lengths when you are combining data from multiple files.

12.3 Changing the Case of Characters

A common programming problem is matching values where the case of the two values may not be the same. Suppose you are presented with two SAS data sets: one has names in mixed case, and the other has names in uppercase. You want to merge the two data sets.

A listing of data sets Mixed and Upper are shown here.

Note: Data sets are already sorted by Name.

Figure 12.1: Listing of Data Sets Mixed and Upper

Listing of Data Set Learn.Mixed

Name	ID
Daniel Fields	123
Patrice Helms	233
Thomas chien	998

Listing of Data Set Learn.Upper

Name	DOB
DANIEL FIELDS	01/03/1966
PATRICE HELMS	05/23/1988
THOMAS CHIEN	11/12/2000

Here is a program to merge the two data sets:

Program 12.2: Changing Values to Uppercase

```
data Mixed;
   set Learn.Mixed;
   Name = upcase(Name);
run;

data Both;
   merge Mixed Learn.Upper;
   by Name;
run;
```

As you might expect, the UPCASE function converts all letters to uppercase.

There are two other functions, LOWCASE and PROPCASE, that convert letters to lowercase and proper case, respectively. Proper case capitalizes the first letter of every "word" and converts the remaining letters to lowercase. By word, we mean any consecutive letters separated by a delimiter. Default delimiters are blank, forward slash, hyphen, open parenthesis, period, and tab. You can specify delimiters as an optional second argument to the PROPCASE function if you want. Program 12.3 demonstrates the use of the PROPCASE function along with the COMPBL function described in the next section.

12.4 Removing Characters from Strings

The two functions in this category are COMPBL and COMPRESS. The former converts two or more blanks to a single blank; the latter removes blanks (default action) or characters that you specify from a character value.

We demonstrate the COMPBL function (don't try to pronounce COMPBL; you might hurt yourself—just say compress blanks) in a program to help standardize some addresses. In the listing of the addresses here, notice that several of the lines contain multiple blanks and there is inconsistent use of case as well.

Figure 12.2: Listing of Data Set Address

```
Listing of Data Set Address

   Name               Street                City          State      Zip

ron    coDY     1178   HIGHWAY 480     camp    verde       tx       78010
jason Tran      123 lake   view drive  East  Rockaway      ny       11518
```

Note: This is a mono-spaced listing so that you can see multiple blanks.

Here is a program that converts all multiple blanks to a single blank; converts the case for Name, Street, and City to proper case; and converts the state abbreviations to uppercase:

Program 12.3: Converting Multiple Blanks to a Single Blank and Demonstrating the PROPCASE Function

```
data Standard;
   set Learn.Address;
   Name = compbl(propcase(Name));
   Street = compbl(propcase(Street));
   City = compbl(propcase(City));
   State = upcase(State);
run;
```

Notice in this example how you can nest one function within another. The COMPBL function has as its argument the value returned by the PROPCASE function (in this example, the order doesn't matter). The following listing shows the result:

Figure 12.3: Listing of Data Set Standard

Listing of Data Set Standard

Name	Street	City	State	Zip
Ron Cody	1178 Highway 480	Camp Verde	TX	78010
Jason Tran	123 Lake View Drive	East Rockaway	NY	11518

12.5 Joining Two or More Strings Together

Putting strings together is called *concatenation*. For example, if you have separate variables representing a first and last name, you can concatenate them (with a blank in between) to create a variable containing the full name.

The concatenation operator, || (or !!), has always been available to SAS programmers. For example, if One = **ABC** and Two = **DEF**, then One || Two is equal to **ABCDEF**.

When you use the || operator, if you do not define the length of the resulting string beforehand, the length of this string is the sum of the lengths of the individual strings you are concatenating.

Three very useful functions, CAT, CATS, and CATX, make this process much easier. The CAT function takes two or more arguments and concatenates them. It is almost the same as using the || operator, except that the length of the resulting string defaults to 200 if you do not define the length first.

The CATS function strips off leading and trailing blanks before joining the strings (remember S for strip). The CATX function is similar to the CATS function, except you supply a separator as the first argument. When you use this function to concatenate strings, leading and trailing blanks are removed and the separator value is placed between each of the strings.

Note: You can use multiple characters for the separator value.

Here are some examples:

Program 12.4: Demonstrating the Concatenation Functions

```
title "Demonstrating the Concatenation Functions";

data _null_;
   Length Join Name1 - Name4 $ 15;
   First = 'Ron   ';
   Last = 'Cody  ';
   Join = ':' || First || ':';
   Name1 = First || Last;
   Name2 = cat(First,Last);
   Name3 = cats(First,Last);
   Name4 = catx(' ',First,Last);
   file print;
   put Join= /
       Name1= /
       Name2= /
       Name3= /
       Name4= /;
run;
```

First the output, and then an explanation:

Figure 12.4: Output from Program 12.4

```
Demonstrating the Concatenation Functions
Join=:Ron   :
Name1=Ron   Cody
Name2=Ron   Cody
Name3=RonCody
Name4=Ron Cody
```

Once again, monospaced output was used instead of HTML so that you can see multiple blanks more easily. The variable Join is created by concatenating a colon to the front and back of the variable First. Notice that there are three blanks between `Ron` and the final colon. Name1 joins the two variables First and Last, but because there are three trailing blanks in First, there are three blanks between the two names. The CAT function gives you the same result as the || operator. Notice that the variable Name3 has no blanks between the names because the CATS function strips leading and trailing blanks before joining the strings. Finally, Name4 has a single blank between the first and last names because a blank was entered as the separator value.

12.6 Removing Leading or Trailing Blanks

Three functions, TRIMN (remember, the N is for null string), LEFT, and STRIP, enable you to remove trailing blanks, leading blanks, or both from a character value. We will demonstrate these functions with a sample program:

Program 12.5: Demonstrating the TRIMN, LEFT, and STRIP Functions

```
data Blanks;
   String = '   ABC  ';
   ***There are 3 leading and 2 trailing blanks in String;
   JoinLeft = ':' || left(String) || ':';
   JoinTrim = ':' || trimn(String) || ':';
   JoinStrip = ':' || strip(String) || ':';
run;
```

Here is the output:

Figure 12.5: Output from Program 12.5

```
Listing of Data Set Blanks

                              Join          Join
String      JoinLeft          Trim          Strip

   ABC      :ABC     :     :   ABC:      :ABC:
```

The variable String has three leading and two trailing blanks. A good way to know if a value has leading and/or trailing blanks is to concatenate a character (a colon, for example) to the beginning and end of a character value.

> **Note:** It is very hard to see trailing blanks on printed output—only experienced SAS programmers can do this!

Between the colons in the three Join variables, we have JoinLeft - ABC followed by five blanks, JoinTrim - three blanks followed by **ABC**, and JoinStrip - no leading or trailing blanks.

When you assign the result of any of these functions to a variable, the length of that variable does not change. So, if you create three variables as in the following examples, the storage length of each of the three variables Left, Trim, and Strip is equal to 8 (the length of String):

```
Left = left(String);
Trim = trimn(String);
Strip = strip(String);
```

Left is **ABC** followed by five blanks. Trim is three blanks followed by **ABC** followed by two blanks. Even though the TRIMN function returns a value with no trailing blanks, when this value is assigned to a variable with a length of 8, the trailing blanks reappear. Using this same logic, Strip is equal to **ABC** followed by five blanks.

12.7 Using the COMPRESS Function to Remove Characters from a String

Here's an interesting, and fairly common problem: you have a SAS data set (or a raw data file) containing phone numbers. These numbers came from various sources and are in various formats, as shown here:

Figure 12.6: Listing of Data Set Learn.Phone

Listing of Data Set Learn.Phone

Phone
(908)232-4856
210.343.4757
(516) 343 - 9293
9342342345

You want to retain only the digits in each of these phone numbers. To do this, you want to remove blanks, left and close parentheses, periods, and hyphens from each of the values. The COMPRESS function allows you to select which characters you want to remove from a character value. Here is one way to accomplish this task:

Program 12.6: Using the COMPRESS Function to Remove Characters from a String

```
data Phone;
   length PhoneNumber $ 10;
   set Learn.Phone;
   PhoneNumber = compress(Phone,' ()-.');
run;
```

The first argument of the COMPRESS function is the character value from which you want to remove characters. The second argument is a list of characters you want to remove. If you omit the second argument, COMPRESS removes all blanks from the string. A listing of data set Phone is shown next:

Figure 12.7: Listing of Data Set Phone

Listing of Data Set Phone

PhoneNumber	Phone
9082324856	(908)232-4856
2103434757	210.343.4757
5163439293	(516) 343 - 9293
9342342345	9342342345

Notice that the character variable PhoneNumber contains only digits.

There is an optional third argument to the COMPRESS function (introduced in SAS 9). This third argument, called a *modifier*, allows you to add character classes such as digits or punctuation to the characters that are to be deleted. If you also include a second argument, the characters identified by the modifiers are added to the ones you already specified.

One of these modifiers (k) reverses the action of the COMPRESS function (to help remember this, k is for *keep*). When you use the k modifier, the COMPRESS function keeps the selected characters and removes everything else.

Note: This is one of the most useful ways to use the COMPRESS function. It is usually better to specify the characters you wish to keep rather than the ones you want to remove.

This table shows some of the more useful modifiers.

Modifier	Action
d	Adds numerals (digits) to the list of characters to be deleted
a	Adds upper- and lowercase letters to the list of characters to be deleted
i	Ignores case
k	Keeps listed characters instead of removing them
s	Adds blanks, tabs, line-feeds, or carriage returns to the list of characters to be deleted
p	Adds punctuation to the list of characters to be deleted

Here are a few examples: If `String = "X1y2Z3"`

Function	Description	Value
`compress(String,,'a')`	Removes all letters	123
`compress(String,,'kd')`	Keeps digits (deletes everything else)	123
`compress(String,'wxyz','i')`	Removes wxyz and ignores case	123
`compress("A?B C99",,'pd')`	Removes punctuation and digits	AB C

Notice that if you want to use modifiers and you do not specify a second argument, you need to use two commas together to indicate that the modifiers are the third argument.

Here is Program 12.6 written using modifiers:

Program 12.7: Demonstrating the COMPRESS Modifiers

```
data Phone;
   length PhoneNumber $ 10;
   set Learn.Phone;
   PhoneNumber = compress(Phone,,'kd');
   *Keep only digits;
run;
```

The resulting data set is identical to the one listed previously.

As you can see in this example, it is easier to use the k modifier to indicate what you want to keep, than it is to list all of the characters you want to remove.

12.8 Searching for Characters

There are a large number of SAS functions that allow you to search for individual characters or character substrings.

Let's start with a problem where you are given a SAS data set containing measurements in either English or metric units.

First, here is a listing of data set Mixed_Units:

Figure 12.8: Listing of Mixed_Units

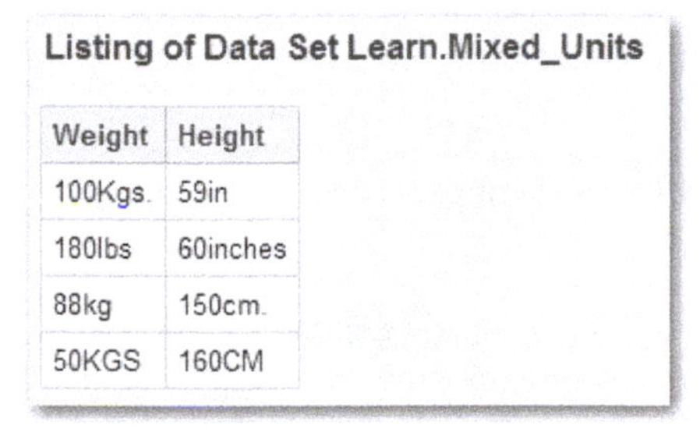

Listing of Data Set Learn.Mixed_Units

Weight	Height
100Kgs.	59in
180lbs	60inches
88kg	150cm.
50KGS	160CM

Notice that the units are not consistently in upper- or lowercase, there may or may not be a period in the units, and they vary slightly. This looks like a difficult problem—but not when you have SAS character functions.

Problems of this type are quite common. Take a look at the program below and then we will discuss the approach:

Program 12.8: Demonstrating the COMPRESS and FIND Functions

```
data Metric;
   set Learn.Mixed_Units;
   Wt_Kg = input(compress(Weight,,'kd'),12.);
   if find(Weight,'lb','i') then Wt_Kg = Wt_Kg/2.2;
   Ht_Cm = input(compress(Height,,'kd'),12.);
   if find(Height,'in','i') then Ht_Cm = Ht_Cm*2.54;
run;
```

The COMPRESS function extracts the digits from the character value Weight, and the INPUT function performs the character to numeric conversion. Even though you called this variable Wt_Kg, you don't know at this point whether or not the weight value you read was in pounds or kilograms.

Next, you use the FIND function to search the value of Weight for the characters `lb`. Here's how this function works:

In its simplest form, you supply the FIND function with two arguments, like this:

```
find(string, find-string)
```

Here, *string* is the string you want to search and *find-string* is the string you are looking for. The function returns the position in *string* where *find-string* is found. If *find-string* is not found, the function returns a zero. For example, the value of the following statement:

```
find("what hat is that","hat")
```

returns a value of 2, the position where the string **hat** begins. An older function, INDEX, was similar to the FIND function as described here. However, the FIND function has some added capabilities; you can add two more arguments to the FIND function. One of these arguments is a modifier that alters how the search works. The most useful modifier is the i (ignore case) modifier. You can also specify a starting position at where to start the search. If you use a negative starting value, the search proceeds from right to left. To summarize, the FIND function has the following syntax:

```
find(string, find-string, modifiers, starting-position)
```

By the way, you can use modifiers, a starting position, or both, and they can be in any order! How is that possible? Because modifiers are always character values and starting positions are always a numeric value, SAS can figure out which argument is a modifier and which argument is a starting position.

In the previous program, the i modifier is used so that the search looks for upper- or lowercase values of **LB** or **IN**.

If you determine the units of Weight contain the characters 'LB' (in upper- or lowercase), you divide the value of Wt_Kg by 2.2 to convert pounds to kilograms—if you determine the units of Heights contain the characters 'IN' (in upper- or lowercase), you multiply the value of Ht_Cm by 2.54 to convert inches to centimeters.

You would most likely drop the original variables Weight and Height, but they were not dropped in this example so that you can see how the conversion worked. A listing of the resulting data set is shown next:

Figure 12.9: Listing of Data Set Metric

Listing of Data Set Metric

Weight	Height	Wt_Kg	Ht_Cm
100Kgs.	59in	100.000	149.86
180lbs	60inches	81.818	152.40
88kg	150cm.	88.000	150.00
50KGS	160CM	50.000	160.00

12.9 Searching for Individual Characters

A companion function to FIND is FINDC. While FIND searches a string for a specific sub-string of characters, FINDC searches a string for the first occurrence of any one of the characters from a list that you supply as the second argument to the function. The table here compares the two functions, FIND and FINDC.

Line	Function	Value Returned
1	`find('XYZCBA','ABC')`	0

Line	Function	Value Returned
2	`findc('XYZCBA','ABC')`	4
3	`find('ABCDE','D')`	4
4	`findc('ABCDE','D')`	4
5	`find('XYZcba','ABC','i')`	0
6	`findc('XYZcba','ABC','i')`	4

In Line 1 of the table, you are looking for the string **`ABC`**. Because this string does not appear in the first argument, the function returns a **`0`**. In line 2, the FINDC function returns a **`4`** (the position of the **`C`**). When you are looking for a single character, as in Lines 3 and 4, both functions return the same value—the position of that character. The last two lines in the table demonstrate the i (ignore case) modifier that you can use with both of these functions.

12.10 Searching for Words in a String

The FINDW function is similar to the FIND function, except that it searches for words (hence the W in the function name).

Words are defined as a series of characters that start and end with a word boundary. The default word boundaries are the beginning of a string, the end of a string, a blank, and the following: ! $ % & ( ) * + , - . / ; < ^ |. The list of default delimiters is slightly different if you are using the EBCDIC character set. You can specify alternative delimiters as the third argument to the FINDW function (you might want to use a single blank instead of the default list of characters).

The syntax for this function is as follows:

```
findw(string, find-string, delimiters, modifiers, starting-position)
```

Here, *string* is the string you want to search, *find-string* is the string you are looking for, and *delimiters* supplies a list of word delimiters. Note that if you want to supply modifiers (most likely the 'i' ignore case modifier), you must also supply one or more delimiters (even if all you want is to use the default values).

The following table shows the difference between the two functions, FIND and FINDW.

Function	Value Returned
`find('there is the dog','the')`	1 (position of the string the)
`findw('there is the dog','the')`	10 (start of the word the)
`findw('there is THE dog','the',' ','i')`	10 (Start of the word THE)
`findw('pear:apple','apple',':')`	6 (start of the word apple)

As an example, suppose you want to look for the name **Roger** in a list of character values (ignoring case). Here is the program:

Program 12.9: Demonstrating the FINDW Function

```
data Look_for_Roger;
   input String $40.;
   if findw(String,'Roger',' ','i') then Match = 'Yes';
   else Match = 'No';
datalines;
Will Rogers
Roger Cody
Was roger here?
Was Roger here?
;
```

Here is a listing:

Figure 12.10: Listing of Data Set Look_for_Roger

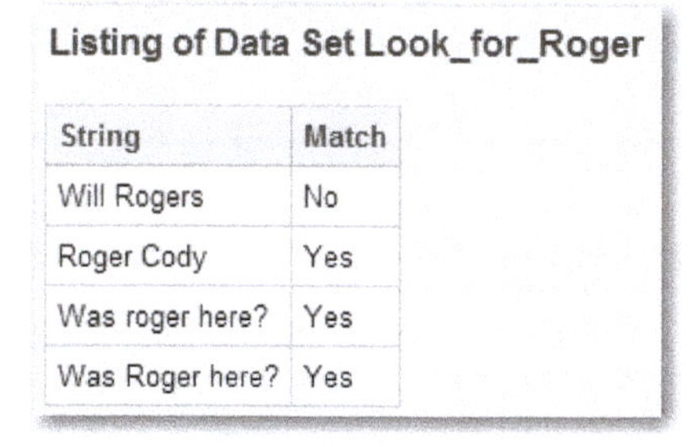

Listing of Data Set Look_for_Roger

String	Match
Will Rogers	No
Roger Cody	Yes
Was roger here?	Yes
Was Roger here?	Yes

This program shows why the FINDW function is useful when you are looking for words that may be part of a longer string.

12.11 Searching for Character Classes

A group of functions that some folks call the *ANY* functions (ANYALNUM, ANYALPHA, ANYDIGIT, ANYPUNCT, and ANYSPACE) allow you to search for alphanumeric values (upper- and lowercase letters and digits), alphas (all letters), digits, punctuation characters, and space characters (blank, tab, line feeds), respectively.

As an example, suppose you have some ID values, some that contain digits and some that do not. You want to search each ID and create a list of all the IDs that contain one or more digits. The IDs are stored in a text file called **ID.txt**.

A listing of the **ID.txt** file is as follows:

```
abcABC
NJ1234
987654
123XYZ
TX78010
```

Here is the program:

Program 12.10: Demonstrating the ANYDIGIT Function

```
data Any_Digit;
   infile 'C:\books\learning\ID.txt';
   input ID : $10.;
   Position = anydigit(ID);
   if Position then output;
run;
```

If the argument of ANYDIGIT contains any digits, the function returns the position of the first one. If not, ANYDIGIT returns a 0. This program added an extra line creating the variable Position so that you can see more clearly how the ANYDIGIT function works. Remember that the value of Position will be nonzero for any ID that contains a digit. Because all values that are not 0 or missing are considered true in logical expressions, these IDs are written out to the data set Any_Digit. Here are the listings:

Figure 12.11: Listing of Data Set Any_Digit

Listing of Data Set Any_Digit

ID	Position
NJ1234	3
987654	1
123XYZ	1
TX78010	3

There is an optional second argument for all of the ANY functions. This argument indicates the starting position of the character value to start looking for the specified characters. If the start value is negative, the search proceeds from right to left, starting at the absolute value of this parameter.

12.12 Using the NOT Functions for Data Cleaning

Another collection of functions, the NOT functions, are similar to the ANY functions. The difference is that these functions return the position of the first character in a string that does **not** belong to the specified class.

That makes this collection of functions especially useful for checking a character value for characters that don't belong. For example, if you expect a string to contain only digits, you can see if the NOTDIGIT function returns any value greater than 0. Let's see some examples.

You have a SAS data set (Cleaning) that contains three variables: Letters, Digits, and Both. As you can probably tell from these variable names, Letters should contain only upper- and lowercase letters; Digits should contain only digits, and Both can contain either letters or digits. You want to write a program to produce an error report. Here is a listing of the Cleaning data set:

Figure 12.12: Listing of Data Set Cleaning

Listing of Data Set Learn.Cleaning

Subject	Letters	Digits	Both
1	Apple	12345	XYZ123
2	Ice9	123X	Abc.123
3	Help!	999	X1Y2Z3

And here is a program that produces the error report:

Program 12.11: Demonstrating the NOT Functions for Data Cleaning

```
title "Data Cleaning Application";

data _null_;
   file print;
   set Learn.Cleaning;
   if notalpha(trimn(Letters)) then put Subject= Letters=;
   if notdigit(trimn(Digits))  then put Subject= Digits=;
   if notalnum(trimn(Both))    then put Subject= Both=;
run;
```

It is very important to understand why you need to use the TRIMN function in this program. Without the TRIMN function, each of the three NOT functions used here would return the position of the first trailing blank in each of the character values. The three NOT functions are used to check for any invalid character type in each of the three variables. Here is the output:

Figure 12.13: Output from Program 12.11

Data Cleaning Application

```
Subject=2 Letters=Ice9
Subject=2 Digits=123X
Subject=2 Both=Abc.123
Subject=3 Letters=Help!
```

12.13 Extracting Part of a String

It is useful to be able to extract one or more characters from a character variable. For example, an ID number such as NJ12M99 might contain a state abbreviation as the first two digits and a gender code in the 5th position. You might even want to extract the digits between the state code and the gender as well. The SUBSTR (stands for substring) function allows you to accomplish these tasks. Here is a program that extracts a state abbreviation, the digits in Positions 3–4 (and creates a numeric variable), the gender code in the 5th position, and the final digit or digits as a character variable:

Program 12.12: Using the SUBSTR Function to Extract Substrings

```
data Extract;
   input ID : $10. @@;
   length State $ 2 Gender $ 1 Last $ 5;
   State = ID;
   Number = input(substr(ID,3,2),3.);
   Gender = substr(ID,5,1);
   Last = substr(ID,6);
datalines;
NJ12M99 NY76F4512 TX91M5
;
```

The first argument to this function is the string you are searching. The second and third arguments of the SUBSTR function specify the starting position and the length of the substring. For example, the variable Gender is extracted from ID, starting in the fifth position for a length of 1. The variable Number is a bit more complicated. The SUBSTR function extracts two digits, starting at Position 3, and then uses the INPUT function to perform a character-to-numeric conversion. Notice that the SUBSTR function used to create the variable Last does not use a third argument. When the third argument is missing, the SUBSTR function extracts characters until the last non-blank character in the string.

Wait! What about State? Why wasn't the SUBSTR function used in that assignment statement? Notice that the variable State is declared as a character variable with a length of 2 in the LENGTH statement. When you set State equal to ID, because the storage length of State is 2, it contains the first two characters of the variable ID. This is a nice "trick" used by many SAS programmers. The reason it is preferred to using the SUBSTR function is that functions use computer resources (CPU time) and this trick avoids using any functions. (Once again, the author's compulsive programming personality shows through.)

Here is a listing of the resulting data set:

Figure 12.14: Listing of Data Set Extract

Listing of Data Set Extract

ID	State	Gender	Last	Number
NJ12M99	NJ	M	99	12
NY76F4512	NY	F	4512	76
TX91M5	TX	M	5	91

In this program, a LENGTH statement is used to define the lengths of State, Gender, and Last. Without this statement, the lengths of these three character variables would be equal to 10, that is the length of ID. Understanding why is very important.

SAS assigns a length to all character variables in the compile stage before any data values are read. In general, the starting position and the length of a substring could be read from data or computed in the DATA step. So, without a LENGTH statement, what length should SAS assign to the variables State, Gender, and Last? The longest substring you can extract from a string of length *n* is a string of length *n*, so if you don't define a length, that's exactly what SAS does.

12.14 Dividing Strings into Words

The SCAN function is typically used to extract words from a string. However, as you will see in a later example, it can also be used to parse (separate) substrings separated by delimiters other than blanks. Let's start with a common example—separating the first and last names from a variable that contains both:

Program 12.13: Demonstrating the SCAN Function

```
data Original;
   input Name $ 30.;
datalines;
Jeffrey Smith
Ron Cody
Alan Wilson
Alfred E. Newman
;

data First_Last;
   set Original;
   length First Last $ 15;
   First = scan(Name,1,' ');
   Last = scan(Name,2,' ');
run;
```

The SCAN function extracts the *n*th word from a string with blanks and most punctuation characters as the default delimiters. (The default delimiters differ in the ASCII and EBCDIC character sets. They are < (+ & ! $ *) ; ^ - / , % | and < (+ | & ! $ *) ; ¬ - / , % | ¢ respectively.) It is usually a good idea to specify the word delimiters you want to use as the third argument to the SCAN function. In this program, a blank is specified as the delimiter.

A LENGTH statement is used to define the length of First and Last. If you don't include this statement, both First and Last will have a length equal to the length of Name (you can't extract a substring longer than the length of the first argument). When this function was first introduced to SAS, a default length of 200 was used. It was changed later to the present value. Here is a listing of data set First_Last:

Figure 12.15: Listing of Data Set First_Last

Listing of Data Set First_Last

Name	First	Last
Jeffrey Smith	Jeffrey	Smith
Ron Cody	Ron	Cody
Alan Wilson	Alan	Wilson
Alfred E. Newman	Alfred	E.

Notice that the middle initial was read as a last name for Mr. Newman. So, how do you solve this problem? You could use the SCAN function to look for a possible third word in this program. If you ask for a third word when there are only two, SCAN returns a missing character value. You could test each Name for a third word. If it is found, you can assume that the second word is an initial. However, there is an easier way.

Here is a program that extracts the last name from the Name value and produces a list of names in alphabetical order:

Program 12.14: Using the SCAN Function to Extract the Last Name

```
data Last;
   set Original;
   length Last_Name $ 15;
   Last_Name = scan(Name,-1,' ');
run;

proc sort data=Last;
   by Last_Name;
run;

title "Alphabetical List of Names";
proc print data=Last noobs;
   var Name Last_Name;
run;
```

The trick is to use a **negative value** as the second argument to the SCAN function. A negative value causes the scan to proceed from right to left, starting at the end of the string. Here is the result:

Figure 12.16: Output from Program 12.14

Alphabetical List of Names

Name	Last_Name
Ron Cody	Cody
Alfred E. Newman	Newman
Jeffrey Smith	Smith
Alan Wilson	Wilson

Notice that the names are in alphabetical order by last name.

12.15 Performing a Fuzzy Match

The SPEDIS function (stands for *spelling distance*) is used for *fuzzy matching*, which is comparing character values that may be spelled differently. The logic is a bit complicated, but using this function is quite easy. As an example, suppose you want to search a list of names to see if the name **`Friedman`** is in the list. You want to look for an exact match or names that are similar. Here is such a program:

Program 12.15: Using the SPEDIS Function to Perform a Fuzzy Match

```
data Fuzzy;
   input Name $20.;
   Value = spedis(Name,'Friedman');
datalines;
Friedman
Freedman
Xriedman
Freidman
Friedmann
Alfred
FRIEDMAN
;
```

Here is a listing of data set Fuzzy:

Figure 12.17: Listing of Data Set Fuzzy

Listing of Data Set Fuzzy

Name	Value
Friedman	0
Freedman	12
Xriedman	25
Freidman	6
Friedmann	3
Alfred	100
FRIEDMAN	87

The SPEDIS function returns a 0 if the two arguments match exactly. The function assigns penalty points for each type of spelling error. For example, getting the first letter wrong is assigned more points than misspelling other letters. Interchanging two letters is a relatively small error, as is adding an extra letter to a word.

Once the total number of penalty points has been computed, the resulting value is computed as a percentage of the length of the first argument. This makes sense because getting one letter wrong in a 3-letter word would be a more serious error than getting one letter wrong in a 10-letter word.

Notice that the two character values evaluated by the SPEDIS function are case-sensitive (look at the last observation in the listing). If case may be a problem, use the UPCASE, LOWCASE, or PROPCASE function before testing the value with SPEDIS.

To identify any name that is similar to **`Friedman`**, you could extract all names where the value returned by the SPEDIS function is less than some predetermined value. In the program here, values less than 15 or 20 would identify some reasonable misspellings of the name.

12.16 Substituting Strings or Words

The last two functions in this chapter are TRANSLATE and TRANWRD. TRANSLATE allows you to substitute one character value for another. TRANWRD lets you substitute one word for another (hence the name, translate word—TRANWRD).

In this first example, you want to substitute the values **A**, **B**, **C**, **D**, and **E** for the character values **1**, **2**, **3**, **4**, and **5**.

Program 12.16: Demonstrating the TRANSLATE function

```
data Trans;
   input Answer : $5.;
      Answer = translate(Answer,'ABCDE','12345');
datalines;
14325
AB123
51492
;
```

The resulting data set Trans contains the following three observations:

Figure 12.18: Listing of Data Set Trans

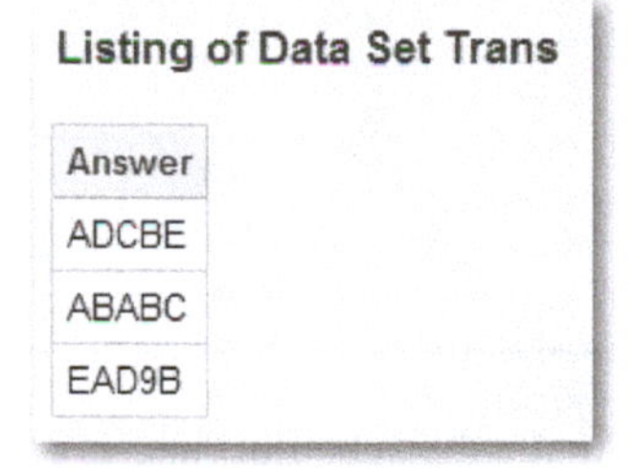

Listing of Data Set Trans

Answer
ADCBE
ABABC
EAD9B

The values **1-5** are replaced by **A-E**. Any other characters are not changed. The TRANSLATE function takes three arguments: the first argument is the character string you want to change; the second and third arguments are the *to-string* and the *from-string*. Each character in the *to-string* is substituted for the corresponding character in the *from-string*. In the program above, a 1 becomes an A, a 2 becomes a B, and so forth.

The last function in this chapter (hurray!) is TRANWRD. This function lets you substitute one word for another.

This function is often used to standardize addresses (changing Street to St., Road to Rd., and so forth). Here is an example:

Program 12.17: Using the TRANWRD Function to Standardize an Address

```
data Address;
   infile datalines dlm=' ,';
   *Blanks or commas are delimiters;
   input #1 Name $30.
         #2 Line1 $40.
         #3 City & $20. State : $2. Zip : $5.;

   Name = tranwrd(Name,'Mr.',' ');
   Name = tranwrd(Name,'Mrs.',' ');
   Name = tranwrd(Name,'Dr.',' ');
   Name = tranwrd(Name,'Ms.',' ');
   Name = left(Name);

   Line1 = tranwrd(Line1,'Street','St.');
   Line1 = tranwrd(Line1,'Road','Rd.');
   Line1 = tranwrd(Line1,'Avenue','Ave.');
datalines;
Dr. Peter Benchley
123 River Road
Oceanside, NY 11518
Mr. Robert Merrill
878 Ocean Avenue
Long Beach, CA 90818
Mrs. Laura Smith
80 Lazy Brook Road
Flemington, NJ 08822
;
```

The first argument to the function is the character value where you want to make the substitutions, and the second and third arguments are the *from-string* and *to-string*, respectively.

Notice the use of the TRANWRD function to remove the titles (Mr., Mrs., Dr., and Ms.) from the names. Here you are substituting a blank for these strings. To remove the leading blank that results, you use the LEFT function to left-align the name.

If the length of the resulting variable is not defined before you use this function, SAS sets the length to 200. The reasoning here is that the resulting string could be longer than the original string after a substitution; therefore, the default length is made quite large. (Two hundred was not a completely arbitrary number—it was the maximum length of a character variable in SAS 6.) In this program, length is not an issue because the length of the variables resulting from the TRANWRD function have already been defined and, because you are substituting shorter strings for longer ones, the original length is long enough to hold the new values. If this were not the case, you would need to place a LENGTH statement **before** your SET statement to define a new length for the character variable you are modifying.

Here is a listing of the Address data set. Notice that all the substitutions have been made and the titles (for example, Mr. and Mrs.) have all been removed.

Figure 12.19: Listing of Data Set Address

Listing of Data Set Address

Name	Line1	City	State	Zip
Peter Benchley	123 River Rd.	Oceanside	NY	11518
Robert Merrill	878 Ocean Ave.	Long Beach	CA	90818
Laura Smith	80 Lazy Brook Rd.	Flemington	NJ	08822

And so ends this rather long chapter. Even though this chapter only covered some of the more useful character functions, you can see that SAS has a tremendously powerful set of functions to manipulate character data.

12.17 Problems

Solutions to odd-numbered problems are located at the back of this book. Solutions to all problems are available to professors. If you are a professor, visit the book's companion website at support.sas.com/cody for information about how to obtain the solutions to all problems.

1. Look at the following program and determine the storage length of each of the variables:

```
data Storage;
   length A $ 4 B $ 4;
   Name = 'Goldstein';
   AandB = A || B;
   Cat = cats(A,B);
   if Name = 'Smith' then Match = 'No';
      else Match = 'Yes';
   Substring = substr(Name,5,2);
run;
```

   A________________
   B________________
   Name________________
   AandB________________
   Cat________________
   Match________________
   Substring________________

2. Using the data set Mixed, create a temporary SAS data set (also called Mixed) with the following new variables:
   a. NameLow – Name in lowercase
   b. NameProp – Name in proper case
   c. (Bonus – difficult) NameHard – Name in proper case without using the PROPCASE function

3. Create a new, temporary SAS data set (Names_And_More) using the permanent SAS data set Names_And_More with the following changes:

a. Name has single blanks between the first and last name.
b. Phone contains only digits (and is still a character value).

4. Data set Names_And_More contains a character variable called Height.

   Here is a listing of data set Names_And_More:

```
Listing of Data Set Learn.Names_and_More

Name                   Phone           Height       Mixed

Roger   Cody         (908)782-1234   5ft. 10in.   50 1/8
Thomas Jefferson     (315) 848-8484  6ft. 1in.    23 1/2
Marco Polo           (800)123-4567   5Ft. 6in.    40
Brian Watson         (518)355-1766   5ft. 10in    89 3/4
Michael DeMarco      (445)232-2233   6ft.         76 1/3
```

   As you can see, the heights are in feet and inches. Assume that these units can be in upper- or lowercase and there may or may not be a period following the units. Create a temporary SAS data set (Height) that contains a numeric variable (HtInches) that is the height in inches.

   Note: One of the Height values is missing an inches value. Be sure that there are no character-to-numeric notes in the SAS log.

   Hints: You can use the KD modifiers to keep the digits and use a single blank in the second argument to keep blanks. You can then use the SCAN function to extract the feet and inches values.

5. Data set Names_And_More contains a character variable called Mixed that is either an integer or a mixed number (such as `50 1/8`). Using this data set, create a new, temporary SAS data set with a numeric variable (Price) that has decimal values. This number should be rounded to the nearest .001.

   Hint: Use the SCAN function with blanks and forward slashes (/) as delimiters.

6. Data set Study (shown here) contains the character variables Group and Dose. Create a new, temporary SAS data set (Study) with a variable called GroupDose by putting these two values together, separated by a hyphen. The length of the resulting variable should be 6 (test this using PROC CONTENTS or the SAS Explorer). Make sure that there are no blanks (except trailing blanks) in this value. Try this problem two ways: first using one of the CAT functions, and second without using any CAT functions.

   Here is the listing:

```
Listing of Data Set Learn.Study

Subj  Group  Dose  Weight   Subgroup

001     A    Low   220lbs.     2
002     A    High  90Kg.       1
003     B    Low   88kg        1
004     B    High  165lbs.     2
005     A    Low   88kG        1
```

7. Data set Study contains a character variable (Group) and a numeric variable (Subgroup). Create a new, temporary SAS data set with these two variables plus a variable consisting of the Group, a dash, and the Subgroup (call it Combined). Try doing this with and without using any of the CAT functions. Be sure there are no conversion messages in your SAS log.
8. Notice in the listing of data set Study in Problem 6 that the variable called Weight contains units (either lbs or kgs). These units are not always consistent in case and may or may not contain a period. Assume an upper- or lowercase **`LB`** indicates pounds and an upper- or lowercase **`KG`** indicates kilograms. Create a new, temporary SAS data set (Study) with a numeric variable also called Weight (careful here) that represents weight in pounds, rounded to the nearest 10th of a pound.

   Note: 1 kilogram = 2.2 pounds.
9. Using the Sales data set, create a temporary SAS data set (Spirited) containing all the observations from Sales where the string (not necessarily the word) **`SPIRIT`** in either upper-, lower-, or mixed case is part of the Customer value (variable name Customer).
10. Data set Errors contains character variables Subj (3 bytes) and PartNumber (8 bytes). (See the partial listing here.) Create a temporary SAS data set (Check1) with any observation in Errors that violates either of the following two rules: first, Subj should contain only digits, and second, PartNumber should contain only the uppercase letters **`L`** and **`S`** and digits.

    Here is a partial listing of Errors:

```
Listing of Data Set Learn.Errors

        Part
Subj   Number      Name

001    L1232    Nichole Brown
0a2    L887X    Fred Beans
003    12321    Alfred 2 Nice
004    abcde    Mary Bumpers
X89    8888S    Gill Sandford
```

11. List the subject number (Subj) for all observations in Errors where the Name contains a digit. (See a listing of the previous data set.)
12. List the subject number (Subj) for any observations in Errors where PartNumber contains an upper- or lowercase **`X`** or **`D`**.
13. Data set Social contains two variables, SS1 and SS2. These variables represent all possible combinations of Social Security numbers from two separate data sets. Using this data set, create two temporary SAS data sets: one, where SS1 is equal to SS2, and two, where SS1 is within a spelling distance of 25 of SS2. Call these data sets Exact and Within25, respectively.

    Hint: You can compute the spelling distance between two Social Security numbers just as you would between two names.
14. List all patients in the Medical data set where the word *antibiotics* is in the comment field (Comment).
15. Using the Names_And_More data set, create a temporary SAS data set containing the phone number (Phone) and the 3-digit area code (AreaCode). Be sure the length of AreaCode is 3. You may need to list a few observations in Names_And_More to see how Phone is stored.

16. Provide a list, in alphabetical order by last name, of the observations in the Names_And_More data set. Set the length of the last name to 15 and remove multiple blanks from Name.

    Note: The variable Name contains a first name, one or more spaces, and then a last name.

17. List the observations in data set Personal. Replace the first 7 digits of the Social Security number (SS) with asterisks and replace the last character (Position 5) of the account number (AcctNumber) with a dash. Do not include the variables Food1–Food8 in the list.

Chapter 13: Working with Arrays

13.1 Introduction

Cody's rule of SAS programming goes something like this: if you are writing a SAS program, and it is becoming very tedious, stop. There is a good chance that there is a SAS tool, perhaps arrays, functions, or macros, that will make your task less tedious.

To the beginning programmer, arrays can be a bit frightening—to an experienced programmer, arrays can be a huge time saver. So, let's get you over the fright and into saving time.

First of all, what is an array? SAS *arrays* are a collection of elements (usually SAS variables) that allow you to write SAS statements referencing this group of variables.

Note: SAS arrays are different from arrays in many other programming languages. They do not hold values, and they allow you to refer to a collection of SAS variables in a convenient manner.

It is much easier to understand what an array is by a few simple examples. So, here goes.

13.2 Setting Values of 999 to a SAS Missing Value for Several Numeric Variables

Typically, arrays are used to perform a similar operation on a group of variables. In this example, you have a SAS data set called SPSS that contains several numeric variables. The folks that created this data set used a value of `999` whenever there was a missing value. (Some statistical packages such as SPSS allow you to substitute a missing value for a specific value. It is common to use values such as `999` or `9999` to represent this missing value.)

The following is a program that solves this problem without using arrays:

Program 13.1: Converting Values of 999 to a SAS Missing Value—Without Using Arrays

```
data New;
   set Learn.SPSS;
   if Height = 999 then Height = .;
   if Weight = 999 then Weight = .;
   if Age    = 999 then Age    = .;
run;
```

Notice that you are writing the same SAS statement several times—the only thing that is changing is the name of the variable. You would like to be able to write something like this:

```
if Height, Weight, or Age = 999 then
   Height, Weight or Age = .;
```

This is pretty much how an array works. Let's first see the program using arrays, and then we'll go through the explanation:

Program 13.2: Converting Values of 999 to a SAS Missing Value—Using Arrays

```
data New;
   set Learn.SPSS;
   array Myvars{3} Height Weight Age;
   do i = 1 to 3;
      if Myvars{i} = 999 then Myvars{i} = .;
   end;
   drop i;
run;
```

The first thing you may notice is that the program with arrays is longer than the one without arrays! However, if you had 50 or 100 variables to process, the program using arrays would be the same length as this program.

The ARRAY statement is used to create the array. Following the keyword ARRAY is the name you choose for your array. Array names follow the same rules you use for SAS variables. In this example, you chose the name MYVARS as the array name. Following the array name, you place the number of elements (in this example, the three variables) in brackets. Finally, you list the variables you want to include in the array. You may use any of the SAS shorthand methods for referring to a list of variables here, such as Var1–Var*n*. This list of variables must be all numeric or all character—you cannot mix them.

You may also use square brackets [] or parentheses () following the array name to specify the number of elements in the array. SAS documentation usually uses curly brackets {}. We recommend that you always use the same type of brackets (either straight or curly) when you use arrays. The reason—SAS function names are always followed by a set of parentheses, so it might be difficult to know if a program is making an array reference or using a function if you use parentheses in your array references.

Once you have defined your array, you can use an array reference in place of a variable name anywhere in the DATA step. For example, MYVARS{2} can be used in place of the variable Weight. The number in the brackets following the array name is called a *subscript*, even though it is not true subscript notation.

By placing the array in a DO loop, you can process each variable in the array. Also, because you do not need or want the DO loop counter included in the SAS data set, you use a DROP statement.

Let's "play computer" and follow the logic of this program. At the top of the DATA step, the SET statement brings in an observation from data set SPSS. Next, the DO loop counter starts at 1. The IF statement references the array element MYVARS{1}, which is equivalent to the variable Height. SAS checks if the value of Height is equal to `999` and, if so, replaces it with a SAS missing value.

The DO loop continues in the same way for all three array elements. When the DO loop finishes, you are at the bottom of the DATA step. An observation is written out to the NEW data set and control returns to the top of the DATA step, where the next observation from data set SPSS is read. This process continues until there are no more observations to read from the SPSS data set.

The best way to get started writing arrays is to first write a few lines of code without an array. Next, write an ARRAY statement where the array elements are all the variables you want to process. Next, use one of the sample lines as a template and write that line, substituting your array name (with a subscript of your choice) for the variable name. Finally, place this line (or lines) of code inside a DO loop. (Don't forget to drop the DO loop counter.) You are done.

13.3 Setting Values of NA and ? to a Missing Character Value

As we mentioned in the introduction, SAS arrays must contain all numeric or all character variables. If the variables you want to include in a character array are already defined as character (for example, they are coming from a SET statement), you can write an ARRAY statement resembling the one in the previous section. However, if you want the array to contain new variables, you need to include a dollar sign ($) and, optionally, a length when you define the array. For example, to create an array of character variables Q1–Q20, each with a length of 2 bytes, you would write the following:

```
array Mychars{20} $ 2 Q1-Q20;
```

It is a good idea to include the dollar sign ($) in every character array, even if the array variables have been previously defined as character.

As an example, the following program uses a character array to convert all values of NA (Not Applicable) or question mark (?) to a SAS missing value. Suppose you are given a SAS data set Chars and you want to create a new data set named Missing with these changes. Here is the program:

Program 13.3: Converting Values of NA and ? to a Missing Character Value

```
data Missing;
   set Learn.Chars;
   array Char_Vars{*} $ _character_;
   do Loop = 1 to dim(Char_Vars);
      if Char_Vars{Loop} in ('NA','?') then
      Char_Vars{Loop} = ' ';
   end;
   drop Loop;
run;
```

This program introduces a number of new features. First, an asterisk is used in place of the number of elements in the array. You can always use an asterisk here if you don't know how many variables are in the array (as may be the case here where you haven't counted the number of character variables in the Chars data set). SAS counts for you and computers are better at counting than people, anyway. Next, the keyword _CHARACTER_ is used as the variable list. Because this statement follows the SET statement, _CHARACTER_ includes all the character variables in the Chars data set. _CHARACTER_ references character variables that are present in the PDV at that point in the DATA step. For example, if you define character variables A, B, and C in the first three lines of a DATA step and then use the reference _CHARACTER_ followed by defining character variables D and E, only variables A, B, and C are referenced by _CHARACTER_. In Program 13.3, if you place the ARRAY statement before the SET statement, the array will not reference any variables.

Because you used an asterisk in place of the number of elements in the array, what value do you use for the upper bound of the DO loop? Luckily, the DIM function comes to the rescue. It returns the number of elements in an array. Finally, an IF statement checks for the values of NA or ? and sets them equal to a SAS character missing value.

13.4 Converting All Character Values to Propercase

If you have data (either a raw data file or a SAS data set) where the data entry folks were careless (or didn't set standards ahead of time), you may have a hodgepodge of upper-, lower-, or propercase values. This section describes a simple program to convert all character values to propercase.

First, here is a listing of data set Careless:

Figure 13.1: Listing of Data Set Careless

Listing of Data Set Learn.Careless

Score	Last_Name	Ans1	Ans2	Ans3
100	COdY	A	b	c
65	sMITH	C	C	d
95	scerbo	D	e	D

The following program converts all the character variable to propercase:

Program 13.4: Converting All Character Values in a SAS Data Set to Propercase

```
data Proper;
   set Learn.Careless;
   array All_Chars{*} _character_;

   do i = 1 to dim(All_Chars);
      All_Chars{i} = propcase(All_Chars{i});
   end;

   drop i;
run;
```

The logic of this program is similar to the previous program. You use the keyword _CHARACTER_ to reference all the character variables in data set Careless, and then use a DO loop to convert all the values to Propercase.

Data set Proper, with all the character values in propercase, is shown here:

Figure 13.2: Listing of Data Set Proper

Listing of Data Set Proper

Score	Last_Name	Ans1	Ans2	Ans3
100	Cody	A	B	C
65	Smith	C	C	D
95	Scerbo	D	E	D

13.5 Using an Array to Create New Variables

You can include variables in an ARRAY statement that do not yet exist in your SAS data set. For example, if your SAS data set had variables Fahren1–Fahren24 containing 24-Fahrenheit temperatures, you could use an array to create 24 new variables (say Celsius1–Celsius24) with the Celsius equivalents. Here is a program that accomplishes this:

Program 13.5: Using an Array to Create New Variables

```
data Temp;
   input Fahren1-Fahren24 @@;
   array Fahren[24];
   array Celsius[24] Celsius1-Celsius24;
   do Hour = 1 to 24;
      Celsius{Hour} = (Fahren{Hour} - 32)/1.8;
   end;
   drop Hour;
datalines;
35 37 40 42 44 48 55 59 62 62 64 66 68 70 72 75 75
72 66 55 53 52 50 45
;
```

The variables Celsius1–Celsius24 are created by the ARRAY statement. Inside the DO loop, you convert each of the 24-Fahrenheit temperatures to Celsius. Data set Temp contains all 24 Fahrenheit and 24-Celsius temperatures. You may wonder where the variable list went in the FAHREN array. If you omit a variable list in an ARRAY statement and you include the number of elements following the array name, SAS automatically creates variable names for you, using the array name as the base and adding the numbers from 1 to *n*, where *n* is the number of elements in the array. In this program, SAS creates the variables Fahren1–Fahren24. You could have used this feature for the Celsius array as well.

13.6 Changing the Array Bounds

By default, SAS numbers the elements of an array starting from 1. There are times when it is useful to specify the beginning and ending values of the array elements. For example, if you have variables Income2010 to Income2017, it would be nice to have the array elements start with 2010 and end with 2017.

The program that follows creates an array of the eight Income values, using the values of 2010 and 2017 as the array bounds, and computes the taxes for each of the eight years:

Program 13.6: Changing the Array Bounds

```
data Account;
   input ID Income2010-Income2017;
   array Income{2010:2017} Income2010-Income2017;
   array Taxes{2010:2017} Taxes2010-Taxes2017;
   do Year = 2010 to 2017;
      Taxes{Year} = .25*Income{Year};
   end;
   format Income2010-Income2017
          Taxes2010-Taxes2017 dollar10.;

datalines;
001 45000 47000 47500 48000 48000 52000 53000 55000
002 67130 68000 72000 70000 65000 52000 49000 40100
;
```

In this program, you specify the lower and upper bounds in the brackets following the array name and separate them with a colon.

13.7 Temporary Arrays

You can create an array that only has elements and no variables! As strange as this sounds, elements of temporary arrays are great places to store values or perform table lookups. If you want, you can assign the array elements initial values when you create the temporary array. Alternatively, you can load values into the temporary array in the DATA step. Either way, the values in the temporary array are automatically retained (that is, they are not set to missing values when the DATA step iterates). Thus, they are useful places to store values that you need during the execution of the DATA step.

We start out with an example that uses a temporary array to store the correct answer for each of 10 questions on a multiple-choice quiz. You can then score the quiz using the temporary array as the answer key. Here is the program:

Program 13.7: Using a Temporary Array to Score a Test

```
data Score;
   array Ans{10} $ 1;
   array Key{10} $ 1 _temporary_
      ('A','B','C','D','E','E','D','C','B','A');
   input ID (Ans1-Ans10)($1.);
   RawScore = 0;
   do Ques = 1 to 10;
      RawScore + (key{Ques} eq Ans{Ques});
   end;
   Percent = 100*RawScore/10;
   keep ID RawScore Percent;
datalines;
123 ABCDEDDDCA
126 ABCDEEDCBA
129 DBCBCEDDEB
;
```

This program uses a temporary array (Key) to hold the answers to the 10 quiz questions. The keyword _TEMPORARY_ tells SAS that this is a temporary array and the 10 values in parentheses are the initial values for each of the elements of this array. It is important to remember that there are no corresponding variables (Key1, Key2, and so on) in this DATA step. Also, because elements of a temporary array are retained, the 10 answer key values are available throughout the DATA step for scoring each of the student tests.

The scoring is done in a DO loop. A "trick" is used to do the scoring: a logical comparison is performed between the student answer and the corresponding answer key. If they match, the logical comparison returns a 1 and this is added to RawScore. If not, the result is a 0 and RawScore is not incremented. Here is a listing of the output:

Figure 13.3: Listing of Data Set Score

Listing of Data Set Score

ID	RawScore	Percent
123	7	70
126	10	100
129	4	40

13.8 Loading the Initial Values of a Temporary Array from a Raw Data File

If you had a long test, you would probably prefer to load the answer key into the array elements by reading the values from a text file (especially if you are scoring the tests using an optical mark-sense reader). In the program that follows, you enter the answer key values on the first line of your data file and the IDs and student answers on the remaining lines (2 through *n*):

Program 13.8: Loading the Initial Values of a Temporary Array from a Raw Data File

```
data Score;
   array Ans{10} $ 1;
   array Key{10} $ 1 _temporary_;

   /* Load the temporary array elements */
   if _n_ = 1 then do Ques = 1 to 10;
      input key{Ques} $1. @;
   end;

   input ID (Ans1-Ans10)($1.);
   RawScore = 0;

   /* Score the test */
   do Ques = 1 to 10;
      RawScore + (key{Ques} eq Ans{Ques});
   end;

   Percent = 100*RawScore/10;
   keep ID RawScore Percent;
datalines;
ABCDEEDCBA
123 ABCDEDDDCA
126 ABCDEEDCBA
129 DBCBCEDDEB
;
```

Because you want to read the first line of the data file differently from the other lines, the value of _N_ (which counts iterations of the DATA step) can be used to ensure that the answer key values are read only once.

The INPUT statement that reads the student ID and student answers uses a form of input that may be new to you. You place the list of variables in a set of parentheses and an informat that you want to use to read these variables in a second set of parentheses. (See Chapter 21 for information about using variable and informat lists.)

The result of running this DATA step is identical to the data set you obtained from Program 13.7.

13.9 Using a Multidimensional Array for Table Lookup

SAS arrays can also be multidimensional. Instead of having a single index to identify elements of an array, you can, for example, use two indices (usually thought of as *row* and *column indices*) to identify an element. This is particularly useful when you want to retrieve a single value based on two selection criteria.

To define a multidimensional array, you specify the number of elements in each dimension in the brackets following the array name, separated by commas. For example, to define an array called MULTI with three elements on the first dimension and five elements on the second dimension, you could use:

```
array Multi{3,5} X1-X15;
```

To determine the number of elements in a multidimensional array, you multiply the number of elements in each dimension. In this example, the 3 by 5 array has 3 times 5 equals 15 elements.

The following table shows the benzene levels for each year (from 1944 to 1949) and the job codes (A through E) at a rubber factory. You want to retrieve a benzene level, given a year and job code. You can see a solution using formats in Chapter 22 of this book. Here is an array solution.

To start, you need to create a two-dimensional array with one index representing the year and the other the job code. To make matters more convenient, you can make the index values for the years range from 1944 to 1949 rather than from 1 to 6. You also need to decide if you want to populate the array with the benzene values as part of the ARRAY statement or if you want to read those values from raw data or, perhaps, a SAS data set. Finally, you can use a regular array or a temporary array.

The solution shown here loads the array from raw data and uses a temporary array to hold the benzene values.

You start with a data set (Expose) that holds a worker ID, the year worked, and the job code. Given this information, you want to look up the worker's benzene exposure.

A listing of data set Expose is shown here:

Figure 13.4: Listing of Data Set Expose

Listing of Data Set Learn.Expose

Worker	Year	JobCode
001	1944	B
002	1948	E
003	1947	C
005	1945	A
006	1948	D

Here is the table of benzene exposures by year and job code.

			Job Code		
Year	**A**	**B**	**C**	**D**	**E**
1944	220	180	210	110	90
1945	202	170	208	100	85
1946	150	110	150	60	50
1947	105	56	88	40	30
1948	60	30	40	20	10
1949	45	22	22	10	8

The first step is to load a temporary array with these values, as follows:

Program 13.9: Loading a Two-Dimensional, Temporary Array with Data Values

```
data Look_Up;
   /*****************************************************
      Create the array, the first index is the year and
      it ranges from 1944 to 1949. The second index is
      the job code (we're using 1-5 to represent job codes
      A through E).
   *****************************************************/

   array Level{1944:1949,5} _temporary_;

   /* Populate the array */
   if _n_ = 1 then do Year = 1944 to 1949;
      do Job = 1 to 5;
         input level{Year,Job} @;
      end;
   end;

   set Learn.Expose;
   /* Compute the job code index from the JobCode value */
   Job = input(translate(Jobcode,'12345','ABCDE'),1.);
   Benzene = level{Year,Job};
   drop Job;
datalines;
220 180 210 110 90
202 170 208 100 85
150 110 150 60 50
105 56 88 40 30
60 30 40 20 10
45 22 22 10 8
;
```

There is a lot going on in this program so let's take it one step at a time. The key is the two-dimensional array (LEVEL). The dimensions of the array are defined by the **comma** in the brackets following the array name. You can think of the first dimension as a row and the second dimension as a column in a table. Each row of raw data following the DATALINES statement represents a different year (starting from 1944) and each of the five columns represents the values for job codes A through E.

Because you want to populate the array only once, you execute the nested DO loops when _N_ is equal to 1. As mentioned earlier, the index values of the first dimension of the array range from 1944 to 1949. This saves you the trouble of computing the correct row value for a given year.

Finally, you use the keyword _TEMPORARY_ to declare the array to be temporary. This has several advantages. First, you don't have to drop (or maintain in the PDV) 30 variables for each of the array elements. Next, these values are automatically retained so they are available for the duration of the DATA step. Finally, using temporary arrays is very efficient. They require less storage than regular variables, and all the values are stored in memory for rapid retrieval.

The only remaining problem is that the JobCode variable is a letter from **A** to **E**. You use the TRANSLATE function (see Chapter 12, section 16 for more details) to convert each of the letters **A** to **E** to the character values **1** to **5**. You then use the INPUT function to do the character-to-numeric conversion.

To look up any benzene level, you simply obtain the array value corresponding to the Year and JobCode (converted to a number) values.

Here is a listing of data set Look_Up:

Figure 13.5: Listing of Data Set Look_Up

Listing of Data Set Look_Up

Year	Worker	JobCode	Benzene
1944	001	B	180
1948	002	E	10
1947	003	C	88
1945	005	A	202
1948	006	D	20

SAS arrays are powerful and flexible. You can change array bounds and create multidimensional arrays. Temporary arrays provide a convenient place to store values for efficient table lookup.

13.10 Problems

Solutions to odd-numbered problems are located at the back of this book. Solutions to all problems are available to professors. If you are a professor, visit the book's companion website at support.sas.com/cody for information about how to obtain the solutions to all problems.

1. Using the permanent SAS data set Survey1, create a new, temporary SAS data set (Survey1) where the values of the variables Ques1–Ques5 are reversed as follows:

   ```
   1 → 5; 2 → 4; 3 → 3; 4 → 2; 5 → 1.
   ```

   Note: Ques1–Ques5 are character variables. Accomplish this using an array.
2. Redo Problem 1, except use data set Survey2.

   Note: Ques1–Ques5 are numeric variables.
3. Using the SAS data set Nines, create a new temporary SAS data set (Nonines) where all values of `999` are replaced by SAS missing values. Do this without explicitly naming the numeric variables in data set Nines (use _NUMERIC_ when you define your array).
4. Data set Survey2 has five numeric variables (Q1–Q5), each with values of `1`, `2`, `3`, `4`, or `5`. You want to determine for each subject (observation) if they responded with a `5` on any of the five questions. This is easily done using the OR or the IN operators. However, for this question, use an array to check each of the five questions. Set variable (ANY5) equal to `Yes` if any of the five questions is a `5` and `No` otherwise.
5. The passing score on each of five tests is 65, 70, 60, 62, and 68. Using the data here, use a temporary array to count the number of tests passed by each student.

ID	Test 1	Test 2	Test 3	Test 4	Test 5
001	90	88	92	95	90
002	64	64	77	72	71
003	68	69	80	75	70
004	88	77	66	77	67

Part 3: Presenting and Summarizing Your Data

Chapter 14: Displaying Your Data

14.1 Introduction

This chapter shows you how to list the observations in a SAS data set using PROC PRINT. In later chapters, you see fancier ways to display and summarize your data.

14.2 The Basics

You have already seen how you can use PROC PRINT to list the observations in a SAS data set. Let's see how you can add some options and statements to this procedure to allow you more control over what is displayed.

You have a SAS data set called Sales that contains the following information on your sales: an employee Name and ID, the region where the sale was made, the name of the company to whom the sale was made, the part number, the quantity, the unit cost of the item, and the total amount of the sale (Quantity times UnitCost). Program 14.1 shows PROC PRINT with all the defaults:

Program 14.1: PROC PRINT Using All the Defaults

```
title "Listing of Data Set Sales";
proc print data=Learn.Sales;
run;
```

You obtain a listing like the one shown in Figure 14.1:

Figure 14.1: Partial Output from Program 14.1

Listing of Data Set Sales

Obs	EmpID	Name	Region	Customer	Item	Quantity	UnitCost	TotalSales
1	1843	George Smith	North	Barco Corporation	144L	50	8.99	449.5
2	1843	George Smith	South	Cost Cutter's	122	100	5.99	599.0
3	1843	George Smith	North	Minimart Inc.	188S	3	5199.00	15597.0
4	1843	George Smith	North	Barco Corporation	908X	1	5129.00	5129.0
5	1843	George Smith	South	Ely Corp.	122L	10	29.95	299.5
6	0177	Glenda Johnson	East	Food Unlimited	188X	100	6.99	699.0
7	0177	Glenda Johnson	East	Shop and Drop	144L	100	8.99	899.0
8	1843	George Smith	South	Cost Cutter's	855W	1	9109.00	9109.0
9	9888	[illegible]	West	Cost Cutter's	122	50	5.99	299.5

14.3 Changing the Appearance of Your Listing

You can control which variables appear in your listing, as well as the order of these variables, by supplying a VAR statement. You place the variables you would like to see, following the keyword VAR. The order of this list also controls the order the variables appear in the listing.

Program 14.2: Controlling Which Variables Appear in the Listing

```
title "Listing of Sales";
proc print data=Learn.Sales;
   var EmpID Customer TotalSales;
run;
```

If you run the code in Program 14.2, you obtain the following.

Figure 14.2: Output from Program 14.2

Listing of Sales

Obs	EmpID	Customer	TotalSales
1	1843	Barco Corporation	449.5
2	1843	Cost Cutter's	599.0
3	1843	Minimart Inc.	15597.0
4	1843	Barco Corporation	5129.0
5	1843	Ely Corp.	299.5
6	0177	Food Unlimited	699.0
7	0177	Shop and Drop	899.0
8	1843	Cost Cutt	

Your next step is to omit the Obs column and replace it with the employee ID. Use an ID statement to do this, as follows:

Program 14.3: Using an ID Statement to Omit the Obs Column

```
title "Listing of Sales";
proc print data=Learn.Sales;
   id EmpID;
   var Customer TotalSales;
run;
```

Notice how the listing has changed.

Figure 14.3: Partial Output from Program 14.3

Listing of Sales

EmpID	Customer	TotalSales
1843	Barco Corporation	449.5
1843	Cost Cutter's	599.0
1843	Minimart Inc.	15597.0
1843	Barco Corporation	5129.0
1843	Ely Corp.	299.5
0177	Food Unlimited	699.0
0177	Shop and Drop	899.0

The variable (or variables) you place in the ID statement replaces the Obs column and is printed in the left-most columns of your listing.

Notice that when you place a variable name in the ID statement, you do not also list it in the VAR statement. If you do, that variable appears twice in the listing.

An alternative way to omit the Obs column is to use PROC PRINT with the option NOOBS. The advantage of using an ID statement is that the ID variable begins each page if you have more variables than will fit across a single page.

14.4 Changing the Appearance of Values

Suppose you would like the TotalSales values to appear with dollar signs and commas. You can change the appearance of values in your listing by associating a format with one or more variables. SAS has many built-in formats that can add commas or dollar signs to numbers or display dates in different ways. Program 14.4 adds a FORMAT statement to list TotalSales with dollar signs and commas and Quantity with commas:

Program 14.4: Adding a FORMAT Statement to PROC PRINT

```
title "Listing of Sales";
proc print data=Learn.Sales;
   id EmpID;
   var Customer Quantity TotalSales;
   format TotalSales dollar10.2 Quantity comma7.;
run;
```

Notice the change in the Quantity and TotalSales columns.

Figure 14.4: Partial Listing from Program 14.4

Listing of Sales

EmpID	Customer	Quantity	TotalSales
1843	Barco Corporation	50	$449.50
1843	Cost Cutter's	100	$599.00
1843	Minimart Inc.	3	$15,597.00
1843	Barco Corporation	1	$5,129.00
1843	Ely Corp.	10	$299.50
0177	Food Unlimited	100	$699.00
0177	Shop and Drop	100	$899.00
1843	Cost Cutter's	1	$9,109.00
9888	Cost Cutter's	50	$299.50
9888	Pet's are Us	1,000	$1,990.00

14.5 Controlling the Observations That Appear in Your Listing

You can also control which observations appear in a listing by including a WHERE statement in the procedure. For example, suppose you want your listing to contain only observations where the Quantity is greater than 400. The following program would do the trick:

Program 14.5: Controlling Which Observations Appear in the Listing (WHERE Statement)

```
title "Listing of Sales with Quantities Greater than 400";
proc print data=Learn.Sales;
   where Quantity gt 400;
   id EmpID;
   var Customer Quantity TotalSales;
   format TotalSales dollar10.2 Quantity comma7.;
run;
```

Figure 14.5: Output from Program 14.5

Listing of Sales with Quantities Greater than 400

EmpID	Customer	Quantity	TotalSales
9888	Pet's are Us	1,000	$1,990.00
0017	Roger's Spirits	500	$19,995.00

Suppose you want to see the sales for two employees: 1843 and 0177. Using an IN operator along with a WHERE statement is a convenient way to do this, as shown in Program 14.6.

Program 14.6: Using the IN Operator in a WHERE Statement

```
title "Listing of Sales from Employees 1843 and 0177";
proc print data=Learn.Sales;
   where EmpID in ('1843','0177');
   id EmpID;
   var Customer Quantity TotalSales;
   format TotalSales dollar10.2 Quantity comma7.;
run;
```

You can use the IN operator with either numeric or character values. You also have the choice of separating the values in the list with spaces or commas. Here is the output from Program 14.6:

Figure 14.6: Output from Program 14.6

Listing of Sales from Employees 1843 and 0177

EmpID	Customer	Quantity	TotalSales
1843	Barco Corporation	50	$449.50
1843	Cost Cutter's	100	$599.00
1843	Minimart Inc.	3	$15,597.00
1843	Barco Corporation	1	$5,129.00
1843	Ely Corp.	10	$299.50
0177	Food Unlimited	100	$699.00
0177	Shop and Drop	100	$899.00
1843	Cost Cutter's	1	$9,109.00
0177	Minimart Inc.	5	$52.50
0177	Barco Corporation	2	$20,000.00
1843	Minimart Inc.	3	$15,597.00

14.6 Adding Titles and Footnotes to Your Listing

You can make your output more meaningful by adding additional title lines and by adding one or more footnotes to your listing. The TITLE*n* statement (where *n* is a number from 1 to 10) allows you to specify multiple title lines. Note that TITLE1 and TITLE are equivalent. The FOOTNOTE*n* statement allows you to specify from 1 to 10 footnotes, lines that appear at the bottom of the page. It is important to remember that once you issue a TITLE or FOOTNOTE statement, the titles or footnotes print on every page of your output until you change or cancel them.

The short program here lists several variables from the Sales data set and adds several title and footnote lines:

Program 14.7: Adding Titles and Footnotes to Your Listing

```
title1 "The XYZ Company";
title3 "Sales Figures for Fiscal 2017";
title4 "Prepared by Roger Rabbit";
title5 "-----------------------------";
footnote "All sales Figures are Confidential";

proc print data=Learn.Sales;
   id EmpID;
   var Customer Quantity TotalSales;
   format TotalSales dollar10.2 Quantity comma7.;
run;
```

First, we show you the listing, followed by an explanation:

Figure 14.7: Partial Output from Program 14.7

The XYZ Company

Sales Figures for Fiscal 2017
Prepared by Roger Rabbit

EmpID	Customer	Quantity	TotalSales
1843	Barco Corporation	50	$449.50
1843	Cost Cutter's	100	$599.00

0177	Minimart Inc.	5	$52.50
0177	Barco Corporation	2	$20,000.00
1843	Minimart Inc.	3	$15,597.00

All sales Figures are Confidential

Because the TITLE2 statement is missing, there is a blank line between the first and third title lines.

If you submit a new TITLE*n* statement, it replaces the current TITLE*n* statement and all TITLE lines with higher values of *n*. For example, suppose you change the TITLE3 line in Program 14.7 to read:

```
title3 "New Sales Figures for 2017";
```

Now you submit a PROC PRINT without any of the previous TITLE lines and your output now read as follows:

Figure 14.8: New Title Lines

The XYZ Company

New Sales Figures for 2017

EmpID	Customer	Quantity	TotalSales
1843	Barco Corporation	50	$449.50

Line 3 has been replaced by the new text and Lines 4 and 5 are removed. The FOOTNOTE*n* statements work the same way.

By the way, to cancel all title statements use the following:

```
title;
```

In a similar manner, use the following to cancel all footnote lines:

```
footnote;
```

14.7 Changing the Order of Your Listing

If you want to see your list in a particular sorted order, you can precede PROC PRINT with a PROC SORT step. If you want to see your listing in order of TotalSales, the following program could be used:

Program 14.8: Using PROC SORT to Change the Order of Your Observations

```
proc sort data=Learn.Sales;
   by TotalSales;
run;

title "Listing of Sales - Sorted by Total Sales";
proc print data=Learn.Sales;
   id EmpID;
   var Customer Quantity TotalSales;
   format TotalSales dollar10.2 Quantity comma7.;
run;
```

Your listing is now in order of TotalSales, starting from the lowest to the highest.

Figure 14.9: Output from Program 14.8

Listing of Sales – Sorted by Total Sales

EmpID	Customer	Quantity	TotalSales
0177	Minimart Inc.	5	$52.50
1843	Ely Corp.	10	$299.50
9888	Cost Cutter's	50	$299.50
1843	Barco Corporation	50	$449.50
1843	Cost Cutter's	100	$599.00
0177	Food Unlimited	100	$699.00
0177	Shop and Drop	100	$899.00
9888	Pet's are Us	1,000	$1,990.00
0017	Spirited Spirits	100	$1,995.00
1843	Barco Corporation	1	$5,129.00
1843	Cost Cutter's	1	$9,109.00
1843	Minimart Inc.	3	$15,597.00
1843	Minimart Inc.	3	$15,597.00
0017	Roger's Spirits	500	$19,995.00
0177	Barco Corporation	2	$20,000.00

To see your listing in order from highest to lowest, precede the variable name (TotalSales) with the keyword DESCENDING, like this:

Program 14.9: Demonstrating the DESCENDING Option of PROC SORT

```
proc sort data=Learn.Sales;
   by descending TotalSales;
run;
```

In Program 14.8, you replaced the original SAS data set with one sorted in order of TotalSales. In some cases, this is fine. However, if you do not want the original data set changed, add an OUT= option to your sort to specify an output data set. This is especially important if you subset the data when you are performing your sort. As an example, the following program creates a temporary SAS data set (Sales) in descending order of TotalSales:

Program 14.10: Sorting the Permanent Data Set and Creating a Temporary Output Data Set

```
proc sort data=Learn.Sales out=Sales;
   by descending TotalSales;
run;
```

Notice that SAS is reading your permanent SAS data set (Sales) and creating a temporary SAS data set (Sales). It is fine (and sometimes convenient) to use the same name for each of the two data sets, as long as they are in different libraries.

14.8 Sorting by More Than One Variable

You can sort your data set by more than one variable. This is called a *multi-level sort*. As an example, the following PROC SORT statements sort your data by employee ID and, within each ID, in decreasing value of total sales:

Program 14.11: Sorting by More than One Variable

```
proc sort data=Learn.Sales out=Sales;
   by EmpID descending TotalSales;
run;

title "Sorting by More than One Variable";
proc print data=sales;
   id EmpID;
   var TotalSales Quantity;
   format TotalSales dollar10.2 Quantity comma7.;
run;
```

As you can see in the listing below, the temporary data set Sales is now in EmpID order and decreasing values of TotalSales for each employee:

Figure 14.10: Output from Program 14.11

Sorting by More than One Variable

EmpID	TotalSales	Quantity
0017	$19,995.00	500
0017	$1,995.00	100
0177	$20,000.00	2
0177	$899.00	100
0177	$699.00	100
0177	$52.50	5
1843	$15,597.00	3
1843	$15,597.00	3
1843	$9,109.00	1
1843	$5,129.00	1
1843	$599.00	100
1843	$449.50	50
1843	$299.50	10
9888	$1,990.00	1,000
9888	$299.50	50

You can see how a multi-level sort works by looking at this listing. Notice that the observations are in order of increasing EmpID (the default order) and, within each EmpID, in decreasing order of Quantity.

14.9 Labeling Your Column Headings

If you want to make your listing a bit more readable, at least for non-programmer types, you may want to use variable labels instead of variable names as your column headings. You need to do two things to make this happen. First, you need to use a LABEL statement, either in your DATA step or as a statement following your PROC PRINT statement. (LABEL statement act somewhat like FORMAT statements—if you place a LABEL statement in the DATA step, it remains associated with the variable it is labeling—if you place the LABEL statement in the PROC step, it is associated only for that procedure.) Next, you need to add a LABEL option to your PROC PRINT statement. We add labels to the Sales listing to demonstrate this.

Program 14.12: Using Labels as Column Headings with PROC PRINT

```
title "Using Labels as Column Headings";
proc print data=Sales label;
   id EmpID;
   var TotalSales Quantity;
   label EmpID = "Employee ID"
         TotalSales = "Total Sales"
         Quantity = "Number Sold";
   format TotalSales dollar10.2 Quantity comma7.;
run;
```

Notice that we have added a LABEL statement to associate a label with each of the variable names, as well as a LABEL option to tell PROC PRINT to use the labels as column headings. If you forget the LABEL option, PROC PRINT will not use labels as column headings even if you have included a LABEL statement. The listing here is more readable than the previous listings that used variable names to head the columns:

Figure 14.11: Partial Output from Program 14.12

Using Labels as Column Headings

Employee ID	Total Sales	Number Sold
0017	$19,995.00	500
0017	$1,995.00	100
0177	$20,000.00	2

14.10 Adding Subtotals and Totals to Your Listing

You can add subtotals and totals to your listing by including SUM and BY statements. In order to include a BY statement in PROC PRINT, remember that you need to have previously sorted your data set in the same order. For example, if you want to break down your listing by region, you first sort your data set by region and include a BY statement in PROC PRINT.

Program 14.13: Using a BY Statement in PROC PRINT

```
proc sort data=Learn.Sales out=Sales;
   by Region;
run;

title "Demonstrating the BY Statement in PROC PRINT";
proc print data=sales label;
   by Region;
   id EmpID;
   var TotalSales Quantity;
   label EmpID = "Employee ID"
         TotalSales = "Total Sales"
         Quantity = "Number Sold";
   format TotalSales dollar10.2 Quantity comma7.;
run;
```

The listing from this program is shown next:

Figure 14.12: Partial Output from Program 14.13

Demonstrating the BY Statement in PROC PRINT

Region=East

Employee ID	Total Sales	Number Sold
0177	$699.00	100
0177	$899.00	100
0017	$19,995.00	500
0177	$20,000.00	2

Region=North

Employee ID	Total Sales	Number Sold
0177	$52.50	5
1843	$449.50	50
1843	$5,129.00	1
1843	$15,597.00	3
1843	$15,597.00	3

Region=South

Employee ID	Total Sales	Number Sold
1843	[illegible]9.50	10

If you want to see each region on a separate page, include a PAGEBY statement as well as a BY statement in PROC PRINT. Make sure the variables following the PAGEBY keyword are the same as the variables following the BY keyword in PROC PRINT.

You can easily modify Program 14.13 to include subtotals and totals by including a SUM statement to your program, like this:

Program 14.14: Adding Totals and Subtotals to Your Listing

```
proc sort data=Learn.Sales out=Sales;
   by Region;
run;

title "Adding Totals and Subtotals to Your Listing";
proc print data=Sales label;
   by Region;
   id EmpID;
   var TotalSales Quantity;
   sum Quantity TotalSales;
   label EmpID = "Employee ID"
         TotalSales = "Total Sales"
         Quantity = "Number Sold";
   format TotalSales dollar10.2 Quantity comma7.;
run;
```

A partial listing is shown next:

Figure 14.13: Partial Output from Program 14.14

Adding Totals and Subtotals to Your Listing

Region=East

Employee ID	Total Sales	Number Sold
0177	$699.00	100
0177	$899.00	100
0017	$19,995.00	500
0177	$20,000.00	2
Region	$41,593.00	702

Region=North

Employee ID	Total Sales	Number Sold
0177	$52.50	5
1843	$449.50	50

The last row for each Region shows totals for Total Sales and Number Sold.

14.11 Making Your Listing Easier to Read

If you include a BY statement and an ID statement, each with the same variables, PROC PRINT does not repeat the variable in the first column if the value has not changed. If you want to see a listing in EmpID order, the listing will look better if you use **EmpID as an ID variable and a BY variable**. Here is the program:

Program 14.15: Using an ID Statement and a BY Statement in PROC PRINT

```
proc sort data=Learn.Sales out=Sales;
   by EmpID;
run;

title "Using the Same Variable in an ID and BY Statement";
proc print data=sales label;
   by EmpID;
   id EmpID;
   var Customer TotalSales Quantity;
   label EmpID = "Employee ID"
         TotalSales = "Total Sales"
         Quantity = "Number Sold";
   format TotalSales dollar10.2 Quantity comma7.;
run;
```

Notice the improved appearance in the listing:

Figure 14.14: Partial Output from Program 14.15

Using the Same Variable in an ID and BY Statement

Employee ID	Customer	Total Sales	Number Sold
0017	Spirited Spirits	$1,995.00	100
	Roger's Spirits	$19,995.00	500

Employee ID	Customer	Total Sales	Number Sold
0177	Minimart Inc.	$52.50	5
	Food Unlimited	$699.00	100
	Shop and Drop	$899.00	100
	Barco Corporation	$20,000.00	2

Employee ID	Customer	Total Sales	Number Sold
1843	Ely Corp.	$299.50	10

14.12 Adding the Number of Observations to Your Listing

By including the N= option of PROC PRINT, the total number of observations in your data set prints at the bottom of the listing. If you want to be even fancier, you can use the form N="*your-label*" to label this number with a label of your own choosing. Here is an example:

Program 14.16: Demonstrating the N= Option with PROC PRINT

```
title "Demonstrating the N= option of PROC PRINT";
proc print data=Sales n="Total number of Observations:";
   id EmpID;
   var TotalSales Quantity;
   label EmpID = "Employee ID"
         TotalSales = "Total Sales"
         Quantity = "Number Sold";
   format TotalSales dollar10.2 Quantity comma7.;
run;
```

The resulting listing is shown next:

Figure 14.15: Partial Output from Program 14.15

Demonstrating the N= option of PROC PRINT

EmpID	TotalSales	Quantity
0017	$1,995.00	100
0017	$19,995.00	500
1843	$15,597.00	3
9888	$299.50	50
9888	$1,990.00	1,000
Total number of Observations:15		

You may wonder why the variable labels do not show up in this listing. Even though there is a LABEL statement, the LABEL option is not used so SAS uses the variable names to head the columns.

14.13 Listing the First *n* Observations of Your Data Set

A very useful data set option, OBS=*n*, allows you to list the first *n* observations of your data set. This is particularly useful when you have a very large data set and want to see just a few observations to be sure your program is running correctly. To demonstrate this, the program here lists the first five observations from the permanent SAS data set Sales:

Program 14.17: Listing the First Five Observations of Your Data Set

```
title "First Five Observations from Sales";
proc print data=Learn.Sales(obs=5);
run;
```

`OBS=n` is a data set option. You can tell this because it is placed in parentheses following the data set name. There are many other data set options that we will discuss later in this book. Notice the listing here stops at Observation 5:

Figure 14.16: Output from Program 14.17

First Five Observations from Sales

Obs	EmpID	Name	Region	Customer	Item	Quantity	UnitCost	TotalSales
1	0177	Glenda Johnson	North	Minimart Inc.	777	5	10.50	52.5
2	1843	George Smith	South	Ely Corp.	122L	10	29.95	299.5
3	9888	Sharon Lu	West	Cost Cutter's	122	50	5.99	299.5
4	1843	George Smith	North	Barco Corporation	144L	50	8.99	449.5
5	1843	George Smith	South	Cost Cutter's	122	100	5.99	599.0

You can combine another data set option, FIRSTOBS=, with OBS= to print observations starting from any point in the data set. For example, if you wrote the following, PROC PRINT would list Observations 4 through 7:

```
proc print data=learn.sales(firstobs=4 obs=7);
```

Keep in mind that OBS= is not the number of observations you want to print—it is the last observation to be processed. When you use FIRSTOBS= and OBS= together, think of OBS= as *LASTOBS*.

Even though PROC PRINT is one of the simplest procedures in Base SAS, there are quite a few options that allow you some control over the appearance of the output. If you want more control over the appearance of a listing, you should consider using PROC REPORT. This procedure takes more programming effort than the PRINT procedure, but it provides much more control over the appearance of the output.

14.14 Problems

Solutions to odd-numbered problems are located at the back of this book. Solutions to all problems are available to professors. If you are a professor, visit the book's companion website at support.sas.com/cody for information about how to obtain the solutions to all problems.

1. List the first 10 observations in data set Blood. Include only the variables Subject, WBC (white blood cell), RBC (red blood cell), and Chol. Label the last three variables "White Blood Cells," "Red Blood Cells," and "Cholesterol," respectively. Omit the Obs column, and place Subject in the first column. Be sure the column headings are the variable labels, not the variable names.

2. Using the data set Sales, create the report shown here:

Sales Figures from the Sales Data Set

Region	Quantity	TotalSales
East	100	699.0
	100	899.0
	500	19995.0
	2	20000.0
East	702	41593.0

Region	Quantity	TotalSales
North	5	52.5

3. Use PROC PRINT (without any DATA steps) to create a listing like the one here.
   Note: The variables in the Hosp data set are Subject, AdmitDate (Admission Date), DischrDate (Discharge Date), and DOB (Date of Birth).

Selected Patients from Hosp Data Set
Admitted in September of 2004
Older than 83 years of age

Subject	Date of Birth	Admission Date	Discharge Date
401	03/21/1921	09/13/2004	09/22/2004
407	08/26/1920	09/13/2004	09/18/2004
409	01/01/1921	09/13/2004	10/02/2004
2577	04/30/1920	09/27/2004	09/27/2004
6889	10/26/1920	09/17/2004	09/22/2004
7495	02/11/1921	09/21/2004	09/22/2004
Number of Patients = 6			

   Hints: Variable labels replace variable names. The number of observations is printed at the bottom of the listing. There are four title lines (the last one being dashes).

4. List the first five observations from data set Blood. Print only variables Subject, Gender, and BloodType. Omit the Obs column.

Chapter 15: Creating Customized Reports

15.1 Introduction

Although you can customize the output produced by PROC PRINT, there are times when you need a bit more control over the appearance of your report. PROC REPORT was developed to fit this need. Not only can you control the appearance of every column of your report, you can produce summary reports as well as detail listings. Two very useful features of PROC REPORT, multiple-panel reports and text wrapping within a column, are often the deciding factor in choosing PROC REPORT over PROC PRINT.

First, let's look at a listing of a data set called Medical using PROC PRINT. You can then see how this listing can be enhanced using PROC REPORT:

Program 15.1: Listing of Medical Using PROC PRINT

```
title "Listing of Data Set Medical from PROC PRINT";
proc print data=Learn.Medical;
   id Patno;
run;
```

This program produces the following listing:

Figure 15.1: Output from Program 15.1

Listing of Data Set Medical from PROC PRINT

Patno	Clinic	VisitDate	Weight	HR	DX	Comment
001	Mayo Clinic	10/21/2006	120	78	7	Patient has had a persistent cough for 3 weeks
003	HMC	09/01/2006	166	58	8	Patient placed on beta-blockers on 7/1/2006
002	Mayo Clinic	10/01/2006	210	68	9	Patient has been on antibiotics for 10 days
004	HMC	11/11/2006	288	88	9	Patient advised to lose some weight
007	Mayo Clinic	05/01/2006	180	54	7	This patient is always under high stress
050	HMC	07/06/2006	199	60	123	Refer this patient to mental health for evaluation

15.2 Using PROC REPORT

Here is a report, using the same data set, produced by PROC REPORT:

Program 15.2: Using PROC REPORT (All Defaults)

```
title "Using the REPORT Procedure";
proc report data=Learn.Medical;
run;
```

Here is the output:

Figure 15.2: Output from Program 15.2

Using the REPORT Procedure

Patno	Clinic	Visit Date	Weight	Heart Rate	DX	Comment
001	Mayo Clinic	10/21/2006	120	78	7	Patient has had a persistent cough for 3 weeks
003	HMC	09/01/2006	166	58	8	Patient placed on beta-blockers on 7/1/2006
002	Mayo Clinic	10/01/2006	210	68	9	Patient has been on antibiotics for 10 days
004	HMC	11/11/2006	288	88	9	Patient advised to lose some weight
007	Mayo Clinic	05/01/2006	180	54	7	This patient is always under high stress
050	HMC	07/06/2006	199	60	123	Refer this patient to mental health for evaluation

One thing to notice about PROC REPORT is that if a variable is associated with a label, the default action is to use the variable label in the report rather than the variable name (which is the default in PROC PRINT). Another difference between PROC REPORT and PROC PRINT relates to column widths. Default column widths with PROC REPORT are computed as follows:

For character variables, the column width is either the length of the character variable or the length of the formatted value (if the variable has a format). For numeric variables, the default column width is 9 or the width of a format (if the variable is formatted). The bottom line is that when you use PROC REPORT instead of PROC PRINT, you will usually be defining the width of each column and not be bothered about what the default widths are. You will see in a moment how to control the width of each column in the report.

15.3 Selecting the Variables to Include in Your Report

To specify which variables you want to include in your report, use a COLUMN statement. The COLUMN statement serves a similar function to the VAR statement of PROC PRINT—it allows you to select which variables you want in your report and the order that they appear. In addition, you need to list variables that you create in COMPUTE blocks (discussed later in this chapter). As an example, look at Program 15.3.

Program 15.3: Adding a COLUMN Statement to PROC REPORT

```
title "Adding a COLUMN Statement";
proc report data=Learn.Medical;
   column Patno DX HR Weight;
run;
```

Here you are selecting four variables (Patno, DX, HR, and Weight) to be included in the report (in that order). Here is the output:

Figure 15.3: Output from Program 15.3

Adding a COLUMN Statement

Patno	DX	Heart Rate	Weight
001	7	78	120
003	8	58	166
002	9	68	210
004	9	88	288
007	7	54	180
050	123	60	199

Because the variable HR was associated with a label when the data set was created, you see the heading "Heart Rate" instead of the variable name HR. As a reminder, PROC REPORT uses variable labels (if they exist) as column headings.

15.4 Comparing Detail and Summary Reports

Unlike PROC PRINT, PROC REPORT is capable of producing both detail reports (listings of all observations) and summary reports (reporting statistics such as sums and means).

By default, PROC REPORT produces detail reports for character variables and summary reports for numeric variables. However, when you include a mix of both types of variables in a single report (as in Program 15.3), you obtain a detail listing showing all observations. To help understand this somewhat complex idea, look at what happens when you include only numeric variables in a report.

Program 15.4: Using PROC REPORT with Only Numeric Variables

```
title "Report with Only Numeric Variables";
proc report data=Learn.Medical;
   column HR Weight;
run;
```

The resulting summary report is as follows:

Figure 15.4: Output from Program 15.4

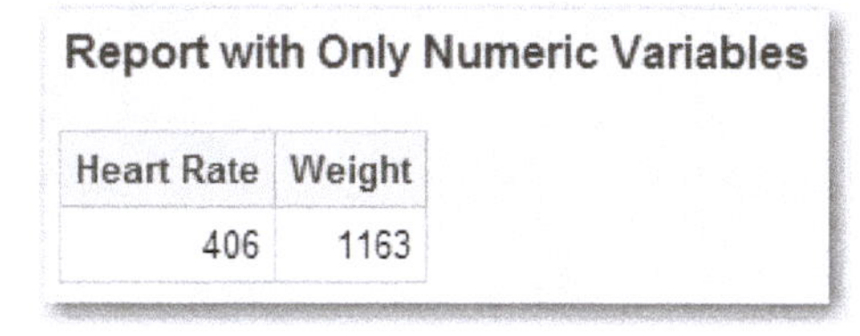
Report with Only Numeric Variables

Heart Rate	Weight
406	1163

These numbers represent the SUM of the heart rates and weights for all the observations in data set Medical. You have learned two things here: first, the default usage for numeric variables is ANALYSIS (which produces a summary report), and second, the default summary statistic is SUM.

Learning how to control whether to produce a detail listing or a summary report and what statistics to produce leads you to the DEFINE statement. You can use a DEFINE statement to specify the usage for each variable; use DISPLAY to create a listing of all the observations, and use ANALYSIS to create a summary report. Suppose you want a detail listing of every person in the Medical data set instead of a summary report. You could use the following program:

Program 15.5: Using DEFINE Statements to Define a Display Usage

```
title "Display Usage for Numeric Variables";
proc report data=Learn.Medical;
   column HR Weight;
   define HR / display "Heart Rate" width=5;
   define Weight / display width=6;
run;
```

Several things have been added to this program. First, notice that there is now a DEFINE statement for each of the two numeric variables. Next, attributes for the variables are entered as options in the DEFINE statement (thus, they follow a slash after the variable name). The DEFINE option DISPLAY is an instruction to produce a detailed listing of all observations. Besides defining the usage as DISPLAY, the

DEFINE statement for HR adds a label (placed in quotes) and a column width. Look at the output here to see the effect of these two DEFINE statements:

Figure 15.5: Output from Program 15.5

Display Usage for Numeric Variables

Heart Rate	Weight
78	120
58	166
68	210
88	288
54	180
60	199

You now see a detailed listing of every observation rather than a single summary statistic.

15.5 Producing a Summary Report

For this example, you want to list the mean heart rate and mean weight for each clinic in the Medical data set. To do this, you need to use the GROUP usage for the variable Clinic. In addition, you need to specify MEAN as the statistic for heart rate and weight. Program 15.6 does the trick:

Program 15.6: Specifying a GROUP Usage to Create a Summary Report

```
title "Demonstrating a GROUP Usage";
proc report data=learn.medical;
   column Clinic HR Weight;
   define Clinic / group width=11;
   define HR / analysis mean "Average Heart Rate" width=12
               format=5.;
   define Weight / analysis mean "Average Weight" width=12
                   format=6.;
run;
```

The MEAN statistic, along with a format, is specified for each of the numeric variables. The keyword ANALYSIS is optional because this is the default usage for numeric variables. You may place these options (label, width, format, and so on) in any order you like. The report is shown next:

Figure 15.6: Output from Program 15.6

Demonstrating a GROUP Usage

Clinic	Average Heart Rate	Average Weight
HMC	69	218
Mayo Clinic	67	170

The figures in the report show the average heart rate and average weight for each of the two clinics.

15.6 Demonstrating the FLOW Option of PROC REPORT

One of the nice features of PROC REPORT is its ability to wrap lines of text within a column when you have long values. To demonstrate this feature, here is a report that includes a variable called Comment that is 50 characters long. Because a value this long would take up most of the width of a page, you can use the FLOW option to improve the appearance of your report.

Note: This option affects only the listing output. It has no effect on other ODS destinations such as HTML or PDF. For other ODS destinations, long text fields will wrap automatically.

Here is the code to accomplish this:

Program 15.7: Demonstrating the FLOW Option with PROC REPORT

```
title "Demonstrating the FLOW Option";
proc report data=Learn.Medical headline
            split=' ' ls=74;
   column Patno VisitDate DX HR Weight Comment;
   define Patno     / "Patient Number" width=7;
   define VisitDate / "Visit Date" width=9 format=date9.;
   define DX        / "DX Code" width=4 right;
   define HR        / "Heart Rate" width=6;
   define Weight    / width=6;
   define Comment   / width=30 flow;
run;
```

This program has several new features. First, the FLOW option was added to the DEFINE statement for the Comment variable. This wraps the comment field within the defined column width of 30. The SPLIT= option is required to tell the program that you want to split the comments between words (blanks). Without this option, PROC REPORT would use other characters such as the slashes in the dates as possible line breaks. The LS= option allows you to select a line size.

The option RIGHT, used for the DX variable, right-aligns the DX values. The default alignment for character variables is LEFT. Alignment options are LEFT, RIGHT, and CENTER. Here is the report:

Figure 15.7: Output from Program 15.7

```
Demonstrating the FLOW Option

 Patient        Visit    DX   Heart
 Number          Date  Code    Rate  Weight  Comment
 ----------------------------------------------------------------------
 001        21OCT2006     7      78     120  Patient has had a persistent
                                             cough for 3 weeks.
 003        01SEP2006     8      58     166  Patient placed on
                                             beta-blockers on 7/1/2006
 002        01OCT2006     9      68     210  Patient has been on
                                             antibiotics for 10 days
 004        11NOV2006     9      88     288  Patient advised to lose some
                                             weight
 007        01MAY2006     7      54     180  This patient is always under
                                             high stress
 050        06JUL2006   123      60     199  Refer this patient to mental
                                             health for evaluation
```

This is a detail report, showing all the observations in the data set. You may wonder why this is so, when the default usage for numeric variables is ANALYSIS (with SUM as the default statistic). Because patient number (Patno) is a character variable (with a default usage of DISPLAY) and it is included in the report, the usage for the numeric variables in the report also has to be DISPLAY. Having one DISPLAY usage variable in the report forces the usage to be DISPLAY for the other variables. Some programmers prefer to explicitly code the usage for every variable in a report. Thus, Program 15.7 could be written like this:

Program 15.8: Explicitly Defining Usage for Every Variable

```
title "Demonstrating the FLOW Option";
proc report data=Learn.Medical headline
            split=' ' ls=74;
   column Patno VisitDate DX HR Weight Comment;
   define Patno     / display "Patient Number" width=7;
   define VisitDate / display "Visit Date" width=9
                      format=date9.;
   define DX        / display "DX Code" width=4 right;
   define HR        / display "Heart Rate" width=6;
   define Weight    / display width=6;
   define Comment   / display width=30 flow;
run;
```

15.7 Using Two Grouping Variables

To demonstrate how you can nest one group within another, we use a data set called Bicycles. This data set contains the country where the bicycles were sold, the model of bicycle (road, mountain, or hybrid), the manufacturer, the number of units sold, and the total sales. The goal here is to see the sum of total sales broken down by country and model. To do this, both the Country and Model variables are defined with a GROUP usage. The order of these variables in the COLUMN statement (Country first, followed by Model) specifies that Country comes first and that Model is nested within Country. Here is the program:

Program 15.9: Demonstrating the Effect of Two Variables with GROUP Usage

```
title "Multiple GROUP Usages";
proc report data=Learn.Bicycles headline ls=80;
   column Country Model Units TotalSales;
   define Country / group width=14;
   define Model   / group width=13;
   define Units   / sum "Number of Units" width=8
                    format=comma8.;
   define TotalSales / sum "Total Sales (in thousands)"
                       width=15 format=dollar10.;
run;
```

The option HEADLINE was added to PROC REPORT. This option places a line between the column headings and the remainder of the report (only for listing output). Notice that Country comes before Model in the COLUMN statement. Output from this program is shown next:

Figure 15.8: Output from Program 15.9

Multiple GROUP Usages

Country	Model	Number of Units	Total Sales (in thousands)
France	Hybrid	1,100	$594
	Mountain Bike	6,400	$8,799
	Road Bike	4,300	$11,830
Italy	Hybrid	700	$483
	Mountain Bike	3,400	$6,382
	Road Bike	4,500	$13,005
USA	Hybrid	4,500	$2,925
	Mountain Bike	10,000	$18,000
	Road Bike	7,000	$15,200
United Kingdom	Hybrid	1,300	$832
	Mountain Bike	1,211	$1,358
	Road Bike	3,644	$7,680

The values of both Country and Model are printed only when the value changes (thus making for a more readable report). The values in the last two columns are the sums of the two variables (Units and TotalSales) for each combination of Country and Model.

15.8 Changing the Order of Variables in the COLUMN Statement

Here is the same report, except that the order of the two variables Country and Model is reversed in the COLUMN statement and the manufacturer (`Manuf`) is added.

Program 15.10: Reversing the Order of Variables in the COLUMN Statement

```
title "Multiple GROUP Usages";
proc report data=Learn.Bicycles headline ls=80;
   column Model Country Manuf Units TotalSales;
   define Country / group width=14;
   define Model   / group width=13;
   define Manuf   / width=12;
   define Units   / sum "Number of Units" width=8
                    format=comma8.;
   define TotalSales / sum "Total Sales (in thousands)"
                      width=15 format=dollar10.;
run;
```

Notice that Model now comes first in the report and that Country is nested within Model.

Figure 15.9: Output from Program 15.10 (Partial Listing)

Multiple GROUP Usages

Model	Country	Manufacturer	Number of Units	Total Sales (in thousands)
Hybrid	France	Trek	1,100	$594
	Italy	Trek	700	$483
	USA	Trek	4,500	$2,925
	United Kingdom	Trek	800	$392
		Cannondale	500	$440
Mountain Bike	France	Trek	5,600	$7,280
		Cannondale	800	$1,519
	Italy	Trek	3,400	$6,382
	USA	Trek	6,000	$7,200
		Cannondale	4,000	$10,800
	United Kingdom	Trek	1,211	$1,358
Road Bike	France	Trek	3,400	$8,500

15.9 Changing the Order of Rows in a Report

If you want a listing in sorted order using PROC PRINT, you must first sort the data set with PROC SORT. PROC REPORT allows you to request a report in sorted order within the procedure itself. This is accomplished by requesting the ORDER usage as an option in your DEFINE statement.

As an example, suppose you want a listing of your Sales data set with the variables EmpID, Quantity, and TotalSales. In addition, you want this listing to be arranged in EmpID order. Here is the program:

Program 15.11: Demonstrating the ORDER Usage of PROC REPORT

```
title "Listing from SALES in EmpID Order";
```

```
proc report data=Learn.Sales headline;
   column EmpID Quantity TotalSales;
   define EmpID / order "Employee ID" width=11;
   define Quantity / width=8 format=comma8.;
   define TotalSales / "Total Sales" width=9
                        format=dollar9.;
run;
```

The keyword ORDER in the DEFINE statement for EmpID produces the report in order of ascending EmpID. This saves you the trouble of having to sort your data set prior to requesting the report. The resulting report is shown next:

Figure 15.10: Output from Program 15.11

Listing from SALES in EmpID Order

Employee ID	Quantity	Total Sales
0017	100	$1,995
	500	$19,995
0177	5	$53
	100	$699
	100	$899
	2	$20,000
1843	10	$300
	50	$450
	100	$599
	1	$5,129
	1	$9,109
	3	$15,597
	3	$15,597
9888	50	$300
	1,000	$1,990

15.10 Applying the ORDER Usage to Two Variables

You can apply the ORDER usage to several variables. For example, suppose you want the same report as the one produced by Program 15.11, but you want to see the total sales for each employee in decreasing order of total sales. Here is the program:

Program 15.12: Applying the ORDER Usage for Two Variables

```
title "Applying the ORDER Usage for Two Variables";
proc report data=Learn.Sales headline;
   column EmpID Quantity TotalSales;
   define EmpID / order "Employee ID" width=11;
   define TotalSales / descending order "Total Sales"
                       width=9 format=dollar9.;
   define Quantity / width=8 format=comma8.;
run;
```

Because you want the report to show total sales for each employee in descending order, you precede ORDER with the keyword DESCENDING. Remember, the order of the ORDER variables in the report is controlled by their order in the COLUMN statement.

Here is the report:

Figure 15.11: Output from Program 15.12 (partial listing)

Applying the ORDER Usage for Two Variables

Employee ID	Quantity	Total Sales
0017	500	$19,995
	100	$1,995
0177	2	$20,000
	100	$899
	100	$699
	5	$53
1843	3	$15,597
	3	
	1	$9,109

The report is in ascending order of employee ID and decreasing order of total sales.

15.11 Creating a Multi-Column Report

PROC REPORT can create telephone book style, multi-column reports. This is especially useful when you have only a few variables to report and you want to save paper. Data set Assign contains subject numbers (Subject) and groups (A, B, or C). Using the PANELS= option of PROC REPORT, you can print this report with multiple columns. If you specify a large number for the number of panels, PROC REPORT fits as many panels as possible for a given line size.

Note: As with the FLOW option, this option affects only the listing output. It has no effect on other ODS destinations such as HTML or PDF. If your output destination is PRINTER, PDF, or RTF, you can use a COLUMNS= option in the ODS statement.

Here is the program:

Program 15.13: Creating a Multi-column Report

```
title "Random Assignment - Three Groups";
proc report data=Learn.Assign panels=99
            headline ps=16;
   columns Subject Group;
   define Subject / display width=7;
   define Group / width=5;
run;
```

The page size (PS) is set to `16` (which limits the number of lines per page to 16) so that you can see the effect of the PANELS= option. In this program, a DISPLAY usage was used for Subject (a numeric variable). This was not necessary because the default usage for Group (a character variable) is DISPLAY, and you would have obtained a detail report anyway. However, it is fine to include it. Here is the report:

Figure 15.12: Output from Program 15.13

```
Random Assignment - Three Groups

 Subject  Group       Subject  Group       Subject  Group
 ---------------      ---------------      ---------------
       1      C            13      A            25      B
       2      B            14      A            26      C
       3      B            15      C            27      A
       4      A            16      A            28      A
       5      B            17      C            29      B
       6      A            18      A            30      B
       7      C            19      B            31      B
       8      B            20      B            32      B
       9      B            21      C            33      C
      10      A            22      A            34      C
      11      C            23      C            35      A
      12      C            24      C            36      A
```

15.12 Producing Report Breaks

PROC REPORT can produce totals and subtotals on ANALYSIS variables by using BREAK and RBREAK statements. RBREAK, which stands for *report break*, is used to report summary statistics (typically sums or means) at the top or bottom of a report. The BREAK statement is used to report summary statistics each time a GROUP or ORDER variable changes value. Let's look at a few examples.

Using the Sales data set, you want to see totals for Quantity and TotalSales at the bottom of the report. The following program produces this report:

Program 15.14: Requesting a Report Break (RBREAK Statement)

```
title "Producing Report Breaks";
proc report data=Learn.Sales;
   column Region Quantity TotalSales;
   define Region / order width=6;
   define Quantity / sum width=8 format=comma8.;
   define TotalSales / sum "Total Sales" width=9
                       format=dollar9.;
   rbreak after / summarize;
run;
```

Following the keyword RBREAK is a location, either BEFORE or AFTER. BEFORE places the summary statistics at the beginning of the report; AFTER places them at the end. The option SUMMARIZE prints a statistic (the one requested in the DEFINE statements) at the top or bottom of the report (depending on whether you request the break before or after). The report is shown below:

Figure 15.13: Output from Program 15.14

Producing Report Breaks

Region	Quantity	Total Sales
East	2	$20,000
	500	$19,995
	100	$899
	100	$699
North	3	$15,597
	3	$15,597
	100	$599
	10	$300
West	1,000	$1,990
	50	$300
	2,025	$92,710

To display summary statistics for each value of one or more GROUP or ORDER variables, use the BREAK statement. For example, if you want to see the total Quantity and TotalSales for each region in the Sales data set, use the following program:

Program 15.15: Demonstrating the BREAK Statement of PROC REPORT

```
title "Producing Report Breaks";
proc report data=Learn.Sales;
   column Region Quantity TotalSales;
   define Region / order width=6;
   define Quantity / sum width=8 format=comma8.;
   define TotalSales / sum "Total Sales" width=9
                     format=dollar9.;
   break after region / summarize;
run;
```

The BREAK statement uses the same location keywords as the RBREAK statement. With the BREAK statement, you also need to specify one or more GROUP or ORDER variables that determine where to print the summary statistics. The option SUMMARIZE prints the appropriate summary statistic after the break. In Program 15.15, a line showing the sum of Quantity and TotalSales is printed for each change in the Region value. The output from this program is as follows:

Figure 15.14: Output from Program 15.15

Producing Report Breaks

Region	Quantity	Total Sales
East	100	$699
	100	$899
	500	$19,995
	2	$20,000
East	702	$41,593
West	50	$300
	1,000	$1,990
West	1,050	$2,290

If you don't want the summary line to contain the values of the BREAK variable(s), use the SUPPRESS option, like this:

```
break after region / summarize suppress;
```

This results in a cleaner looking report:

Figure 15.15: Demonstrating the SUPPRESS Option

	211	$12,003
West	50	$300
	1,000	$1,990
	1,050	$2,290

West	50	$300
	1,000	$1,990
	1,050	$2,290

15.13 Using a Nonprinting Variable to Order a Report

PROC REPORT can use a variable to order the rows of a report without including the variable in the report. For example, in data set Sales, the Name variable stores names as first name followed by last name. Now, if you want a report showing this variable but you want to arrange the rows alphabetically by last name, you can run the following program:

Program 15.16: Using a Nonprinting Variable to Order the Rows of a Report

```
data Temp;
   set Learn.Sales;
   length LastName $ 10;
   LastName = scan(Name,-1,' ');
run;

title "Listing Ordered by Last Name";
proc report data=Temp;
   column LastName Name EmpID TotalSales;
   define LastName / order group noprint;
   define Name / group width=15;
   define EmpID / "Employee ID" group width=11;
   define TotalSales / sum "Total Sales" width=9
                       format=dollar9.;
run;
```

In the short DATA step, the SCAN function extracts the last name from the Name variable. The second argument to the SCAN function tells the system which "word" to extract from the first argument. A negative value says to start scanning from the right; the third argument defines the word delimiters—in this case, a space.

The LastName variable must be listed in the COLUMN statement, even if you are not planning to include it in the report. In the DEFINE statement for LastName, you need to use the NOPRINT option. This option allows the report to be sorted by last name but removes this variable from the report. In the listing from Program 15.16, notice that the rows of the report are ordered by the last name:

Figure 15.16: Output from Program 15.16

Listing Ordered by Last Name

Name	Employee ID	Total Sales
Glenda Johnson	0177	$21,651
Sharon Lu	9888	$2,290
Jason Nguyen	0017	$21,990
George Smith	1843	$46,780

15.14 Computing a New Variable with PROC REPORT

One powerful feature of PROC REPORT is its ability to compute new variables. This makes PROC REPORT somewhat unique among SAS procedures. With most SAS procedures, you need to run a DATA step first if you want to perform any calculations.

Suppose you want to report the weights of the patients in your Medical data set, but instead of reporting the weights in pounds (the units used in the data set), you want to see the weights in kilograms. Program 15.17 uses a compute block to accomplish this:

Program 15.17: Computing a New Variable with PROC REPORT

```
title "Computing a New Variable";
proc report data=Learn.Medical;
   column Patno Weight WtKg;
   define Patno / display "Patient Number" width=7;
   define Weight / display noprint width=6;
   define WtKg / computed "Weight in Kg"
                 width=6 format=6.1;
   compute WtKg;
      WtKg = Weight / 2.2;
   endcomp;

run;
```

Notice several things about this program:

- The new variable (WtKg) and the variable used to compute it (Weight) are both listed in the COLUMN statement. It is important that you list Weight before WtKg in this statement—a computed value must follow the variable or variables used to define it. If you don't want to include the original value of weight in the report, you use the keyword NOPRINT in the DEFINE statement for Weight.
- Use a usage of COMPUTED in the DEFINE statement for your new variable.

- Use COMPUTE and ENDCOMP statements to create a COMPUTE block. You define your new variable inside this block.

As you will see in the next example, this block can contain programming logic as well as arithmetic computations.

Here is the output:

Figure 15.17: Output from Program 15.17

Computing a New Variable

Patient Number	Weight in Kg
001	54.5
003	75.5
002	95.5
004	130.9
007	81.8
050	90.5

15.15 Computing a Character Variable in a COMPUTE Block

As we mentioned in the previous example, you can include programming logic within a COMPUTE block. In this example, you want to create a new character variable (Rate) that has a value of `Fast`, `Normal`, or `Slow`, based on the heart rate (HR). Here is the program:

Program 15.18: Computing a Character Variable in a COMPUTE Block

```
title "Creating a Character Variable in a COMPUTE Block";
proc report data=Learn.Medical;
   column Patno HR Weight Rate;
   define Patno / display "Patient Number" width=7;
   define HR / display "Heart Rate" width=5;
   define Weight / display width=6;
   define Rate / computed width=6;

   compute Rate / character length=6;
     if HR gt 75 then Rate = 'Fast';
     else if HR gt 55 then Rate = 'Normal';
     else if not missing(HR) then Rate='Slow';
   endcomp;

run;
```

As before, you include the variable you are computing in the COLUMN statement. The COMPUTE statement now includes the keyword CHARACTER to tell SAS that Rate is a character variable. The option **LENGTH=6** defines the storage length for this variable.

Next, the logical statements to compute Rate are sandwiched between the COMPUTE and ENDCOMP statements. Here is the output:

Figure 15.18: Output from Program 15.18

Creating a Character Variable in a COMPUTE Block

Patient Number	Heart Rate	Weight	Rate
001	78	120	Fast
003	58	166	Normal
002	68	210	Normal
004	88	288	Fast
007	54	180	Slow
050	60	199	Normal

The new variable (Rate) is now included in the report.

15.16 Creating an ACROSS Variable with PROC REPORT

Besides creating simple column reports, PROC REPORT can also create a tabular style report with each unique value of a variable forming a new column in your report.

This is accomplished by using an ACROSS usage for your variable. In the example that follows, each type of bicycle (hybrid, mountain, and road) in the Bicycle data set is used to create a separate column in the report:

Program 15.19: Demonstrating an ACROSS Usage in PROC REPORT

```
***Demonstrating an Across Usage;
title "Demonstrating an ACROSS Usage";
proc report data=Learn.Bicycles;
   column Model,Units Country;
   define Country  / group width=14;
   define Model    / across "Model";
   define Units    / sum "# of Units" width=14
                     format=comma8.;
run;
```

Besides defining Model as an Across variable, you need to tell the procedure what value you want to display for each row of the table. Notice the **Model,Units** term in the COLUMN statement. This is an instruction to PROC REPORT to report the number of units within each category of Model. The listing here should help clarify how this works:

Figure 15.19: Output from Program 15.19

Demonstrating an ACROSS Usage

Model			
Hybrid	Mountain Bike	Road Bike	
# of Units	# of Units	# of Units	Country
1,100	6,400	4,300	France
700	3,400	4,500	Italy
4,500	10,000	7,000	USA
1,300	1,211	3,644	United Kingdom

15.17 Using an ACROSS Usage to Display Statistics

You can use an ACROSS usage to create a report showing a summary statistic for each level of the ACROSS variable. Here is an example.

You want to see the average white and red blood cell counts for each combination of gender, blood type, and age group in the Blood data set. You also want the age group variable to be displayed in separate columns of the report. Here is the program:

Program 15.20: Using ACROSS Usage to Display Statistics

```
title "Average Blood Counts by Age Group";
proc report data=Learn.Blood;
   column Gender BloodType AgeGroup,WBC AgeGroup,RBC;
   define Gender    / group width=8;
   define BloodType / group width=8 "Blood Group";
   define AgeGroup  / across "Age Group";
   define WBC       / analysis mean format=comma8.;
   define RBC       / analysis mean format=8.2;
run;
```

The COLUMN statement shows that the mean value of WBC (white blood cells) and RBC (red blood cells) should be displayed for each value of AgeGroup. You also need to define WBC and RBC with an ANALYSIS usage with MEAN as the desired statistic. Here is the output:

Figure 15.20: Output from Program 15.20

Average Blood Counts by Age Group

		Age Group		Age Group	
		Old	Young	Old	Young
Gender	Blood Group	WBC	WBC	RBC	RBC
Female	A	7,162	7,310	5.40	5.58
	AB	7,556	7,251	5.05	5.96
	B	6,931	6,501	5.69	5.30
	O	7,033	7,071	5.56	5.48
Male	A	6,995	7,138	5.38	5.59
	AB	6,769	7,079	5.82	5.36
	B	7,082	6,807	5.41	5.43
	O	6,853	7,041	5.47	5.48

You should be aware that PROC TABULATE is well-suited for creating tables such as this. Many programmers, including this author, find PROC TABULATE easier to use when creating more complicated two-way tables.

This chapter is only the "tip of the iceberg" as far as PROC REPORT is concerned. If you would like to learn more, there are excellent SAS Press books and SAS online Docs devoted to PROC REPORT.

15.18 Problems

Solutions to odd-numbered problems are located at the back of this book. Solutions to all problems are available to professors. If you are a professor, visit the book's companion website at support.sas.com/cody for information about how to obtain the solutions to all problems.

1. Use PROC REPORT to create a report, as shown here:

   Note: The data set is Blood, and the variables to be listed are Subject, WBC, and RBC. All three variables are numeric (be careful).

First 5 Observations from Blood Data Set

Subject Number	White Blood Cells	Red Blood Cells
1	7,710	7.40
2	6,560	4.70
3	5,690	7.53
4	6,680	6.85
5	.	7.72

2. Using the Blood data set, produce a summary report showing the average WBC and RBC count for each value of Gender as well as an overall average. Your report should look like this:

Statistics from BLOOD by Gender

Gender	Average WBC	Average RBC
Female	7,112	5.50
Male	6,988	5.47
	7,043	5.48

3. Using the Hosp data set, create the report shown here. Age should be computed using the YRDIF function and rounded to the nearest integer:

Demonstrating a Compute Block

Subject	Admission Date	DOB	Age at Admission
1	03/28/2003	09/15/1926	77
2	03/28/2003	07/08/1950	53
3	03/28/2003	12/30/1981	21
4	03/28/2003	06/11/1942	61
5	08/03/2003	06/28/1928	75

4. Using the SAS data set BloodPressure, compute a new variable in your report. This variable (Hypertensive) is defined as `Yes` for females (`Gender=F`) if the SBP is greater than 138 or the DBP is greater than 88 and `No` otherwise. For males (`Gender=M`), Hypertensive is defined as `Yes` if the SBP is over 140 or the DBP is over 90 and `No` otherwise. Your report should look like this:

Hypertensive Patients

Gender	SBP	DBP	Hypertensive?
M	144	90	Yes
F	110	62	No
M	130	80	No
F	120	70	No
M	142	82	Yes
M	150	96	Yes
F	138	88	No
F	132	76	No

5. Using the SAS data set BloodPressure, produce a report showing Gender and Age, along with a new variable called AgeGroup. Values of AgeGroup are `<= 50` or `> 50` depending on the value of Age. Label this variable "Age Group." Your report should look like this:

Patient Age Groups

Gender	Age	Age Group
M	23	<= 50
F	68	> 50
M	55	> 50
F	28	<= 50
M	35	<= 50
M	45	<= 50
F	48	<= 50
F	78	> 50

6. Using the SAS data set BloodPressure, produce a report showing Gender, Age, SBP, and DBP. Order the report in Gender and Age order as shown here:

Subjects in Gender and Age Order

Gender	Age	Systolic Blood Pressure	Diastolic Blood Pressure
F	28	120	70
	48	138	88
	68	110	62
	78	132	76
M	23	144	90
	35	142	82
	45	150	96
	55	130	80

7. Using the SAS data set Blood, produce a report showing the mean cholesterol (Chol) for each combination of Gender and blood type (BloodType). Your report should look like this:

Mean Cholesterol by Gender and Blood Type

Gender	Blood Type	Mean Cholesterol
Female	A	201.4
	AB	166.5
	B	208.1
	O	205.6
Male	A	201.5
	AB	191.2
	B	211.1
	O	197.9

8. Using the data set Blood, produce a report like the one here. The numbers in the table are the average WBC and RBC counts for each combination of blood type and gender.

Report on the Survey Data Set

Subject ID	Age as of 1/1/2006	Gender	Yearly Salary	Average Response
001	Less than 30	Male	$28,000	1.8
002	51+	Female	$76,123	2.6
003	30 to 50	Male	$36,500	1.8
004	51+	Female	$128,000	3.2
005	Less than 30	Male	$23,060	3.0
006	51+	Male	$90,000	3.4
007	30 to 50	Female	$76,100	3.6

9. Using the SAS data set Survey, produce a report showing the ID, Gender, Age, and Salary variables and the average of the five variables Ques1–Ques5. Your report should look like this:

Report on the Survey Data Set

Subject ID	Age as of 1/1/2006	Gender	Yearly Salary	Average Response
001	Less than 30	Male	$28,000	1.8
002	51+	Female	$76,123	2.6
003	30 to 50	Male	$36,500	1.8
004	51+	Female	$128,000	3.2
005	Less than 30	Male	$23,060	3.0
006	51+	Male	$90,000	3.4
007	30 to 50	Female	$76,100	3.6

Chapter 16: Summarizing Your Data

16.1 Introduction

You may have thought of PROC MEANS (or PROC SUMMARY) primarily as a way to generate summary reports, reporting the sums and means of your numeric variables. However, these procedures are much more versatile and can be used to create summary data sets that can then be analyzed with more DATA or PROC steps.

All the examples in this chapter use PROC MEANS rather than PROC SUMMARY, even when all you want is an output data set. The reason for this is that using PROC MEANS with a NOPRINT option is identical to using PROC SUMMARY when you are creating an output data set.

16.2 PROC MEANS—Starting from the Beginning

You can begin by running PROC MEANS with all the defaults, using the permanent SAS data set Blood.

Here it is:

Program 16.1: PROC MEANS with All the Defaults

```
title "PROC MEANS With All the Defaults";
proc means data=Learn.Blood;
run;
```

Here is the resulting output:

Figure 16.1: Output from Program 16.1

PROC MEANS With All the Defaults

Variable	Label	N	Mean	Std Dev	Minimum	Maximum
Subject		1000	500.5000000	288.8194361	1.0000000	1000.00
WBC		908	7042.97	1003.37	4070.00	10550.00
RBC		916	5.4835262	0.9841158	1.7100000	8.7500000
Chol	Cholesterol	795	201.4352201	49.8867157	17.0000000	331.0000000

By default, PROC MEANS produces statistics on all the numeric variables in the input SAS data set. Looking at the output, you see the default statistics produced are N (number of nonmissing values), Mean (average), Std Dev (standard deviation), Minimum, and Maximum. The next step is to gain some control over this process.

You can control which variables to include in the report by supplying a VAR statement. Statistics are chosen by selecting options in the PROC MEANS statement. Here is a partial list of some of the more commonly used options.

PROC MEANS Option	Statistic Produced
N	Number of nonmissing values
NMISS	Number of missing values
MEAN	Mean or Average
CLM	Confidence Limit for the Mean
SUM	Sum of the values
MIN	Minimum (nonmissing) value
MAX	Maximum value
MEDIAN	Median value
STD	Standard deviation
VAR	Variance
CLM	95% confidence interval for the mean
Q1	Value of the first quartile (25th percentile)
Q3	Value of the third quartile (75th percentile)
QRANGE	Interquartile range (equal to Q3–Q1)

Besides these statistics, the option MAXDEC=*value* is especially useful. This value controls the number of places to the right of the decimal point that are printed in the output.

Let's use some of these options to customize the output. You will compute the number of missing and nonmissing values, the mean, median, minimum, and maximum for the variables RBC (red blood cells) and WBC (white blood cells) in the Blood data set. Finally, you will report all statistics to the nearest 10th. Here is the program:

Program 16.2: Adding a VAR Statement and Requesting Specific Statistics with PROC MEANS

```
title "Selected Statistics Using PROC MEANS";
proc means data=Learn.Blood n nmiss mean median
                            min max maxdec=1;
   var RBC WBC;
run;
```

The output follows:

Figure 16.2: Output from Program 16.2

Selected Statistics Using PROC MEANS

Variable	N	N Miss	Mean	Median	Minimum	Maximum
RBC	916	84	5.5	5.5	1.7	8.8
WBC	908	92	7043.0	7040.0	4070.0	10550.0

You now have only the statistics you requested on the variables listed in the VAR list. Notice also that all the statistics are reported to one decimal place, due to the **MAXDEC=1** option. (Unfortunately, the number of decimal places you choose is applied to all variables. If you need more control over the number of printed decimal places for each variable, you can use PROC TABULATE.) Values for N and NMISS are always integers.

16.3 Adding a BY Statement to PROC MEANS

If you want to see descriptive statistics for each level of another variable, you can include a BY statement, listing one or more BY variables. Remember that you have to sort your data set first by the same variable or variables you list in the BY statement. Here are the same statistics displayed by Figure 16.2, broken down by gender:

Program 16.3: Adding a BY Statement to PROC MEANS

```
proc sort data=Learn.Blood out=Blood;
   by Gender;
run;

title "Adding a BY Statement to PROC MEANS";
proc means data=Blood n nmiss mean median
                      min max maxdec=1;
   by Gender;
   var RBC WBC;
run;
```

You now have your descriptive statistics for males and females separately, as shown next:

Figure 16.3: Output from Program 16.3

Adding a BY Statement to PROC MEANS

Gender=Female

Variable	N	N Miss	Mean	Median	Minimum	Maximum
RBC	409	31	5.5	5.6	1.7	8.8
WBC	403	37	7112.4	7150.0	4620.0	10260.0

Gender=Male

Variable	N	N Miss	Mean	Median	Minimum	Maximum
RBC	507	53	5.5	5.5	2.3	8.4
WBC	505	55	6987.5	6930.0	4070.0	10550.0

16.4 Using a CLASS Statement with PROC MEANS

PROC MEANS lets you use a CLASS statement in place of a BY statement. The CLASS statement performs a similar function to the BY statement, with some significant differences. If you are using PROC MEANS to print a report and are not creating a summary output data set, the differences in the printed output between a BY and CLASS statement are basically cosmetic. The main difference, from a programmer's perspective, is that you do not have to sort your data set before using a CLASS statement. To demonstrate this, run Program 16.3 again, substituting the CLASS statement for the BY statement, as follows:

Program 16.4: Using a CLASS Statement with PROC MEANS

```
title "Using a CLASS Statement with PROC MEANS";
proc means data=Learn.Blood n nmiss mean median
                            min max maxdec=1;
   class Gender;
   var RBC WBC;
run;
```

Notice that you are using the permanent SAS data set (that may not be sorted) instead of the temporary sorted data set used in Program 16.3. There are some minor differences in the appearance of the printed output, as shown here:

Figure 16.4: Output from Program 16.4

Using a CLASS Statement with PROC MEANS

Gender	N Obs	Variable	N	N Miss	Mean	Median	Minimum	Maximum
Female	440	RBC	409	31	5.5	5.6	1.7	8.8
		WBC	403	37	7112.4	7150.0	4620.0	10260.0
Male	560	RBC	507	53	5.5	5.5	2.3	8.4
		WBC	505	55	6987.5	6930.0	4070.0	10550.0

16.5 Applying a Format to a CLASS Variable

One very nice feature of using a CLASS statement (besides not having to sort your data) is that SAS uses formatted values of the CLASS variable(s). You can use this to your advantage by adding a FORMAT statement to the procedure and changing how the CLASS variable groups your data, all without having to modify your original data set.

Suppose you want to see some basic statistics (mean and median) for the two blood count variables (RBC and WBC), broken down by subjects with cholesterol levels below 200 versus subjects with cholesterol levels at or above 200. See how easy this is to do with a CLASS statement and a FORMAT statement:

Program 16.5: Demonstrating the Effect of a Formatted CLASS Variable

```
proc format;
   value Chol_Group
    low -< 200 = 'Low'
    200 - high = 'High';
run;

title "Using a CLASS Statement with PROC MEANS";
proc means data=Learn.Blood n nmiss mean median maxdec=1;
   class Chol;
   format Chol Chol_Group.;
   var RBC WBC;
run;
```

Do you see the tremendous power in this method? You can try different groupings of CLASS variables by supplying a new format. Here is the output:

Figure 16.5: Output from Program 16.5

Using a CLASS Statement with PROC MEANS

Cholesterol	N Obs	Variable	N	N Miss	Mean	Median
Low	384	RBC	352	32	5.5	5.5
		WBC	351	33	6938.2	6910.0
High	411	RBC	376	35	5.5	5.5
		WBC	374	37	7138.9	7130.0

Even though the variable Chol (Cholesterol) is a continuous value, the output displays statistics for the formatted values of the variable.

Note: The technique of using formats with CLASS variables is very powerful and efficient. You can create summary reports for different groupings of a continuous variable without having to make modifications to your original data set.

16.6 Deciding between a BY Statement and a CLASS Statement

There are a number of considerations in deciding whether to use a BY or a CLASS statement when you want to use PROC MEANS to produce printed output (the differences are more important if you are using PROC MEANS to create summary output data sets).

First, if you have a very large data set that is not sorted, you may want to use a CLASS statement. However, if the data set is already in the correct sorted order, you can use either a CLASS or a BY statement. If you have a large number of CLASS variables and there are many distinct values for each of these variables, you may need considerable computer memory to run the procedure. If you have relatively small data sets, choose the statement that produces the style of printed output that you prefer.

16.7 Creating Summary Data Sets Using PROC MEANS

You can use PROC MEANS (or SUMMARY) to create a new data set that contains summary information such as sums and means. This data set can then be used for further analysis. This first example shows how to compute two means and output them to a SAS data set.

Program 16.6: Creating a Summary Data Set Using PROC MEANS

```
proc means data=Learn.Blood noprint;
   var RBC WBC;
   output out = My_Summary
          mean = MeanRBC MeanWBC;
run;

title "Listing of My_Summary";
proc print data=My_Summary noobs;
run;
```

An OUTPUT statement tells PROC MEANS that you want to create a summary SAS data set. The keyword OUT= is used to name the new data set. In this example, you are going to name your output data set My_Summary. Next, you can specify exactly what statistics you want in this data set. You can output most of the statistics listed earlier in this chapter. Following the keyword and an equal sign, you provide a list of the variable names that correspond to these values. In this program, the variable MeanRBC is the mean of all the RBC values, and MeanWBC is the mean of all the WBC values. You can name these variables anything you want—they are arbitrary. The order of the variables here corresponds to the order of the variables in the VAR list. It is convenient to name these variables in a way that helps you remember what they stand for. Shortly, you will see that SAS can name these variables for you in a very convenient way. Finally, if you want to create a SAS data set but do not want any printed output from PROC MEANS, use the NOPRINT option. As we mentioned earlier, if you choose to use PROC SUMMARY instead of PROC MEANS, you do not need a NOPRINT option—it is the default for that procedure.

Program 16.6 also includes a PROC PRINT statement so that you can see the contents of the output data set. Here is the listing of the summary data set My_Summary:

Figure 16.6: Output from Program 16.6

Listing of My_Summary

TYPE	_FREQ_	MeanRBC	MeanWBC
0	1000	5.48353	7042.97

This data set has only one observation. It includes the two means plus two additional variables created by PROC MEANS. We will discuss these two variables later when we run the procedure with a CLASS statement.

When you are outputting only a single statistic (such as a mean in the previous example), you can omit the variable list following the keyword MEAN= if you want. If you do, SAS gives these summary variables the same names as the variables listed in the VAR statement. **It is strongly recommended that you do not do this**. It is easy to make a mistake when you have the same variable name in two data sets, with one representing individual values and the other representing a summary statistic.

16.8 Outputting Other Descriptive Statistics with PROC MEANS

As we mentioned earlier, you can output more than one statistic in your output data set. As an example, let's output the mean, the number of nonmissing observations, the number of missing observations, and the median. Here is the program:

Program 16.7: Outputting More Than One Statistic with PROC MEANS

```
proc means data=Learn.Blood noprint;
   var RBC WBC;
   output out      = Many_Stats
          mean     = M_RBC M_WBC
          n        = N_RBC N_WBC
          nmiss    = Miss_RBC Miss_WBC
          median   = Med_RBC Med_WBC;
run;
```

So, you get the idea. You can output as many statistics as you want and name them anything you want. A listing of this data set, using PROC PRINT, is shown here:

Figure 16.7: Listing of Data Set Many_Stats

Listing of Data Set Many_Stats

TYPE	_FREQ_	M_RBC	M_WBC	N_RBC	N_WBC	Miss_RBC	Miss_WBC	Med_RBC	Med_WBC
0	1000	5.48353	7042.97	916	908	84	92	5.52	7040

The two variables M_RBC and M_WBC represent the means of RBC and WBC, respectively. The two variables Med_RBC and Med_WBC are the medians of the two variables. N_RBC and N_WBC represent the number of nonmissing values; Miss_RBC and Miss_WBC represent the number of missing values for these two variables. As a check, notice that N_RBC plus Miss_RBC equals 1,000, the total number of observations in data set Blood.

16.9 Asking SAS to Name the Variables in the Output Data Set

You can ask SAS to create variable names for any of the summary statistics produced by PROC MEANS. The OUTPUT option AUTONAME causes PROC MEANS to create variable names for the selected statistics by using the variable names in the VAR statement and adding an underscore character followed by the name of the statistic. The best way to see this is to look at a program and a listing of the output data set:

Program 16.8: Demonstrating the OUTPUT Option AUTONAME

```
proc means data=Learn.Blood noprint;
   var RBC WBC;
   output out = Many_Stats
                mean    =
                n       =
                nmiss   =
                median  = / autoname;
run;
```

Because AUTONAME is a statement option, it follows a slash in the OUTPUT statement. Take a look at a listing of this data set to see how SAS named these variables:

Figure 16.8: Listing of Data Set Many_Stats

Listing of Data Set Many_Stats

TYPE	_FREQ_	RBC_Mean	WBC_Mean	RBC_N	WBC_N	RBC_NMiss	WBC_NMiss	RBC_Median	WBC_Median
0	1000	5.48353	7042.97	916	908	84	92	5.52	7040

The OUTPUT option AUTONAME is quite useful, and it creates consistent and easy-to-understand variable names. Be aware that the suffix used for standard deviation is StdDev and not the abbreviation Std that is commonly used. Also remember that SAS variable names cannot be longer than 32 characters—this includes the suffix that PROC MEANS adds. Although it is unlikely, if adding the suffix causes the variable name to exceed 32 characters, the original variable name will be truncated.

16.10 Outputting a Summary Data Set: Including a BY Statement

Although there are many uses for a summary data set containing summary statistics on an entire data set, there are times when you would like to output summary statistics for each level of one or more classification variables. For example, you might want to see n's and means of RBC and WBC broken down by gender. Remember that to use a BY statement, the data set must be sorted in the same order. Program 16.9 creates a summary data set containing the mean values of RBC and WBC for males and females.

Program 16.9: Adding a BY Statement to PROC MEANS

```
proc sort data=Learn.Blood out=Blood;
   by Gender;
run;

proc means data=Blood noprint;
   by Gender;
   var RBC WBC;
   output out   = By_Gender
          mean =
          n    =  / autoname;
run;
```

A listing of the output data set follows:

Figure 16.9: Listing of Data Set By_Gender

Listing of Data Set By_Gender

Gender	_TYPE_	_FREQ_	RBC_Mean	WBC_Mean	RBC_N	WBC_N
Female	0	440	5.49848	7112.43	409	403
Male	0	560	5.47146	6987.54	507	505

In this data set, _FREQ_ represents the number of observations for each value of gender (it doesn't matter if there are missing values for RBC or WBC). For example, you see that there are 440 females in the data set and that the means for RBC and WBC are 5.49848 and 7112.43, respectively. The variables RBC_N and WBC_N represent the number of nonmissing values that were used to compute the two means.

16.11 Outputting a Summary Data Set: Using a CLASS Statement

As you saw earlier in this chapter, you can use a CLASS statement with PROC MEANS to obtain results similar to those obtained using a BY statement. The difference between the two becomes more apparent when you are creating output data sets. The program here is a repeat of Program 16.9, with a CLASS statement replacing the BY statement (and the PROC SORT omitted):

Program 16.10: Adding a CLASS Statement to PROC MEANS

```
proc means data=Learn.Blood noprint;
  class Gender;
  var RBC WBC;
  output out  = With_Class
         mean =
         n    = / autoname;
run;
```

The resulting data set is listed here:

Figure 16.10: Listing of Data Set With_Class

Listing of Data Set With_Class

Gender	_TYPE_	_FREQ_	RBC_Mean	WBC_Mean	RBC_N	WBC_N
	0	1000	5.48353	7042.97	916	908
Female	1	440	5.49848	7112.43	409	403
Male	1	560	5.47146	6987.54	507	505

Notice that you now have three observations instead of just two. The first observation in this data set, with _TYPE_ equal to `0`, is the mean for both males and females. Statisticians call this the *grand mean*. The other two observations with _TYPE_ equal to `1` represent the means for females and males. We will discuss the interpretation of the _TYPE_ variable in the next section, which demonstrates the use of more than one CLASS variable.

If you only want the means broken down by gender and do not want an observation with the grand mean in your output data set, use the NWAY option of PROC MEANS. Using this option makes the output data set using a CLASS statement identical (except for the value of _TYPE_) to an output data set using a BY statement. Program 16.11 demonstrates the effect of the NWAY option:

Program 16.11: Adding the NWAY Option to PROC MEANS

```
proc means data=Learn.Blood noprint nway;
   class Gender;
   var RBC WBC;
   output out  = With_Class
          mean =
          n    =  / autoname;
run;
```

The resulting data set is shown below:

Figure 16.11: Listing of Data Set With_Class

Listing of Data Set With_Class

Gender	_TYPE_	_FREQ_	RBC_Mean	WBC_Mean	RBC_N	WBC_N
Female	1	440	5.49848	7112.43	409	403
Male	1	560	5.47146	6987.54	507	505

Notice that the extra observation containing the grand mean is no longer present.

Note: It is very important to remember to use the NWAY option if you only want to see your descriptive statistics at each level of the CLASS variable.

16.12 Using Two CLASS Variables with PROC MEANS

Things start to get more complicated when you have more than one CLASS variable. Suppose you want to compute the mean and the number of nonmissing values for RBC and WBC for each combination of AgeGroup and Gender. You place these two variables in the CLASS statement and create an output data set just as you did earlier with only one CLASS variable. Here are the PROC statements to create such an output data set:

Program 16.12: Using Two CLASS Variables with PROC MEANS

```
proc means data=Learn.Blood noprint;
   class Gender AgeGroup;
   var RBC WBC;
   output out  = Summary
          mean =
          n    = / autoname;
run;
```

And here is a listing of the output data set:

Figure 16.12: Listing of Data Set Summary

Listing of Data Set Summary

Gender	AgeGroup	_TYPE_	_FREQ_	RBC_Mean	WBC_Mean	RBC_N	WBC_N
		0	1000	5.48353	7042.97	916	908
	Old	1	598	5.45779	7011.56	551	540
	Young	1	402	5.52238	7089.08	365	368
Female		2	440	5.49848	7112.43	409	403
Male		2	560	5.47146	6987.54	507	505
Female	Old	3	258	5.47921	7105.98	242	234
Female	Young	3	182	5.52641	7121.36	167	169
Male	Old	3	340	5.44100	6939.35	309	306
Male	Young	3	220	5.51899	7061.66	198	199

Notice that things are getting more complicated. Even though there are only two levels of Gender and two levels of AgeGroup, there are nine observations in the output data set with four values of _TYPE_. The _TYPE_ = `0` observation is the same as you saw earlier—it represents the grand mean, the mean of all nonmissing values in the data set. To understand the other observations in this data set, you have to know either how to count in binary or how to use another PROC MEANS option—CHARTYPE. Let's take the easy way out and rerun this program with the CHARTYPE procedure option as follows:

Program 16.13: Adding the CHARTYPE Procedure Option to PROC MEANS

```
proc means data=Learn.Blood noprint chartype;
   class Gender AgeGroup;
   var RBC WBC;
   output out   = Summary
          mean =
          n    = / autoname;
run;
```

A listing of the output now looks like this:

Figure 16.13: Listing of Data Set Summary (with CHARTYPE Procedure Option)

Listing of Data Set Summary

Gender	AgeGroup	_TYPE_	_FREQ_	RBC_Mean	WBC_Mean	RBC_N	WBC_N
		00	1000	5.48353	7042.97	916	908
	Old	01	598	5.45779	7011.56	551	540
	Young	01	402	5.52238	7089.08	365	368
Female		10	440	5.49848	7112.43	409	403
Male		10	560	5.47146	6987.54	507	505
Female	Old	11	258	5.47921	7105.98	242	234
Female	Young	11	182	5.52641	7121.36	167	169
Male	Old	11	340	5.44100	6939.35	309	306
Male	Young	11	220	5.51899	7061.66	198	199

This may not look any simpler, but it is now easier to explain what the variable _TYPE_ represents. First of all, the option CHARTYPE (this stands for *character type*) causes the _TYPE_ variable to be a character string of 1's and 0's. If you are familiar with counting in binary, you will observe that the character strings under the _TYPE_ column represent the previous _TYPE_ values, except they are now in a binary representation. (It's OK if this last sentence is meaningless to you—that's why SAS created the CHARTYPE option in the first place.)

If you look at the CLASS statement, you see the two variables Gender and AgeGroup. If you imagine writing the left-most character of _TYPE_ under Gender and the other character of _TYPE_ under AgeGroup, there is a simple rule for interpreting the summary statistic for any value of _TYPE_. The following table displays the CLASS variables with the values of _TYPE_ written underneath (we will use the mean as an example):

	Gender	AgeGroup	Interpretation
TYPE	0	0	Mean for all Genders and all AgeGroups
	0	1	Mean for each level of AgeGroup
	1	0	Mean for each level of Gender
	1	1	Mean for each combination of Gender and AgeGroup

Here is the rule: if there is a `1` under the CLASS variable, the statistics in the output data set represent the statistics computed on each level of that variable. For example, if you are computing means, the mean with a value of _TYPE_ equal to `10` is a mean computed at each level of Gender (for all age groups). If you only want the statistics broken down by each level of the CLASS variables, you can use the NWAY option. This selects only the largest value of _TYPE_. (If you use the CHARTYPE option, NWAY selects the value of _TYPE_ containing all 1's.)

You can use the _TYPE_ variable to select the appropriate breakdown. For example, you can use the _TYPE_ values in a WHERE statement in a PROC PRINT statement, like this:

Program 16.14: Using the _TYPE_ Variable to Select CLASS Variable Breakdowns

```
title "Statistics Broken Down by Gender";
proc print data=Summary(drop = _freq_) noobs;
   where _TYPE_ = '10';
run;
```

Figure 16.14: Output from Program 16.14

Statistics Broken Down by Gender

Gender	AgeGroup	_TYPE_	RBC_Mean	WBC_Mean	RBC_N	WBC_N
Female		10	5.49848	7112.43	409	403
Male		10	5.47146	6987.54	507	505

Because the values are broken down by Gender, AgeGroup has a missing value in each observation (it was left in the listing for demonstration purposes).

You can also use this summary data set to create separate data sets: one for the grand mean, one for means broken down by Gender, one broken down by AgeGroup, and one containing cell means. You can do this in one DATA step, like this:

Program 16.15: Using a DATA Step to Create Separate Summary Data Sets

```
data Grand(drop = Gender AgeGroup)
     By_gender(drop = AgeGroup)
     By_Age(drop = Gender)
     Cellmeans;
  set Summary;
  drop _type_;
  rename _freq_ = Number;
  if _type_ = '00' then output Grand;
  else if _type_ = '01' then output By_Age;
  else if _type_ = '10' then output By_Gender;
  else if _type_ = '11' then output Cellmeans;
run;
```

Because you want a different selection of variables in each data set, you can use a KEEP= or DROP= data set option to make your selection. Because you want to drop _TYPE_ from every data set, it is easier to do this with a DROP statement in the DATA step. A RENAME statement renames _FREQ_ to Number. Again, this change applies to all four data sets.

TYPE is a character variable because of the CHARTYPE option, and you can use it to route each observation to the proper output data set.

Below are listings of the four data sets produced by Program 16.15:

Figure 16.15: Listing of the Four Data Sets Produced by Program 16.15

Listing of Data Set Grand

Number	RBC_Mean	WBC_Mean	RBC_N	WBC_N
1000	5.48353	7042.97	916	908

Listing of Data Set By_Gender

Gender	Number	RBC_Mean	WBC_Mean	RBC_N	WBC_N
Female	440	5.49848	7112.43	409	403
Male	560	5.47146	6987.54	507	505

Listing of Data Set By_Age

AgeGroup	Number	RBC_Mean	WBC_Mean	RBC_N	WBC_N
Old	598	5.45779	7011.56	551	540
Young	402	5.52238	7089.08	365	368

Listing of Data Set Cellmeans

Gender	AgeGroup	Number	RBC_Mean	WBC_Mean	RBC_N	WBC_N
Female	Old	258	5.47921	7105.98	242	234
Female	Young	182	5.52641	7121.36	167	169
Male	Old	340	5.44100	6939.35	309	306
Male	Young	220	5.51899	7061.66	198	199

16.13 Selecting Different Statistics for Each Variable

There is an alternative way of selecting which variables and statistics you want in your summary data set, as shown in Program 16.16.

Program 16.16: Selecting Different Statistics for Each Variable Using PROC MEANS

```
proc means data=Learn.Blood noprint nway;
   class Gender AgeGroup;
   output out = Summary(drop = _:)
          mean(RBC WBC)     =
          n(RBC WBC Chol) =
          median(Chol)      = / autoname;
run;
```

We have added several new features to this program. Each statistic keyword is followed by a list of variables (placed in parentheses) for which you want to compute this statistic. For example, this program computes means for RBC and WBC, the number of nonmissing values for all three variables (RBC, WBC, and Chol), and the median of Chol. Although you can name the newly created statistic following the equal sign as before, this program uses the AUTONAME output option to name them.

Finally, you may wonder about the rather strange-looking DROP= data set option in the SUMMARY data set. This DROP= option uses the colon wildcard notation. All variables beginning with an underscore character will be dropped. In general, **`name:`** used in any location where you would place a variable list refers to all variables beginning with the letters *name*. You can think of the colon the same way you use an asterisk as a DOS wildcard (if anyone still remembers DOS).

A listing of the Summary data set follows:

Figure 16.16: Listing of Data Set Summary

Listing of Data Set Summary

Gender	AgeGroup	RBC_Mean	WBC_Mean	RBC_N	WBC_N	Chol_N	Chol_Median
Female	Old	5.47921	7105.98	242	234	208	198
Female	Young	5.52641	7121.36	167	169	141	211
Male	Old	5.44100	6939.35	309	306	279	195
Male	Young	5.51899	7061.66	198	199	167	208

As you saw in the examples in this chapter, PROC MEANS can do a lot more than print summary statistics—it can create summary data sets broken down by one or more CLASS variables. By using options such as CHARTYPE and AUTONAME, you can simplify your programs and let SAS name your summary variables for you.

16.14 Printing all Possible Combinations of Your Class Variables

A very useful PROC MEANS option called PRINTALLTYPES prints summaries of every combination of your class variables. To demonstrate this, take a look at Program 16.17:

Program 16.17: Demonstrating the PRINTALLTYPES PROC MEANS Option

```
title "Demonstrating the PRINTALLTYPES Option";
proc means data=Learn.Blood printalltypes;
   class Gender AgeGroup;
   var RBC WBC;
run;
```

Adding this single option to PROC MEANS results in the following:

Figure 16.17: Output from Program 16.17

Demonstrating the PRINTALLTYPES Option

N Obs	Variable	N	Mean	Std Dev	Minimum	Maximum
1000	RBC	916	5.4835262	0.9841158	1.7100000	8.7500000
	WBC	908	7042.97	1003.37	4070.00	10550.00

Age Group	N Obs	Variable	N	Mean	Std Dev	Minimum	Maximum
Old	598	RBC	551	5.4577858	0.9900724	1.7100000	8.4300000
		WBC	540	7011.56	1000.86	4290.00	10260.00
Young	402	RBC	365	5.5223836	0.9751199	2.9200000	8.7500000
		WBC	368	7089.08	1006.63	4070.00	10550.00

Gender	N Obs	Variable	N	Mean	Std Dev	Minimum	Maximum
Female	440	RBC	409	5.4984841	0.9823118	1.7100000	8.7500000
		WBC	403	7112.43	997.8255175	4620.00	10260.00
Male	560	RBC	507	5.4714596	0.9863729	2.3300000	8.4300000
		WBC	505	6987.54	1005.31	4070.00	10550.00

Gender	Age Group	N Obs	Variable	N	Mean	Std Dev	Minimum	Maximum
Female	Old	258	RBC	242	5.4792149	0.9802759	1.7100000	7.9900000
			WBC	234	7105.98	1001.16	4620.00	10260.00
	Young	182	RBC	167	5.5264072	0.9875345	2.9200000	8.7500000
			WBC	169	7121.36	996.0893384	4640.00	9820.00
Male	Old	340	RBC	309	5.4410032	0.9989426	2.3300000	8.4300000
			WBC	306	6939.35	996.2198541	4290.00	9360.00
	Young	220	RBC	198	5.5189899	0.9670206	3.5800000	7.7200000
			WBC	199	7061.66	1017.20	4070.00	10550.00

This really useful option was not available when I wrote the last edition of this book. I'm happy to share it with you.

16.15 Problems

Solutions to odd-numbered problems are located at the back of this book. Solutions to all problems are available to professors. If you are a professor, visit the book's companion website at support.sas.com/cody for information about how to obtain the solutions to all problems.

1. Using the SAS data set College, compute the mean, median, minimum, and maximum and the number of both missing and nonmissing values for the variables ClassRank and GPA. Report the statistics to two decimal places.

Repeat Problem 1, except compute the desired statistics for each combination of Gender and SchoolSize. Do this twice, once using a BY statement, and once using a CLASS statement. **Note:** The data set College has permanent formats for Gender, SchoolSize, and Scholarship. Either make this format catalog available to SAS (see Chapter 5), run the PROC FORMAT statements that follow, or use the system option NOFMTERR (no format error) that allows you to access SAS data sets that have permanent user-defined formats without causing an error.

```
proc format;
   value $Yesno  'Y','1' = 'Yes'
                 'N','0' = 'No'
                 ' '     = 'Not Given';
   value $Size   'S' = 'Small'
                 'M' = 'Medium'
                 'L' = 'Large'
                 ' ' = 'Missing';
   value $Gender 'F' = 'Female'
                 'M' = 'Male'
                 ' ' = 'Not Given';
run;
```

2. Using the SAS data set College, report the mean and median GPA and ClassRank broken down by school size (SchoolSize). Do this twice, once using a BY statement, and once using a CLASS statement.
3. Repeat Problem 3 (CLASS statement only), except group small and medium school sizes together. Do this by writing a new format for SchoolSize (values are **S**, **M**, and **L**). Do not use any DATA steps.
4. Using the SAS data set College, report the mean GPA for the following categories of ClassRank: 0–50 = **bottom half**, 51–74 = **3rd quartile**, and 75 to 100 = **top quarter**. Do this by creating an appropriate format. Do not use a DATA step.
5. Using the SAS data set College, create a summary data set (call it Class_Summary) containing the n, mean, and median of ClassRank and GPA for each value of SchoolSize. Use a CLASS statement and be sure that the summary data set only contains statistics for each level of SchoolSize. Use the AUTONAME option to name the variables in this data set.
6. Using the SAS data set College, create four summary data sets containing the number of nonmissing and missing values and the mean, minimum, and maximum for ClassRank and GPA, broken down by Gender and SchoolSize. The first data set (Grand) should contain the statistics for all subjects, the second data set (ByGender) should contain the statistics broken down by Gender,

the third data set (BySize) should contain the statistics broken down by SchoolSize, and the fourth data set (Cell) should contain the statistics broken down by Gender and SchoolSize. Do this by using PROC MEANS (with a CLASS statement) and one DATA step.

Hint: Use the CHARTYPE procedure option.

7. Using the SAS data set SASHELP.Cars, compute the mean and standard deviation for Horsepower broken down by Make and Origin. Include the number of nonmissing and missing values in the output and print all statistics to one decimal place. Use the PRINTALLTYPES option to see all possible combinations of the CLASS variables.

Use PROC PRINT (without any DATA steps) to create a listing like the one here.

Note: The variables in the Hosp data set are Subject, AdmitDate (Admission Date), DischrDate (Discharge Date), and DOB (Date of Birth).

Selected Patients from Hosp Data Set
Admitted in September of 2004
Older than 83 years of age

Subject	Date of Birth	Admission Date	Discharge Date
401	03/21/1921	09/13/2004	09/22/2004
407	08/26/1920	09/13/2004	09/18/2004
409	01/01/1921	09/13/2004	10/02/2004
2577	04/30/1920	09/27/2004	09/27/2004
6889	10/26/1920	09/17/2004	09/22/2004
7495	02/11/1921	09/21/2004	09/22/2004

Number of Patients = 6

Hints: Variable labels replace variable names. The number of observations is printed at the bottom of the listing. There are four title lines (the last one being a dashed line).

Chapter 17: Counting Frequencies

17.1 Introduction

PROC FREQ can be used to count frequencies of both character and numeric variables, in one-way, two-way, and three-way tables. In addition, you can use PROC FREQ to create output data sets containing counts and percentages. Finally, if you are statistically inclined, you can use this procedure to compute various statistics such as chi-square, odds ratio, and relative risk. (See Ron Cody and Jeffrey K. Smith, *Applied Statistics and the Programming Language*, *5th ed. (Englewood Cliffs, NJ: Prentice Hall*, 2005 or *SAS Statistics by Example,* published by SAS Press (Cody, 2011)

17.2 Counting Frequencies

Let's start out by running PROC FREQ with all the defaults, as shown in Program 17.1:

Program 17.1: Counting Frequencies: One-Way Tables Using PROC FREQ

```
title "PROC FREQ with all the Defaults";
proc freq data=Learn.Survey;
run;
```

The default action of PROC FREQ is to compute frequencies on every variable in your data set, both character and numeric. Here is a partial listing of the output:

Figure 17.1: Output from Program 17.1

PROC FREQ with all the Defaults

Subject ID				
ID	Frequency	Percent	Cumulative Frequency	Cumulative Percent
001	1	14.29	1	14.29
002	1	14.29	2	28.57
003	1	14.29	3	42.86
004	1	14.29	4	57.14
005	1	14.29	5	71.43
006	1	14.29	6	85.71
007	1	14.29	7	100.00

Gender				
Gender	Frequency	Percent	Cumulative Frequency	Cumulative Percent
F	3	42.86	3	42.86
M	4	57.14	7	100.00

Age as of 1/1/2006				
Age	Frequency	Percent	Cumulative Frequency	Cumulative Percent
22	1	14.29	1	14.29
23	1	14.29	2	28.57
38	1	14.29	3	42.86
45	1	14.29	4	57.14
55	1	14.29	5	71.43
63	1	14.29	6	85.71
67	1	14.29	7	100.00

Yearly Salary				
Salary	Frequency	Percent	Cumulative Frequency	Cumulative Percent
$23,060	1	14.29	1	14.29
$28,000	1	14.29	2	28.57
$36,500	1	14.29	3	42.86
$76,100	1	14.29	4	57.14
$76,123	1	14.29	5	71.43
$90,000	1	14.29	6	85.71
$128,000	1	14.29	7	100.00

The governor doing a good job?				
Ques1	Frequency	Percent	Cumulative Frequency	Cumulative Percent
1	1	14.29	1	14.29
2	2	28.57	3	42.86
3	1	14.29	4	57.14
4	1	14.29	5	71.43
5	2	28.57	7	100.00

The property tax should be lowered				
Ques2	Frequency	Percent	Cumulative Frequency	Cumulative Percent
2	2	28.57	2	28.57
3	4	57.14	6	85.71
5	1	14.29	7	100.00

Guns should be banned				
Ques3	Frequency	Percent	Cumulative Frequency	Cumulative Percent
1	1	14.29	1	14.29
2	3	42.86	4	57.14
3	1	14.29	5	71.43

The school needs to be expanded				
Ques5	Frequency	Percent	Cumulative Frequency	Cumulative Percent
1	2	28.57	2	28.57
2	1	14.29	3	42.86
3	3	42.86	6	85.71
4	1	14.29	7	100.00

Take a moment to look at the frequencies for Gender. The column labeled Frequency tells you that there are three females and four males in the data set. Expressed as a percentage, this corresponds to 42.86% and 57.14%, respectively.

The two columns labeled Cumulative Frequency and Cumulative Percent are a cumulative count of frequencies and percentages.

PROC FREQ computes counts for each unique value of a variable. For example, if you look at the variable Age, you see how many subjects are 22 years old, 23 years old, and so forth. You would probably prefer to see frequencies based on age groups. We'll get to that in a minute.

17.3 Selecting Variables for PROC FREQ

Because you will rarely want to compute frequencies on every variable in a data set, you need to include a TABLES statement to list the variables for which you want to compute frequencies. You may also want to eliminate the cumulative columns because they are not usually needed. The following program selects Gender and Ques1–Ques3 and eliminates the cumulative statistics as well.

Program 17.2: Adding a TABLES Statement to PROC FREQ

```
title "Adding a TABLES Statement and the NOCUM Tables Option";
proc freq data=learn.survey;
   tables Gender Ques1-Ques3 / nocum;
run;
```

You use a TABLES statement (TABLE, singular, works as well) to list the variables you want PROC FREQ to include. NOCUM is a TABLES option that tells PROC FREQ not to include the two cumulative statistics columns in the output. Because NOCUM is an option in the TABLES statement, it follows a slash (this is the syntax for all statement options within a procedure). In this program, you obtain counts and percentages for the four variables Gender, Ques1, Ques2, and Ques3. Here is the output:

Figure 17.2: Output from Program 17.2

Adding a TABLES Statement and the NOCUM Option

Gender		
Gender	Frequency	Percent
F	3	42.86
M	4	57.14

The governor doing a good job?		
Ques1	Frequency	Percent
1	1	14.29
2	2	28.57
3	1	14.29
4	1	14.29
5	2	28.57

The property tax should be lowered		
Ques2	Frequency	Percent
2	2	28.57
3	4	57.14
5	1	14.29

Guns should be banned		
Ques3	Frequency	Percent
1	1	14.29
2	3	42.86
3	1	14.29
4	1	14.29
5	1	14.29

By the way, if you don't want PROC FREQ to compute percentages, you can add the NOPERCENT option in the TABLES statement.

17.4 Using Formats to Label the Output

It is easy enough to realize that **F** stands for Female and **M** for Male in the listing for Gender. However, it would improve the appearance of the report if you used a format to label these values. You might also want to supply a format for the Ques variables. Here is Program 17.2 with formats added:

Program 17.3: Adding Formats to Program 17.2

```
proc format;
   value $Gender
      'F' = 'Female'
      'M' = 'Male';
   value $Likert
      '1' = 'Strongly disagree'
      '2' = 'Disagree'
      '3' = 'No opinion'
      '4' = 'Agree'
      '5' = 'Strongly agree';
run;

title "Adding Formats";
proc freq data=Learn.Survey;
   tables Gender Ques1-Ques3 / nocum;
   format Gender $Gender.
          Ques1-Ques3 $Likert.;
run;
```

Adding formats to these variables greatly improves the readability of the output as shown here:

Figure 17.3: Partial Output from Program 17.4

Adding Formats

Gender		
Gender	**Frequency**	**Percent**
Female	3	42.86
Male	4	57.14

The governor doing a good job?		
Ques1	**Frequency**	**Percent**
Strongly disagree	1	14.29
Disagree	2	28.57
No opinion	1	14.29
Agree	1	14.29
Strongly agree	2	28.57

The property tax s

Strongly agree ... 14.29

Guns should be banned		
Ques3	**Frequency**	**Percent**
Strongly disagree	1	14.29
Disagree	3	42.86
No opinion	1	14.29
Agree	1	14.29
Strongly agree	1	14.29

Notice that the format labels are displayed in the frequency tables.

17.5 Using Formats to Group Values

Because PROC FREQ computes frequencies on formatted values, you can use formats to group values together into larger categories. Data set Survey contains an Age variable. Suppose you want to generate frequencies for age, broken down into three age groups. You also want to look at question 5 (Ques5) with the values `1` and `2` combined into a **`Generally disagree`** category and values `4` and `5` combined into a **`Generally agree`** category. You can use formats to accomplish these tasks, as shown in the next program:

Program 17.4: Using Formats to Group Values

```
proc format;
   value AgeGroup
      low-<30   = 'Less than 30'
      30-<60    = '30 to 59'
      60-high   = '60 and higher';

   value $Agree_Disagree
      '1','2' = 'Generally disagree'
      '3'     = 'No opinion'
      '4','5' = 'Generally agree';
run;

title "Using Formats to Create Groups";
proc freq data=Learn.Survey;
   tables Age Ques5 / nocum nopercent;
   format Age AgeGroup.
          Ques5 $Agree_Disagree.;
run;
```

This is a useful technique because you don't have to create a new data set. If you want to see frequencies for different groupings, you only need to make a new format. Here is the output:

Figure 17.4: Output from Program 17.4

Using Formats to Create Groups

Age as of 1/1/2006	
Age	Frequency
Less than 30	2
30 to 59	3
60 and higher	2

The school needs to be expanded	
Ques5	Frequency
Generally disagree	3
No opinion	3
Generally agree	1

Age frequencies are now grouped into three categories and the Likert scale for Ques5 has only three categories.

17.6 Problems Grouping Values with PROC FREQ

A problem can occur when PROC FREQ uses formatted values to create groups. When you use the keyword OTHER as a range when you create a format, all values that do not match a format range are grouped together. PROC FREQ assigns all of them to the value of the variable with the lowest value.

As an example, you have a data set (Grouping) with the following values:

Figure 17.5: Data Values from Data Set Grouping

Obs	X	Obs	X	Obs	X
1	2	5	4	9	5
2	2	6	4	10	5
3	3	7	4	11	5
4	3	8	missing	12	6

You write the following program:

Program 17.5: Demonstrating a Problem in How PROC FREQ Groups Values

```
proc format;
   value Two
      low-3 = 'Group 1'
      4-5   = 'Group 2'
      other = 'Other values';
run;

title "Grouping Values (First Try)";
proc freq data=Learn.Grouping;
   tables X / nocum nopercent;
   format X Two.;
run;
```

Looking at the values in data set Grouping, you would expect to see four values in Group 1 and six values in Group 2, one value for Other values and one missing value. Here is what you get:

Figure 17.6: Output from Program 17.5

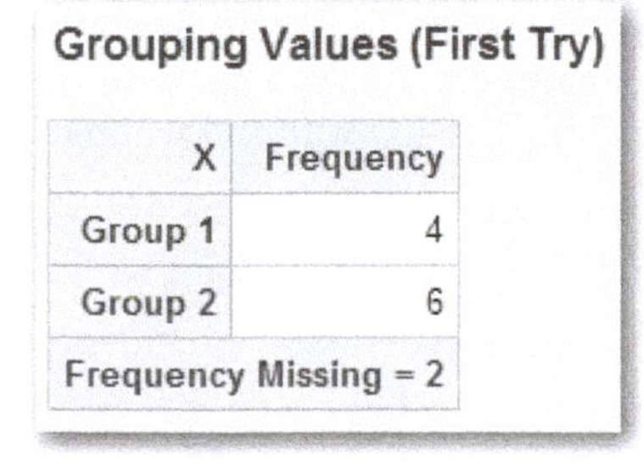

Grouping Values (First Try)

X	Frequency
Group 1	4
Group 2	6
Frequency Missing = 2	

Because the values of **missing** and **6** both fall into the OTHER category, they are assigned the smallest value of the two (**missing**), resulting in this output.

To fix this problem, all you need to do is assign a separate category for missing values when you create your format, as in the following example:

Program 17.6: Fixing the Grouping Problem

```
proc format;
   value Two
      low-3 = 'Group 1'
      4-5   = 'Group 2'
      .     = 'Missing'
      other = 'Other values';
run;
```

Now, missing values and other values are not placed into the same group. Here is the result:

Figure 17.7: Adding a Label for Missing Values in the Format

Grouping Values (Adding a Category for Missing)

X	Frequency
Group 1	4
Group 2	6
Other values	1
Frequency Missing = 1	

By adding a separate label for missing values, the value of 6 (observation 12) is placed in the Other values category.

17.7 Displaying Missing Values in the Frequency Table

SAS normally tells you the frequency of missing values in a separate listing below the frequency table. You can ask SAS to treat missing values just as any other values and include them in the frequency table by including the TABLES option MISSING. Not only does this option bring the count of missing values up into the main table, it also changes how SAS reports percentages. Without the MISSING option, percentages are computed as the frequency of each category divided by the number of nonmissing values; with the MISSING option, SAS computes frequencies by dividing the frequencies by the number of missing and nonmissing observations. To be sure this is clear, let's run PROC FREQ with and without the MISSING option:

Program 17.7: Demonstrating the Effect of the MISSING Option of PROC FREQ

```
 title "PROC FREQ Without the MISSING Option";
 proc freq data=Learn.Grouping;
    format X two.;
    tables X;
 run;

title "PROC FREQ With the MISSING Option";
 proc freq data=Learn.Grouping;
    tables X / missing;
    format X Two.;
 run;
```

Here is the output:

Figure 17.8: Output from Program 17.7

PROC FREQ Without the MISSING Option

X	Frequency	Percent	Cumulative Frequency	Cumulative Percent
Group 1	4	36.36	4	36.36
Group 2	6	54.55	10	90.91
Other values	1	9.09	11	100.00
Frequency Missing = 1				

PROC FREQ With the MISSING Option

X	Frequency	Percent	Cumulative Frequency	Cumulative Percent
Missing	1	8.33	1	8.33
Group 1	4	33.33	5	41.67
Group 2	6	50.00	11	91.67
Other values	1	8.33	12	100.00

Look at the Percent column for Group 1 in both of these tables. Without the MISSING option, the percent, `36.36%`, is obtained by dividing the frequency for Group 1 (`4`) by the number of nonmissing values (`11`). With the MISSING option, the value, `33.33%`, is obtained by dividing the frequency for Group 1 (`4`) by the total number of observations (`12`).

Note: It is important to remember that when you use the MISSING option in the TABLES statement in PROC FREQ, the values in the percent column (and cumulative percent column) are not usually the value you want.

17.8 Changing the Order of Values in PROC FREQ

By default, PROC FREQ orders the values based on internal values (even if a variable has a format). You can change the order of the frequency listings with the ORDER= procedure option. Valid values for the ORDER= Option in PROC FREQ are as follows:

Order= Option	Effect
Internal (default)	Orders values by their internal value
Formatted	Orders values by their formatted value
Freq	Orders values from the most frequent to the least frequent
Data	Orders values based on their order in the input data set

Note that these options do not affect missing values, which are always listed first.

To demonstrate how these options work, we start with a data set (Test) with values of **1**, **2**, **3**, **4**, and **missing**, with frequencies as shown here:

Internal Value	Formatted Value	Frequency
1	Yellow	2
2	Blue	3
3	Red	4
4	Green	1

Using these values, you create a format that corresponds to the values in the table and then run PROC FREQ like this:

Program 17.8: Demonstrating the ORDER= Option of PROC FREQ

```
proc format;
   value Colors
      1 = 'Yellow'
      2 = 'Blue'
      3 = 'Red'
      4 = 'Green'
      . = 'Missing';
run;

data Test;
   input Color @@;
datalines;
3 4 1 2 3 3 3 1 2 2
;

title "Default Order (Internal)";
proc freq data=Test;
   tables Color / nocum nopercent missing;
   format Color Colors.;
run;
```

Here is the output:

Figure 17.9: Output from Program 17.8

Default Order (Internal)

Color	Frequency
Yellow	2
Blue	3
Red	4
Green	1

Notice that the values are ordered by the internal values of Color.

The next program demonstrates how the other ORDER= options work:

Program 17.9: Demonstrating the ORDER= Formatted, Data, and Freq Options

```
title "ORDER = Formatted";
proc freq data=Test order=formatted;
   tables Color / nocum nopercent;
   format Color Colors.;
run;

title "ORDER = Data";
proc freq data=Test order=data;
   tables Color / nocum nopercent;
   format Color Colors.;
run;

title "ORDER = Freq";
proc freq data=test order=freq;
   tables Color / nocum nopercent;
   format Color Colors.;
run;
```

Here is the output:

Figure 17.10: Output from Program 17.9

ORDER = Formatted

Color	Frequency
Blue	3
Green	1
Red	4
Yellow	2

ORDER = Data

Color	Frequency
Red	4
Green	1
Yellow	2
Blue	3

ORDER = Freq

Color	Frequency
Red	4
Blue	3
Yellow	2
Green	1

ORDER=formatted alphabetizes the list based on the formatted values.

To understand the **ORDER=data** option, notice that the first four observations in the data set Test are 3, 4, 1, and 2. This order controls the order of the frequencies in the table. Keep in mind that **ORDER=data** produces tables that may change if you run the program with different data or data in a different order. Using the **ORDER=data** option is usually a bad idea because having a program work differently with different data values is not consistent with good program practices.

Finally, the **ORDER=freq** option lists the values from the most frequent to the least frequent. This last option is especially useful when you have a large number of categories and you want to see which values are most frequent.

17.9 Producing Two-Way Tables

You can easily produce two-way tables (also called *crosstab tables*) by specifying the row and column variables in a TABLES statement, separated by an asterisk (*). For example, to see a table of gender by blood type in the Blood data set, you would write the following:

Program 17.10: Requesting a Two-Way Table

```
title "A Two-way Table of Gender by Blood Type";
proc freq data=Learn.Blood;
   tables Gender * BloodType;
run;
```

The asterisk between Gender and BloodType tells PROC FREQ that you want a two-way table with Gender forming the rows of the table and BloodType forming the columns.

Here is the table:

Figure 17.11: Output from Program 17.10

A Two-way Table of Gender by Blood Type

Frequency Percent Row Pct Col Pct	Table of Gender by BloodType					
		BloodType(Blood Type)				
	Gender(Gender)	A	AB	B	O	Total
	Female	178 17.80 40.45 43.20	20 2.00 4.55 45.45	34 3.40 7.73 35.42	208 20.80 47.27 46.43	440 44.00
	Male	234 23.40 41.79 56.80	24 2.40 4.29 54.55	62 6.20 11.07 64.58	240 24.00 42.86 53.57	560 56.00
	Total	412 41.20	44 4.40	96 9.60	448 44.80	1000 100.00

The upper left corner of the table is the key to the four numbers in each box. For example, looking at females with blood type A, the first number (Frequency = 178) tells you there are 178 females with type A blood. The second number (17.80) is the percentage of all subjects who were females with type A blood. The third number in the box is a row percentage. The 40.45 in this cell tells you that of all the females (which form a row in the table), 40.45% have type A blood. Finally, the fourth number in each cell is a column percentage. In this same cell, 43.20% of those subjects with type A blood are female.

17.10 Requesting Multiple Two-Way Tables

You can request multiple two-way tables in several ways. For example, if you want to see one row variable broken down by several column variables, you can use a TABLES statement like this:

```
tables A * (B C D);
```

This statement generates three tables: A by B, A by C, and A by D. You can supply a list of variables (in parentheses) for both the row and column variables like this:

```
tables (A B) * (C D);
```

This request generates four tables: A by C, A by D, B by C, and B by D. Remember that you can use any one of the short-hand notations that SAS allows for referring to a group of variables. To review, you have the following:

Notation	Result
Ques1 – Ques10	`Ques1, Ques2, … Ques10`
VarA -- VarB	`All variables from VarA to VarB in the order in which they appear in the SAS data set`
Ques:	`All variables that begin with 'Ques:'`
numeric	`All numeric variables in the SAS data set`
character	`All character variables in the SAS data set`
all	`All variables in the SAS data set`

Of course, you couldn't use **`_all_`** in a two-way table request; it was included in the table for completeness.

17.11 Producing Three-Way Tables

You may wonder, "How can I display three dimensions on a piece of paper?" You can do so when you request a three-way table in the form:

```
tables Page * Row * Column;
```

SAS uses separate pages for each value of Page and displays a table of Row by Column on each page. Unless you have a relative in the paper business, you should be very cautious when submitting three-way tables because the output can become quite large.

As an example of a three-way table, Program 17.11 produces a table of AgeGroup by BloodType for each value of Gender.

Program 17.11: Requesting a Three-Way Table with PROC FREQ

```
title "Example of a Three-way Table";
proc freq data=learn.blood;
   tables Gender * AgeGroup * BloodType /
          nocol norow nopercent;
run;
```

Three table options, NOCOL, NOROW, and NOPERCENT, were added to this program. They eliminate the column percentage, row percentage, and overall percentage figures from the table. Here is the output:

Figure 17.12: Output from Program 17.11

Example of a Three-way Table

Frequency	Table 1 of AgeGroup by BloodType				
	Controlling for Gender=Female				
AgeGroup(Age Group)	BloodType(Blood Type)				
	A	AB	B	O	Total
Old	110	11	18	119	258
Young	68	9	16	89	182
Total	178	20	34	208	440

Frequency	Table 2 of AgeGroup by BloodType				
	Controlling for Gender=Male				
AgeGroup(Age Group)	BloodType(Blood Type)				
	A	AB	B	O	Total
Old	143	15	41	141	340
Young	91	9	21	99	220
Total	234	24	62	240	560

For a discussion of how to create a data set containing frequency counts using PROC FREQ, please refer to Chapter 24.

17.12 Problems

Solutions to odd-numbered problems are located at the back of this book. Solutions to all problems are available to professors. If you are a professor, visit the book's companion website at support.sas.com/cody for information about how to obtain the solutions to all problems.

1. Using the SAS data set Blood, generate one-way frequencies for the variables Gender, BloodType, and AgeGroup. Use the appropriate options to omit the cumulative statistics and percentages.
2. Using the SAS data set BloodPressure, generate frequencies for the variable Age. Use a user-defined format to group ages into three categories: 40 and younger, 41 to 60, and 61 and older. Use the appropriate options to omit the cumulative statistics and percentages.
3. Using the data set Blood, produce frequencies for the variable Chol (cholesterol). Use a format to group the frequencies into three groups: low to 200 (normal), 201 and higher (high), and missing. Run PROC FREQ twice, once using the MISSING option, and once without. Compare the percentages in both listings.
4. Using the SAS data set Voter, produce two-way tables for Party by each of the four questions (Ques1–Ques4).

5. Using the SAS data set College, create a two-way table of Scholarship (rows) by ClassRank (columns). Use a user-defined format to group class rank into two groups: 70 and lower, and 71 and higher. (Please see the note in Chapter 16, Problem 2, about the permanent formats used in this data set.)
6. Using the SAS data set College, produce a three-way table of Gender (page) by Scholarship (row) by SchoolSize (column).
7. Using the SAS data set Blood, produce a table of frequencies for BloodType, in frequency order.

Chapter 18: Creating Tabular Reports

18.1 Introduction

PROC TABULATE is an underused, underappreciated procedure that can create a wide variety of tabular reports, displaying frequencies, percentages, and descriptive statistics (such as sums and means) broken down by one or more CLASS variables. There are several excellent books devoted to PROC TABULATE. I recommend Lauren E. Haworth's book, *PROC TABULATE By Example* (Cary, NC: SAS Institute Inc., 1999), for one of the best and most complete books on this topic.

This chapter introduces you to this procedure and, we hope, piques your interest.

To demonstrate many of the features of PROC TABULATE, we will use a data set called Blood that has information on blood types, genders, age groups, red and white blood cell counts, and cholesterol levels. A listing showing the first 10 observations of this data set follows:

Figure 18.1: Listing of Data Set Learn.Blood (First 10 Observations)

Listing of Data Set Learn.Blood
First 10 Observations

Subject	Gender	BloodType	AgeGroup	WBC	RBC	Chol
1	Female	AB	Young	7710	7.40	258
2	Male	AB	Old	6560	4.70	.
3	Male	A	Young	5690	7.53	184
4	Male	B	Old	6680	6.85	.
5	Male	A	Young	.	7.72	187
6	Male	A	Old	6140	3.69	142
7	Female	A	Young	6550	4.78	290
8	Male	O	Old	5200	4.96	151
9	Male	O	Young	.	5.66	311
10	Female	O	Young	7710	5.55	.

18.2 A Simple PROC TABULATE Table

Although PROC TABULATE can create complex tables, there are only three operators that control a table's appearance. Let's start out by running PROC TABULATE with all the defaults and selecting a single CLASS variable, like this:

Program 18.1: PROC TABULATE with All the Defaults and a Single CLASS Variable

```
title "All Defaults with One CLASS Variable";
proc tabulate data=Learn.Blood;
   class Gender;
   table Gender;
run;
```

PROC TABULATE uses a CLASS statement where you specify variables that represent categories (often character variables) to compute frequencies or percentages. In Program 18.1, Gender is selected as the CLASS variable. The TABLE statement specifies the table's appearance. Any variable listed in a TABLE statement must be listed in a CLASS statement or a VAR statement (to be discussed shortly). Here is the output:

Figure 18.2: Output from Program 18.1

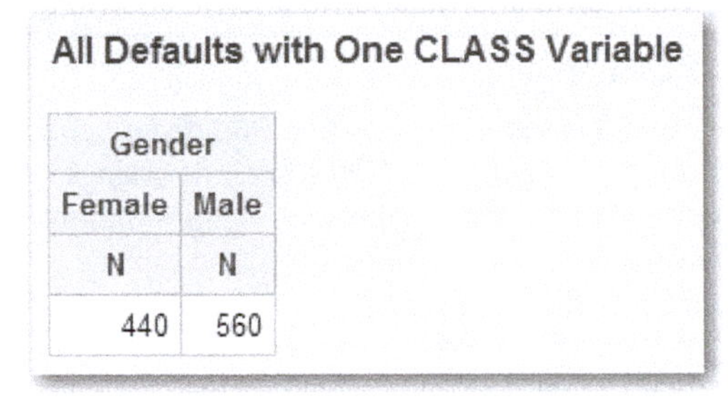

All Defaults with One CLASS Variable

Gender	
Female	Male
N	N
440	560

With only a single CLASS variable listed in the TABLE statement, PROC TABULATE computes frequencies for each value of the selected variable. It also displays these frequencies as columns. You can see here that there are 440 females and 560 males in the data set.

18.3 Describing the Three PROC TABULATE Operators

18.3.1 Concatenation

The first tabulate operator that we will discuss is concatenation. That is, you can place information about several variables next to each other in a table. A space between variables in a TABLE statement is used to concatenate table values. If you want to see Gender and BloodType as columns in a table, you can submit the following tabulate statements:

Program 18.2: Demonstrating Concatenation with PROC TABULATE

```
title "Demonstrating Concatenation";
proc tabulate data=Learn.Blood format=6.;
   class Gender BloodType;
   table Gender BloodType;
run;
```

The space between Gender and BloodType in the TABLE statement represents concatenation. The procedure option, FORMAT=, was included in this program. If you specify this option, the selected format is used for every cell in the table. Later, you will see how to control the format of every individual cell of a PROC TABULATE listing. Here is the table:

Figure 18.3: Output from Program 18.2

Demonstrating Concatenation

Gender		Blood Type			
Female	Male	A	AB	B	O
N	N	N	N	N	N
440	560	412	44	96	448

Notice that the frequencies for BloodType have been placed next to the frequencies for Gender.

18.3.2 Table Dimensions (Page, Row, and Column)

The comma used in a TABLE statement controls the table's dimensions. If you do not include any commas in the TABLE statement, all of your table information is displayed as columns. If you include a single comma, the specification following the comma generates columns in the table; the specification before the comma generates rows in the table. To demonstrate, Program 18.3 shows a request for blood type frequencies to be displayed as columns and gender values to be displayed as rows.

Program 18.3: Demonstrating Table Dimensions with PROC TABULATE

```
title "Demonstrating Table Dimensions";
proc tabulate data=Learn.Blood format=6.;
   class Gender BloodType;
   table Gender,
         BloodType;
run;
```

The resulting table is shown next:

Figure 18.4: Output from Program 18.3

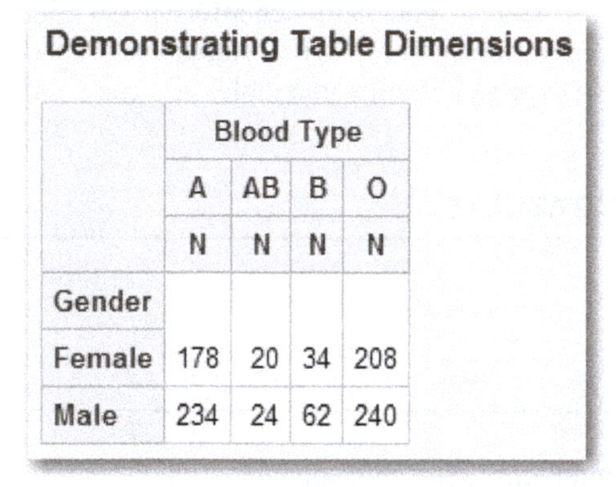

Demonstrating Table Dimensions

	Blood Type			
	A	AB	B	O
	N	N	N	N
Gender				
Female	178	20	34	208
Male	234	24	62	240

If you use two commas in a table request, the first specification represents pages, the second, rows, and the last, columns.

18.3.3 Nesting

The third TABLE operator is the asterisk (*). If you specify variable 1 * variable 2, variable 2 is nested within variable 1. To demonstrate, here is a request to nest BloodType within Gender:

Program 18.4: Demonstrating the Nesting Operator with PROC TABULATE

```
title "Demonstrating Nesting";
proc tabulate data=Learn.Blood format=6.;
   class Gender BloodType;
   table Gender * BloodType;
run;
```

Here is the resulting table:

Figure 18.5: Output from Program 18.4

Demonstrating Nesting

Gender							
Female				Male			
Blood Type				Blood Type			
A	AB	B	O	A	AB	B	O
N	N	N	N	N	N	N	N
178	20	34	208	234	24	62	240

For each value of Gender, you see the frequencies of each of the values of BloodType.

Notice that you can add spaces between any of the operators (for example, there are spaces around the asterisk in the previous program) to make your TABLE statements more readable. If you do not place a comma or asterisk between two variable names, they are concatenated in the table.

18.4 Using the Keyword ALL

Unlike many other SAS procedures, there are some reserved keywords used by PROC TABULATE (which also means you have to be careful not to name any of your variables the same as a PROC TABULATE keyword). One such keyword is ALL. When you place ALL after a variable name in a table request, PROC TABULATE includes a column representing all levels of the preceding variable. To demonstrate, you can add ALL to both dimensions of the table produced by Program 18.3, as follows:

Program 18.5: Adding the Keyword ALL to Your Table Request

```
title "Adding the Keyword ALL to the TABLE Request";
proc tabulate data=Learn.Blood format=6.;
   class Gender BloodType;
   table Gender ALL,
         BloodType ALL;
run;
```

Here is the resulting table:

Figure 18.6: Output from Program 18.5

Adding the Keyword ALL to the TABLE Request

	Blood Type				All
	A	AB	B	O	
	N	N	N	N	N
Gender					
Female	178	20	34	208	440
Male	234	24	62	240	560
All	412	44	96	448	1000

Because there is a space between ALL and the preceding variable, the frequency associated with ALL is concatenated to the appropriate variable.

18.5 Producing Descriptive Statistics

PROC TABULATE can also produce descriptive statistics such as sums and means. To specify that you want to generate descriptive statistics rather than frequency counts, you list all of your analysis variables (that is, the ones for which you want statistics) in a VAR statement.

Note: Descriptive statistics may appear on any dimension of a table, but you may only define statistics on a single dimension of a table.

Program 18.6 computes descriptive statistics for the RBC (red blood cell) and WBC (white blood cell) counts:

Program 18.6: Using PROC TABULATE to Produce Descriptive Statistics

```
title "Demonstrating Analysis Variables";
proc tabulate data=Learn.Blood;
   var RBC WBC;
   table RBC WBC;
run;
```

The resulting output follows:

Figure 18.7: Output from Program 18.6

Demonstrating Analysis Variables

RBC	WBC
Sum	Sum
5022.91	6395020.00

Because the operator between RBC and WBC was a space, the analysis columns are concatenated. The default descriptive statistic is the sum (not a very useful statistic for blood counts). To specify one or more analyses for each variable, you use keywords (sum, mean, min, max, std, and so on) to specify which statistics you want and you associate these statistics with a variable using the nesting (asterisk) operator. For example, to specify the mean rather than a sum for RBC and WBC, you would use the code in Program 18.7.

Program 18.7: Specifying Statistics on an Analysis Variable with PROC TABULATE

```
title "Specifying Statistics";
proc tabulate data=Learn.Blood;
   var RBC WBC;
   table RBC*mean WBC*mean;
run;
```

To save typing, you can also write the TABLE statement like this:

```
table (RBC WBC)*mean;
```

Either way, you now have a table showing the mean (average) blood counts:

Figure 18.8: Output from Program 18.7

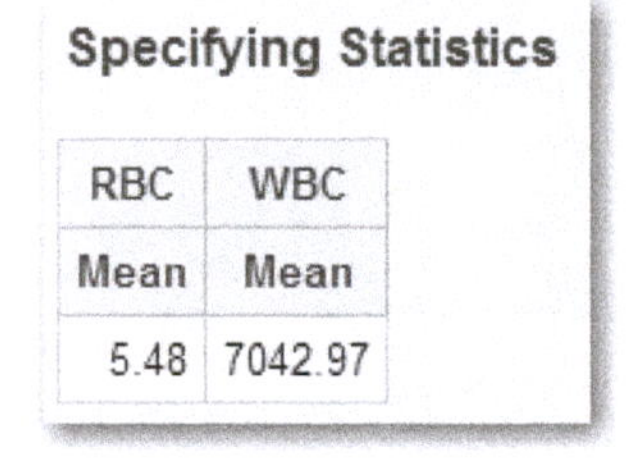

Specifying Statistics

RBC	WBC
Mean	Mean
5.48	7042.97

The output now shows means instead of the default statistic of sums. You can specify several statistics for each variable as well. For example, if you would like the mean, minimum, and maximum values for the blood counts, you could write Program 18.8:

Program 18.8: Specifying More than One Descriptive Statistic with PROC TABULATE

```
title "Specifying More than One Statistic";
proc tabulate data=Learn.Blood format=comma9.2;
   var RBC WBC;
   table (RBC WBC)*(mean min max);
run;
```

Each statistic is now listed for each of the two variables, like this:

Figure 18.9: Output from Program 18.8

Specifying More than One Statistic

RBC			WBC		
Mean	Min	Max	Mean	Min	Max
5.48	1.71	8.75	7,042.97	4,070.00	10,550.00

This program uses the PROC TABULATE option **FORMAT=COMMA9.2** to improve the readability of the output. You now have the mean, minimum, and maximum value for RBC and WBC counts, each displayed with a 9.2 numeric format.

18.6 Combining CLASS and Analysis Variables in a Table

What if you want to see the mean value for RBC and WBC broken down by one or more of the CLASS variables (such as Gender or BloodType). You can combine frequencies and descriptive statistics in a single table. Here is where some people complain that PROC TABULATE is very difficult to use. Actually, the problem is usually not with PROC TABULATE, rather, the problem is more likely that the user did not decide what the table should look like before starting to write the TABULATE statements. It also takes some practice to arrange a table so that columns are not split between pages (making it very hard to read).

The next program is a request for the mean RBC, WBC, and Chol (cholesterol), broken down by Gender and AgeGroup.

Program 18.9: Combining CLASS and Analysis Variables in a Table

```
title "Combining CLASS and Analysis Variables";
proc tabulate data=Learn.Blood format=comma11.2;
   class Gender AgeGroup;
   var RBC WBC Chol;
   table (Gender ALL)*(AgeGroup All),
         (RBC WBC Chol)*mean;
run;
```

This request nests AgeGroup and ALL within each value of Gender and ALL. The means of RBC, WBC, and Chol are displayed as columns of the table, like this:

Figure 18.10: Output from Program 18.9

Combining CLASS and Analysis Variables

		RBC	WBC	Cholesterol
		Mean	Mean	Mean
Gender	Age Group			
Female	Old	5.48	7,105.98	195.88
	Young	5.53	7,121.36	212.27
	All	5.50	7,112.43	202.50
Male	Age Group			
	Old	5.44	6,939.35	199.12
	Young	5.52	7,061.66	203.08
	All	5.47	6,987.54	200.60
All	Age Group			
	Old	5.46	7,011.56	197.73
	Young	5.52	7,089.08	207.29
	All	5.48	7,042.97	201.44

18.7 Customizing Your Table

PROC TABULATE allows you to customize your table in several ways. You can associate a format with a particular value in the table or rename any of the keywords, such as ALL or MEAN.

If you look back at the output produced by Program 18.7, you see that the default format is useful for values of RBC but not for values of WBC (where you would rather not see any decimal values). The program that follows associates a different format with each of these two variables:

Program 18.10: Associating a Different Format with Each Variable in a Table

```
title "Specifying Formats";
proc tabulate data=Learn.Blood;
   var RBC WBC;
   table RBC*mean*f=7.2 WBC*mean*f=comma7.;
run;
```

Here you see another use for the asterisk. It is used to associate a format with a specific value. You can think of the format being nested within the statistic.

Here is the output:

Figure 18.11: Output from Program 18.10

Specifying Formats

RBC	WBC
Mean	Mean
5.48	7,043

To demonstrate how to rename PROC TABULATE keywords, take a look at Program 18.11.

Program 18.11: Renaming Keywords with PROC TABULATE

```
title "Specifying Formats and Renaming Keywords";
proc tabulate data=Learn.Blood;
   class Gender;
   var RBC WBC;
   table Gender ALL,
         RBC*(mean*f=9.1 std*f=9.2)
         WBC*(mean*f=comma9. std*f=comma9.1);
   keylabel ALL  = 'Total'
            mean = 'Average'
            std  = 'Standard Deviation';
run;
```

The KEYLABEL statement allows you to provide a label for any of the keywords used by the procedure. In this program, ALL is replaced with Total, Mean by Average, and Std by Standard Deviation. In addition, different formats are associated with the mean and the standard deviation. This is particularly useful when you are computing statistics like N (the number of nonmissing observations) or NMISS (the number of missing values), where you want integer values, along with means or other statistics, where you may want to see digits to the right of the decimal point. Here is the table:

Figure 18.12: Output from Program 18.11

Specifying Formats and Renaming Keywords

	RBC		WBC	
	Average	Standard Deviation	Average	Standard Deviation
Gender				
Female	5.5	0.98	7,112	997.8
Male	5.5	0.99	6,988	1,005.3
Total	5.5	0.98	7,043	1,003.4

While we are on the topic of improving the appearance of your tables, look back at the output table from Program 18.1. The row of Ns across the table is not necessary. You can eliminate them by providing a null label (a blank) within the TABLE statement request, like this:

Program 18.12: Eliminating the N Column in a PROC TABULATE Table

```
title "Eliminating the 'N' Row from the Table";
proc tabulate data=Learn.Blood format=6.;
   class Gender;
   table Gender*n=' ';
run;
```

The resulting table follows:

Figure 18.13: Output from Program 18.12

Eliminating the 'N' Row from the Table

Gender	
Female	Male
440	560

You could accomplish the same goal by including a KEYLABEL statement and providing a null label for N, like this:

```
keylabel n = ' ';
```

18.8 Demonstrating a More Complex Table

As a final demonstration of combining CLASS and analysis variables in a table and customizing the table, we present the following program:

Program 18.13: Demonstrating a More Complex Table

```
title "Combining CLASS and Analysis Variables";
proc tabulate data=Learn.Blood format=comma9.2;
   class Gender AgeGroup;
   var RBC WBC Chol;
   table (Gender=' ' ALL)*(AgeGroup=' ' All),
          RBC*(n*f=3. mean*f=5.1)
          WBC*(n*f=3. mean*f=comma7.)
          Chol*(n*f=4. mean*f=7.1);
   keylabel ALL = 'Total';
run;
```

Program 18.13 demonstrates several PROC TABULATE features. First, a label is attached to the variables Gender and AgeGroup. This is accomplished by following the variable name with an equal sign, followed by the label. Because a null label is used, the row normally used to display these variable names is eliminated altogether. (Try running this program without this addition to see the difference.) Next, a separate format is attached to each of the statistics in the table (N and MEAN). Often, the formats you use are chosen so that the column headings or labels are not split between rows, rather than the optimum value, to display a particular value. Here is the nice compact table produced by this program:

Figure 18.14: Output from Program 18.13

Combining CLASS and Analysis Variables

		RBC		WBC		Cholesterol	
		N	Mean	N	Mean	N	Mean
Female	Old	242	5.5	234	7,106	208	195.9
	Young	167	5.5	169	7,121	141	212.3
	Total	409	5.5	403	7,112	349	202.5
Male	Old	309	5.4	306	6,939	279	199.1
	Young	198	5.5	199	7,062	167	203.1
	Total	507	5.5	505	6,988	446	200.6
Total	Old	551	5.5	540	7,012	487	197.7
	Young	365	5.5	368	7,089	308	207.3
	Total	916	5.5	908	7,043	795	201.4

18.9 Computing Row and Column Percentages

The statistic PCTN is used with CLASS variables to compute percentages. Program 18.14 demonstrates how this statistic is used. This program computes the numbers and percentages of each of the four blood types in a one-dimensional table.

Program 18.14: Computing Percentages in a One-Dimensional Table

```
title "Counts and Percentages";
proc tabulate data=Learn.Blood format=6.;
   class BloodType;
   table BloodType*(n pctn);
run;
```

The output, while not too pretty, follows:

Figure 18.15: Output from Program 18.14

Counts and Percentages

Blood Type							
A		AB		B		O	
N	PctN	N	PctN	N	PctN	N	PctN
412	41	44	4	96	10	448	45

The numbers in the PctN columns add up to 100%.

It's time to improve on this table's appearance. Let's add an ALL column, rename N and PctN, and add percent signs to the percentage numbers. Here goes:

Program 18.15: Improving the Appearance of the Output from Program 18.14

```
proc format;
   picture Pctfmt low-high='009.9%';
run;

title "Counts and Percentages";
proc tabulate data=Learn.Blood;
   class BloodType;
   table (BloodType ALL)*(n*f=5. pctn*f=Pctfmt7.1);
   keylabel n     = 'Count'
            pctn = 'Percent';
run;
```

Several improvements were made in this program. A user-defined picture format was created with PROC FORMAT. This format allows for values up to three digits before the decimal place, with the right-most digit always being printed (even if it is 0). The 0s in a picture format are place holders; the 9s are also place holders, but they print leading 0s if necessary. The percent sign in the picture definition is printed as is. You can read more about picture formats in SAS Help Center at http://go.documentation.sas.com.

Separate formats were requested for N (5.) and for PctN (Pctfmt7.1). Finally, a KEYLABEL statement was used to rename N and PctN to Count and Percent, respectively. By the way, the format width of 7 for the percentage values was chosen so the Percent column heading would fit. Here is the table:

Figure 18.16: Output from Program 18.15

Counts and Percentages

Blood Type								All	
A		AB		B		O			
Count	Percent	Count	Percent	Count	Percent	Count	Percent	Count	Percent
412	41.2%	44	4.4%	96	9.6%	448	44.8%	1000	100.0%

18.10 Displaying Percentages in a Two-Dimensional Table

Here is where things get tricky. Let's make a table of Gender (rows) by BloodType (columns) and request N and PctN, like this:

Program 18.16: Counts and Percentages in a Two-Dimensional Table

```
proc format;
   picture Pctfmt low-high='009.9%';
run;

title "Counts and Percentages in a Two-way Table";
proc tabulate data=Learn.Blood;
   class Gender BloodType;
   table (BloodType ALL='All Blood Types'),
         (Gender ALL)*(n*f=5. pctn*f=Pctfmt7.1) /RTS=25;
   keylabel ALL   = 'Both Genders'
            n     = 'Count'
            pctn  = 'Percent';
run;
```

The TABLE option, RTS=, increases the row title space (the space for the width of the row title). In the table here, there are 25 spaces from the left-most vertical line to the line preceding the word "Female".

Here is the output:

Figure 18.17: Output from Program 18.16

Counts and Percentages in a Two-way Table

	Gender				Both Genders	
	Female		Male			
	Count	Percent	Count	Percent	Count	Percent
Blood Type						
A	178	17.8%	234	23.4%	412	41.2%
AB	20	2.0%	24	2.4%	44	4.4%
B	34	3.4%	62	6.2%	96	9.6%
O	208	20.8%	240	24.0%	448	44.8%
All Blood Types	440	44.0%	560	56.0%	1000	100.0%

The percentage values in this table represent the number of subjects in a cell (females with blood type A, for example) as a percentage of all subjects. Thus, the number of females who have blood type A (178) represents 17.8% of the total number of subjects (1,000). Suppose you want the percentages to represent the distribution of blood types within each gender value. That is, you want the percentages in each column to add to 100%.

18.11 Computing Column Percentages

If you want the table to include column percentages, you need to modify Program 18.16. The original (*hard*) way to do this was to provide a denominator definition. Lauren Haworth's book, which was mentioned in the introduction to this chapter, discusses denominator definitions. The easy way to do this is to use the keywords COLPCTN and ROWPCTN, which compute column and row percentages respectively. Program 18.17 uses the COLPCTN keyword to compute column percentages.

Program 18.17: Using COLPCTN to Compute Column Percentages

```
title "Percents on the Column Dimension";
proc tabulate data=Learn.Blood;
   class Gender BloodType;
   table (BloodType ALL='All Blood Types'),
         (Gender ALL)*(n*f=5. colpctn*f=Pctfmt7.1) /RTS=25;
   keylabel All     = 'Both Genders'
            n       = 'Count'
            colpctn = 'Percent';
run;
```

Gender (and ALL) form the columns of this table. By substituting COLPCTN for PTCN, the percentages you see in the table are now column percentages. For example, the value of 40.4% for females with blood type A is computed by dividing the number of females with blood type A (178) by the total number of females (440). One other change was made to this program. The KEYLABEL statement assigns the label Both Genders to the keyword ALL. However, you want the label for ALL, associated with BloodType, to read All Blood Types. Another way to assign a label in a PROC TABULATE table is to follow the variable name with an equal sign, followed by the label in quotation marks. That is why the bottom row of the table below reads All Blood Types.

Figure 18.18: Output from Program 18.17

Percents on the Column Dimension

	Gender				Both Genders	
	Female		Male			
	Count	Percent	Count	Percent	Count	Percent
Blood Type						
A	178	40.4%	234	41.7%	412	41.2%
AB	20	4.5%	24	4.2%	44	4.4%
B	34	7.7%	62	11.0%	96	9.6%
O	208	47.2%	240	42.8%	448	44.8%
All Blood Types	440	100.0%	560	100.0%	1000	100.0%

Row percentages are requested in a similar fashion, using the keyword ROWPCTN instead of COLPCTN.

18.12 Computing Percentages on Numeric Variables

There is another type of percentage calculation that is available in PROC TABULATE—you can compute the percentage of the SUM statistic that is represented by a column or row value. For example, the data set Sales contains variables Region and TotalSales. If you want to see the contribution to total sales made by each region, you can use the PCTSUM statistic. If you place Region on the row dimension and TotalSales on the column dimension, you can use the PCTSUM statistic to see the regional contribution of sales as a percentage of total sales for all regions. Here is the program:

Program 18.18: Computing Percentages on a Numeric Value

```
title "Computing Percentages on a Numerical Value";
proc tabulate data=Learn.Sales;
   class Region;
   var TotalSales;
   table (Region ALL),
          TotalSales*(n*f=6. sum*f=dollar8.
                      pctsum*f=pctfmt7.);

             keylabel ALL   = 'All Regions'
             n          = 'Number of Sales'
             sum        = 'Sum'
             pctsum     = 'Percent';
   label TotalSales = 'Total Sales';
run;
```

If you have CLASS variables on both dimensions of a table, you can use the two keywords COLPCTSUM and ROWPCTSUM in a similar manner to COLPCTN and ROWPCTN discussed in the previous section. In the table below, the column labeled Percent represents the sum of the variable TotalSales for each region divided by the sum of TotalSales for all regions.

Figure 18.19: Output from Program 18.18

Computing Percentages on a Numerical Value

	Total Sales		
	Number of Sales	Sum	Percent
Region			
East	4	$41,593	44.8%
North	5	$36,825	39.7%
South	4	$12,003	12.9%
West	2	$2,290	2.4%
All Regions	15	$92,710	100.0%

18.13 Understanding How Missing Values Affect PROC TABULATE Output

Although this is the last topic in this chapter, it is a very important one.

> **Note:** If you have missing values for one or more variables listed in a CLASS statement, any observation with one or more missing values will be eliminated from the table!
>
> This is true even for a variable that is listed in a CLASS statement and is not referenced in a TABLE statement.

To help understand how this works (and later to see how to fix it), look at the small data set (Missing):

Figure 18.20: Listing of Data Set Missing

Listing of Data Set Learn.Missing

A	B	C
X	Y	Z
X	Y	Y
Z	Z	Z
X	X	
Y	Z	
X		

Now, see what happens when you request some simple tables:

Program 18.19: Demonstrating the Effect of Missing Values on CLASS Variables

```
title "The Effect of Missing Values on CLASS variables";
proc tabulate data=Learn.Missing format=4.;
   class A B;
   table A ALL,B ALL;
run;
```

Figure 18.21: Output from Program 18.19

The Effect of Missing Values on CLASS variables

	B			All
	X	Y	Z	
	N	N	N	N
A				
X	1	2	.	3
Y	.	.	1	1
Z	.	.	1	1
All	1	2	2	5

Notice that although there are six observations in the data set, the total shown in the table is five. Now, look what happens if you include the variable C in the CLASS statement, like this:

Program 18.20: Missing Values on a CLASS Variable That Is Not Used in the Table

```
title "The Effect of Missing Values on CLASS variables";
proc tabulate data=Learn.Missing format=4.;
   class A B C;
   table A ALL,B ALL;
run;
```

Here is the table:

Figure 18.22: Output from Program 18.20

The Effect of Missing Values on CLASS variables

	B		All
	Y	Z	
	N	N	N
A			
X	2	.	2
Z	.	1	1
All	2	1	3

Notice that the total number of observations is now three, even though variable C was not used to generate the table. PROC TABULATE chooses the observations to exclude from the table based on any missing values in any of the variables listed in the CLASS statement.

Let's rerun Program 18.19 but include the procedure option MISSING, like this:

Program 18.21: Adding the PROC TABULATE Procedure Option MISSING

```
title "The Effect of Missing Values on CLASS variables";
proc tabulate data=Learn.Missing format=4. missing;
   class A B;
   table A ALL,B ALL;
run;
```

The MISSING option includes missing values as a category in the table. Take a look:

Figure 18.23: Output from Program 18.21

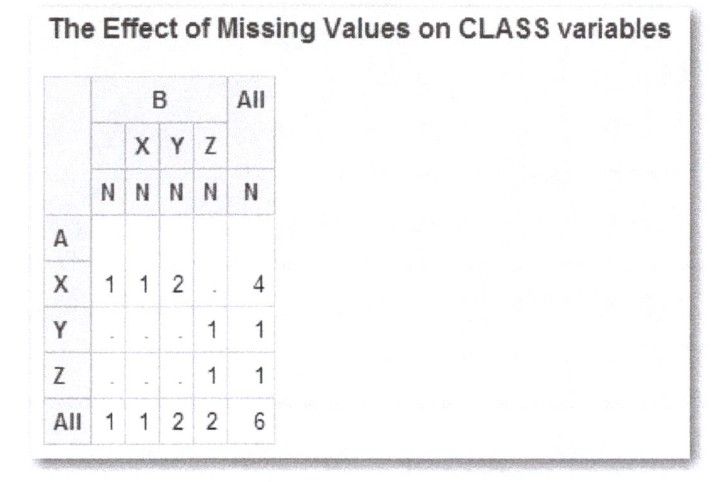

The Effect of Missing Values on CLASS variables

	B				All
		X	Y	Z	
	N	N	N	N	N
A					
X	1	1	2	.	4
Y	.	.	.	1	1
Z	.	.	.	1	1
All	1	1	2	2	6

Notice that variable B now has a column representing missing values and the total number of observations in the table is six.

By the way, if you would prefer to see something other than a period representing missing values in your table, the TABLES option MISSTEXT= can be used to substitute any text you want. For example:

Program 18.22: Demonstrating the MISSTEXT= TABLES Option

```
title "Demonstrating the MISSTEXT TABLES Option";
proc tabulate data=Learn.Missing format=7. missing;
   class A B;
   table A ALL,B ALL / misstext='No Data';
run;
```

The text `No Data` now appears instead of the period in each cell that contains a missing value.

Note: The format width increased to 7 to accommodate the text.

Here is the listing:

Figure 18.24: Output from Program 18.22

Demonstrating the MISSTEXT TABLES Option

	B				All
		X	Y	Z	
	N	N	N	N	N
A					
X	1	1	2	No Data	4
Y	No Data	No Data	No Data	1	1
Z	No Data	No Data	No Data	1	1
All	1	1	2	2	6

We hope that this introduction to PROC TABULATE gives you the courage to give this useful and powerful procedure a try. This author was TABULATE-phobic for a long time until forced to use the procedure for a project that required nice-looking summary tables. Now, he is a convert and tries to get others to use this procedure more often.

18.14 Problems

Solutions to odd-numbered problems are located at the back of this book. Solutions to all problems are available to professors. If you are a professor, visit the book's companion website at support.sas.com/cody for information about how to obtain the solutions to all problems.

All the problems, except Problem 10, use the SAS data set College.

Note: Data set College has permanent formats for Gender, SchoolSize, and Scholarship. Either make this format catalog available to SAS (see Chapter 5), run the PROC FORMAT statements here, or use the system option NOFMTERR (no format error) that allows you to access SAS data sets that have permanent user-defined formats without causing an error.

```
proc format;
   value $Yesno 'Y','1' = 'Yes'
                'N','0' = 'No'
                ' '     = 'Not Given';
   value $Size 'S' = 'Small'
               'M' = 'Medium'
               'L' = 'Large'
                ' ' = 'Missing';
   value $Gender 'F' = 'Female'
                 'M' = 'Male'
                 ' ' = 'Not Given';
run;
```

For each problem, you need to write the appropriate PROC TABULATE statements to produce the given table.

1. Produce the following table. Note that the last row in the table represents all subjects.

Demographics from COLLEGE Data Set

	School Size		
	Large	Medium	Small
Gender			
F	10	23	26
M	8	18	11
Scholarship			
No	16	36	32
Yes	2	5	5
All	18	41	37

2. Produce the following table. Note that the ALL column has been renamed Total.

Demographics from COLLEGE Data Set

	Gender		Scholarship		Total
	F	M	No	Yes	
School Size					
Large	10	8	16	2	18
Medium	23	18	36	5	41
Small	26	11	32	5	37
Total	59	37	84	12	96

3. Produce the following table. Note that the ALL column has been renamed Total and Gender has been formatted.

Demographics from COLLEGE Data Set

		SchoolSize			Total
		Large	Medium	Small	
Gender	Scholarship				
Female	No	9	19	22	50
	Yes	1	4	4	9
	Total	10	23	26	59
Male	Scholarship				
	No	7	17	10	34
	Yes	1	1	1	3
	Total	8	18	11	37
Total	Scholarship				
	No	16	36	32	84
	Yes	2	5	5	12
	Total	18	41	37	96

4. Produce the following table. Note that the keyword ALL has been renamed Total, Gender is formatted, and ClassRank (a continuous numeric variable) has been formatted into two groups (0–70 and 71 and higher).

Demographics from COLLEGE Data Set

	ClassRank					
	Low to 70			71 and higher		
	Gender		Total	Gender		Total
	Female	Male		Female	Male	
Scholarship						
No	20	15	35	23	19	42
Yes	4	.	4	5	2	7
Total	24	15	39	28	21	49

5. Produce the following table. Note that the keywords ALL, N, MIN, and MAX have all been renamed.

Descriptive Statistics

		Gender		Total
		F	M	
GPA	Number	56	38	94
	Average	3.5	3.5	3.5
	Minimum	2.3	2.4	2.3
	Maximum	4.0	4.0	4.0

6. Produce the following table. Note that the keywords ALL, N, and MEAN have all been renamed.

Descriptive Statistics

		Gender		Total
		F	M	
GPA	Number	56	38	94
	Average	3.5	3.5	3.5
Class Rank	Number	52	36	88
	Average	71.3	72.3	71.7

7. Produce the following table. Note that the keywords MIN and MAX have been renamed and the two variables ClassRank and SchoolSize now have labels.

More Descriptive Statistics

	GPA			Class Rank		
	Median	Minimum	Maximum	Median	Minimum	Maximum
School Size						
Large	3.6	3.0	4.0	71	45	98
Medium	3.7	2.4	4.0	71	42	100
Small	3.4	2.3	4.0	79	41	99
Total	3.6	2.3	4.0	73	41	100

8. Produce the following table. Note that the keyword ROWPCTN has been renamed as Percent and Gender has been formatted.

Demonstrating Row Percents

	Scholarship		All
	No	Yes	
	Percent	Percent	Percent
Gender			
Female	83	17	100
Male	93	8	100
All	87	13	100

9. Produce the following table. Note that the ALL column has been renamed Total, COLPCTN has been renamed Percent, Gender has been formatted, and the order of the columns is Yes followed by No. **Hint:** Think of using the procedure option ORDER=data and figure a way to place the data set with `Yes` values before the `No` values.

Demonstrating Column Percents

	Scholarship		Total
	Yes	No	
	Percent	Percent	Percent
Gender			
Female	77	57	60
Male	23	43	40
Total	100	100	100

10. Using the SAS data set Sales, produce the table shown here. The variable TotalSales has been labeled as Total Sales. The percentages shown in the table represent percentages of the total quantities sold and of the total sales.

Demonstrating the Pctsum Statistic

	Quantity	Total Sales
	Percent of Total	Percent of Total
Region		
East	35	45
North	3	40
South	10	13
West	52	2
All	100	100

Chapter 19: Introducing the Output Delivery System

19.1 Introduction

The Output Delivery System (abbreviated ODS) allows you to send SAS output to a variety of destinations, such as HTML, PDF, RTF, and SAS data sets.

Not only can you send your SAS output to all of these formats, you can, if you are brave enough, customize the output's fonts, colors, size, and layout. Finally, you can now capture virtually every piece of SAS output to a SAS data set for further processing.

19.2 Sending SAS Output to an HTML File

Beginning in SAS 9.3 the default ODS Destination for SAS output is HTML. If you don't need your output in a file, you don't need to submit any ODS statements unless you want to capture procedure output to a SAS data set.

As an example, suppose you want to send the output of PROC PRINT and PROC MEANS to an HTML file. It takes only two SAS statement to do this, as shown next:

Program 19.1: Sending SAS Output to an HTML File

```
ods html path='C:\books\learing'
    file='Sample.html';
title "Listing of Test_Scores";
proc print data=Learn.Test_Scores;
   title2 "Sample of HTML Output - all defaults";
   id ID;
   var Score1-Score3;
run;

title "Descriptive Statistics";
proc means data=Learn.Test_Scores n nmiss mean min max maxdec=2;
   var Score1-Score3;
run;

ods html close;
```

All that is required is to place an ODS HTML statement before the procedures whose output you want to capture and an ODS HTML CLOSE statement at the end. You specify a PATH= and a FILE= option on the HTML statement. The HTML extension (or HTM for some environments) is not needed by SAS, but it allows the operating system to recognize that the file contains HTML tags and to open it with the appropriate browser.

The HTML output is shown here:

Figure 19.1: HTML File Created by Program 19.1

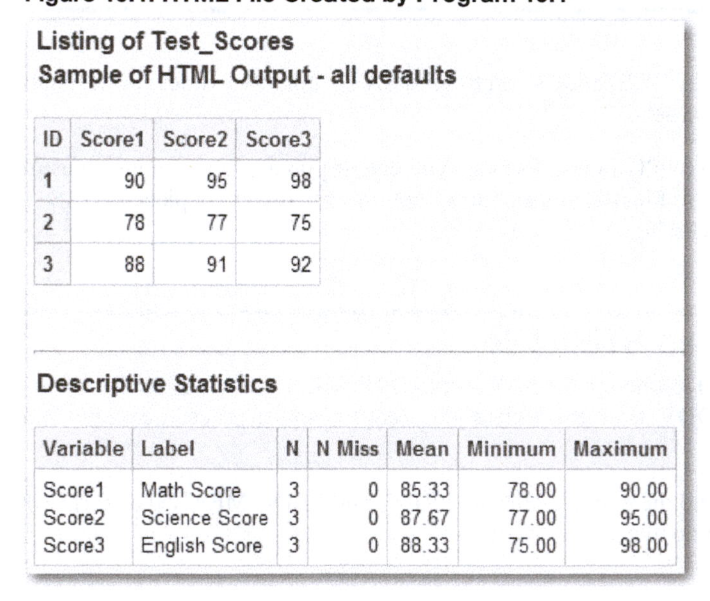

Listing of Test_Scores
Sample of HTML Output - all defaults

ID	Score1	Score2	Score3
1	90	95	98
2	78	77	75
3	88	91	92

Descriptive Statistics

Variable	Label	N	N Miss	Mean	Minimum	Maximum
Score1	Math Score	3	0	85.33	78.00	90.00
Score2	Science Score	3	0	87.67	77.00	95.00
Score3	English Score	3	0	88.33	75.00	98.00

If you are using the SAS windowing system and you have chosen the option of producing listing files (Tools ▸ Options ▸ Preferences ▸ Results tab ▸ (Check LISTING Box), when you run Program 19.1, SAS will produce a listing file in addition to the HTML file on your disk drive.

19.3 Creating a Table of Contents

If your HTML output is large, you may elect to produce a table of contents along with the normal HTML output. Remember that HTML is primarily intended to be viewed with a web browser, not printed. A table of contents embeds links to each separate part of the output, allowing users to click on a link and go directly to different parts of the output.

To create a table of contents, you need to specify three output files:

- A main file that contains the procedure output
- A table of contents file
- A frame file to display the other two

Here is an example, using the same procedures as Program 19.1:

Program 19.2: Creating a Table of Contents for HTML Output

```
ods html body = 'Body_Sample.html'
         contents = 'Contents_Sample.html'
         frame = 'Frame_Sample.html'
         path = 'c:\books\learning';

title "Using ODS to Create a Table of Contents";
proc print data=Learn.Test_Scores;
   id ID;
   var Name Score1-Score3;
run;

title "Descriptive Statistics";
proc means data=Learn.Test_Scores n mean min max;
   var Score1-Score3;
run;

ods html close;
```

The four keywords are BODY=, CONTENTS=, FRAME= and PATH=.

You can name the three files anything you like—the names do not need to include the words *body*, *contents*, or *frame* as in this example. If you are creating a web page from these files, you need to provide a link to the FRAME file. A display of the file Contents_Sample.html is shown next.

Figure 19.2: The Contents_Sample.html File

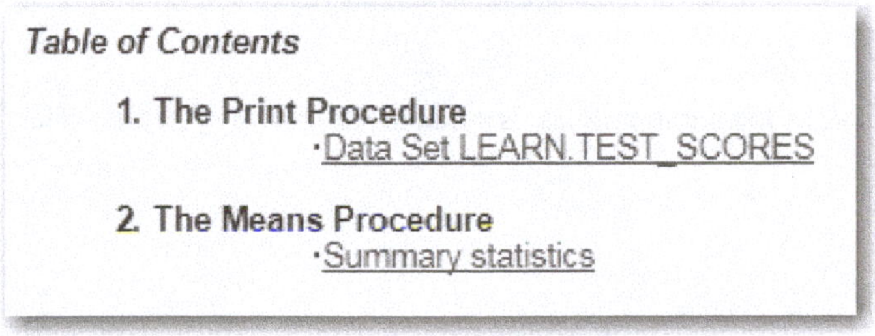

Table of Contents

1. The Print Procedure
   ·Data Set LEARN.TEST_SCORES

2. The Means Procedure
   ·Summary statistics

If you click on the two links in this file, you will see the PROC PRINT and PROC MEANS output.

19.4 Selecting a Different HTML Style

Although you can customize every aspect of the HTML output, it takes time and effort. There are a number of built-in styles that change the appearance of the output without any work on your part. The default style is called HTMLBLUE.

One way to see a list of styles, if you are in a windowing environment, is to select **Tools ▶ Options ▶ Preferences ▶** and then select the Results tab. You will see a long list of built-in styles that you can choose from. For example, if you want black and white output, you could choose the Journal style, like this:

Program 19.3: Choosing a Style for HTML Output

```
ods html path = 'c:\books\learning'
         file = 'Journal_Example.html'
                style=Journal;

title "Listing of Test_Scores";
proc print data=Learn.Test_Scores;
   id ID;
   var Name Score1-Score3;
run;

ods html close;
```

The resulting output follows:

Figure 19.3: Output Using the Journal Style

Listing of Test_Scores

ID	*Score1*	*Score2*	*Score3*
1	90	95	98
2	78	77	75
3	88	91	92

This style does not contain any shading and, as the name implies, is suitable for printing journal-style tables.

19.5 Choosing Other ODS Destinations

You can create RTF (rich text format) or PDF (portable document format) files, which are both readable with Adobe Reader, in the same manner as HTML output. Just substitute the keywords RTF or PDF for HTML in the statements in Program 19.1 and change the file extensions to RTF or PDF, respectively.

Besides providing a better appearance, using RTF or PDF files allows users of software other than SAS to see well-formatted SAS output without having SAS fonts on their machines. RTF is a somewhat universal format that can be incorporated into a Microsoft Office Word document directly. Have you ever seen what a SAS listing file looks like when it is displayed in a font other than a SAS font? Take a look at output from PROC FREQ, which is displayed in Word with a Courier font (it looks even worse when printed with a proportional font such as Arial or Times Roman):

Figure 19.4: Opening a SAS Listing File in Microsoft Notebook

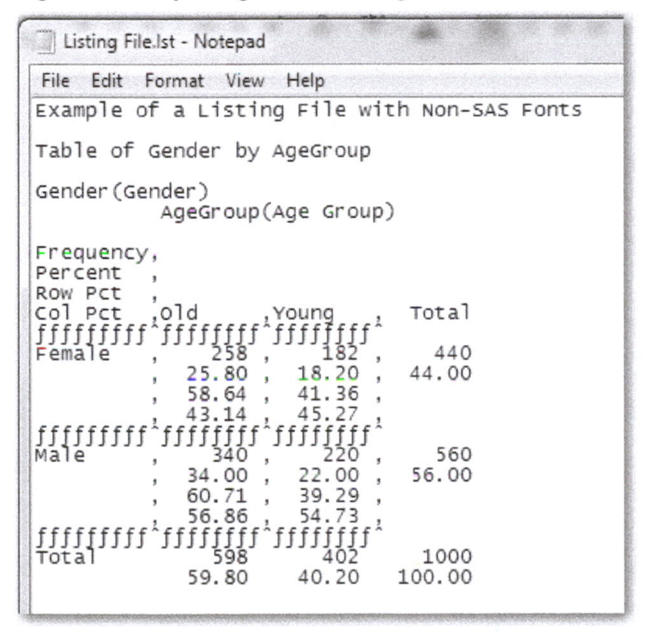


```
Example of a Listing File with Non-SAS Fonts

Table of Gender by AgeGroup

Gender(Gender)
          AgeGroup(Age Group)

Frequency,
Percent  ,
Row Pct  ,
Col Pct  ,Old     ,Young   ,  Total
ƒƒƒƒƒƒƒƒƒˆƒƒƒƒƒƒƒƒˆƒƒƒƒƒƒƒƒˆ
Female   ,    258 ,    182 ,    440
         ,  25.80 ,  18.20 ,  44.00
         ,  58.64 ,  41.36 ,
         ,  43.14 ,  45.27 ,
ƒƒƒƒƒƒƒƒƒˆƒƒƒƒƒƒƒƒˆƒƒƒƒƒƒƒƒˆ
Male     ,    340 ,    220 ,    560
         ,  34.00 ,  22.00 ,  56.00
         ,  60.71 ,  39.29 ,
         ,  56.86 ,  54.73 ,
ƒƒƒƒƒƒƒƒƒˆƒƒƒƒƒƒƒƒˆƒƒƒƒƒƒƒƒˆ
Total         598      402     1000
            59.80    40.20   100.00
```

This output is a strong argument for using RTF or PDF output destinations.

19.6 Selecting or Excluding Portions of SAS Output

You can use an ODS SELECT or ODS EXCLUDE statement before a SAS procedure to control which portions of the output you want.

Suppose you want to use PROC UNIVARIATE to list the five highest and five lowest values of a variable. This is a normal part of the output from PROC UNIVARIATE, but you also get several pages of additional output as well. Program 19.4 uses an ODS SELECT statement to restrict the output.

Program 19.4: Using an ODS SELECT Statement to Restrict PROC UNIVARIATE Output

```
ods select extremeobs;

title "Extreme Values of RBC";
proc Univariate data=Learn.Blood;
   id Subject;
   var RBC;
run;
```

When you run this program, the only output is as follows:

Figure 19.5: Output from Program 19.4

Extreme Values of RBC

Variable: RBC

Extreme Observations					
Lowest			Highest		
Value	Subject	Obs	Value	Subject	Obs
1.71	525	525	7.99	565	565
2.33	440	440	8.12	984	984
2.55	113	113	8.26	288	288
2.92	293	293	8.43	726	726
3.13	635	635	8.75	135	135

This seems simple enough, but how do you determine the name of the output objects (in this case, EXTREMEOBS)? One way is to run the procedure sandwiched between ODS TRACE ON and ODS TRACE OFF statements, like this:

Program 19.5: Using the ODS TRACE Statement to Identify Output Objects

```
ods trace on;

title "Extreme Values of RBC";
proc Univariate data=Learn.Blood;
   id Subject;
   var RBC;
run;

ods trace off;
```

When you run the procedure, the names of the output objects appear in the SAS log as shown next (only selected portions are shown):

Figure 19.6: SAS Log Showing ODS Output Objects

```
8        ods trace on;
9        title "Extreme Values of RBC";
10       proc Univariate data=Learn.Blood;
11          id Subject;
12          var RBC;
13       run;

Output Added:
-------------
Name:         Moments
Label:        Moments
Template:     base.univariate.Moments
Path:         Univariate.RBC.Moments
-------------

Output Added:
-------------
Name:         BasicMeasures
Label:        Basic Measures of Location and Variability
```

```
-------------

Output Added:
-------------
Name:         ExtremeObs
Label:        Extreme Observations
Template:     base.univariate.ExtObs
Path:         Univariate.RBC.ExtremeObs
-------------
```

One way to figure out the correspondence between output objects and portions of the output is to look through the output listing and the list of objects in the SAS log and make an educated guess. It's usually quite obvious which object goes with each portion of output. If you use TRACE ON/LISTING, the information on each output object is placed in the Output window, along with the listing. This is yet another way to know which output object names go with each portion of the output.

For a more systematic approach, look at the Results window in the SAS windowing environment and notice the labels in the list. For example, the following display results from running Program 19.5:

Figure 19.7: The Results Window in the SAS Windowing Environment

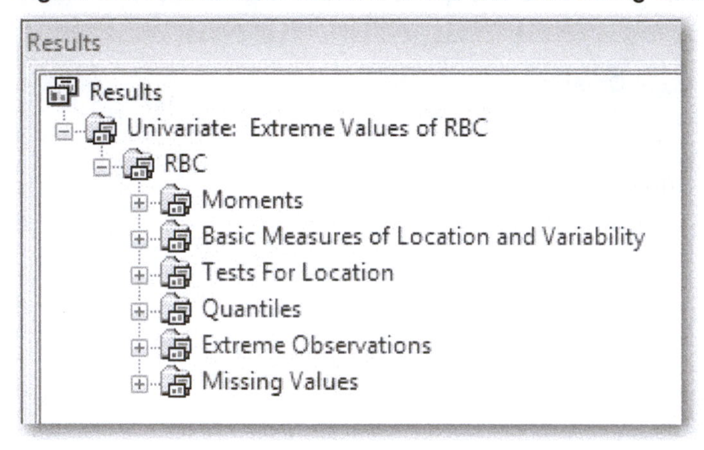

If you then right-click on any one of these labels, you will see more information, including the names of all the output objects.

Below is a screen shot taken after right-clicking on Extreme Observation and selecting Preferences:

Figure 19.8: Properties of Extreme Observations

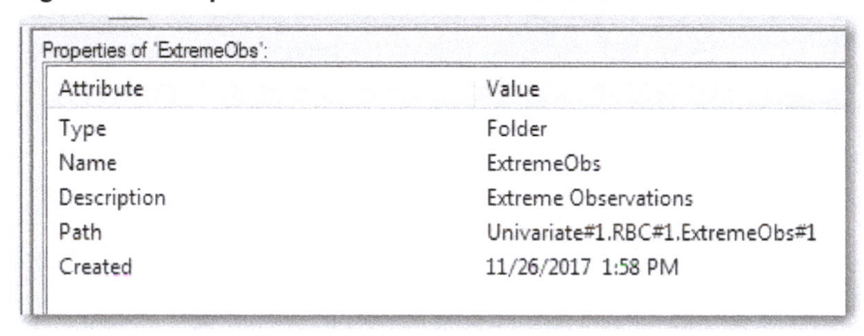
Properties of 'ExtremeObs':

Attribute	Value
Type	Folder
Name	ExtremeObs
Description	Extreme Observations
Path	Univariate#1.RBC#1.ExtremeObs#1
Created	11/26/2017 1:58 PM

You see here that the name of this output object is ExtremeObs.

As an alternative to using the object name in the ODS SELECT statement, you can also select objects by using the labels instead of the object names. For example, you can use the following:

```
ods select "Extreme Observations";
```

in place of this statement:

```
ods select ExtremeObs;
```

If you use the label, be sure to place it in single or double quotes.

You can include a list of output objects following the ODS SELECT statement (separated by spaces). If you want to select most of the output objects from a particular procedure, it might be easier to use an ODS EXCLUDE statement to exclude the ones you don't want.

Remember that ODS SELECT or EXCLUDE statements only operate on the procedure that follows. If you want the selections to remain for other procedures, you can use the PERSIST option, like this:

```
ods select ExtremeObs(persist);
```

19.7 Sending Output to a SAS Data Set

One of the possible ODS output destinations is a SAS data set. This feature of ODS allows you to capture just about any value computed by a procedure and use it in further calculations or a customized report. Prior to the development of ODS, PROC PRINTTO was used to capture SAS output from a procedure and use it as input to a DATA step.

Many procedures already provide you with the ability to capture information in a SAS data set with an OUT= option, usually in an OUTPUT statement. Using ODS to capture output is more general—it lets you capture any value you want. Suppose you want to run a statistical procedure called a *t*-test. This statistical test produces a *t*-value and a *p*-value (the significance of the result). These two values are not available in an output data set from PROC TTEST, but you can use ODS to place these values into a SAS data set. If you are not familiar with PROC TTEST, it should still be clear how to capture SAS procedure output into SAS data sets. Here is a program to capture a portion of the *t*-test output to a data set:

Program 19.6: Using ODS to Send Procedure Output to a SAS Data Set

```
ods listing close;
ods output ttests=T_Test_Data;

proc ttest data=Learn.Blood;
   class Gender;
   var RBC WBC Chol;
run;

ods listing;
title "Listing of T_Test_Data";
proc print data=T_Test_Data;
run;
```

This program first closes the listing destination so that output is not sent to the Output window.

> **Note:** You cannot use a NOPRINT option on procedures that allow it. If you do, the values will not be available to be sent to alternate ODS destinations.

Next, the ODS OUTPUT statement is used to send the output to a data set. The keyword to the left of the equal sign is the name of an output object produced by PROC TTEST.

Following the equal sign is the name of the data set you want to create. You then run the procedure in the usual way. The first time you do this, you should run a PROC PRINT to determine the structure of the output data set.

Here is the output:

Figure 19.9: Output from Program 19.6

Listing of T_Test_Data

Obs	Variable	Method	Variances	tValue	DF	Probt
1	RBC	Pooled	Equal	0.41	914	0.6797
2	RBC	Satterthwaite	Unequal	0.41	874.92	0.6796
3	WBC	Pooled	Equal	1.87	906	0.0624
4	WBC	Satterthwaite	Unequal	1.87	864.56	0.0622
5	Chol	Pooled	Equal	0.53	793	0.5953
6	Chol	Satterthwaite	Unequal	0.54	771.14	0.5918

The structure of these output data sets can be complicated. In this instance, you know that PROC TTEST produces two *t*-values, one under the assumption of equal variance and the other under the assumption of unequal variance. What can you do with this data set?

Sending computed values to a data set enables you to perform additional analyses on them. Also, you might want to present the output from a SAS procedure differently from the way SAS presents it. As an example, suppose you want to see a simple report showing the *t*- and *p*-values from your *t*-test, rather than the more complicated output from PROC TTEST. Once you know the structure of the output data set created by ODS, you can proceed like this:

Program 19.7: Using an Output Data Set to Create a Simplified Report

```
title "T-Test Results - Using Equal Variance Method";
proc report data=T_Test_Data;
   where Variances = "Equal";
   columns Variable tValue ProbT;
   define Variable / width=8;
   define tValue / display "T-Value" width=7 format=7.2;
   define ProbT / display "P-Value" width=7 format=7.5;
run;
```

Inspection of data set T_Test_Data shows that there are separate observations for the two variance assumptions. You can use a WHERE statement to select statistics using the equal variance assumption. The result of the PROC REPORT is a simple listing showing the variable name along with the *t*- and *p*-values. Here it is:

Figure 19.10: Output from Program 19.7

T-Test Results – Using Equal Variance Method

Variable	T-Value	P-Value
RBC	0.41	0.67972
WBC	1.87	0.06237
Chol	0.53	0.59529

This chapter has just touched the surface of the capabilities of the Output Delivery System. However, by routing your output to destinations such as HTML or PDF files or by using ODS SELECT statements, you can create output for web pages or company reports that look much more impressive than a standard listing report.

19.8 Problems

Solutions to odd-numbered problems are located at the back of this book. Solutions to all problems are available to professors. If you are a professor, visit the book's companion website at support.sas.com/cody for information about how to obtain the solutions to all problems.

1. Run the following program, sending the output to an HTML file. See Chapter 16, Problem 2, for the note about creating formats for this data set.

```
title "Sending Output to an HTML File";
proc print data=Learn.College(obs=8) noobs;
run;
proc means data=learn.college n mean maxdec=2;
   var GPA ClassRank;
run;
```

2. Run the same two procedures shown in Problem 1, except create a contents file, a body file, and a frame file.
3. Run the same two procedures as shown in Problem 1, except use the JOURNAL (or FANCYPRINTER) style instead of the default style.
4. Send the results of a PROC PRINT on the data set Survey to an RTF file.
5. Run PROC UNIVARIATE on the variables Age and Salary from the Survey data set. Use the TRACE ON/TRACE OFF statements to display the names of the output objects created by this procedure. Once you have done this, run PROC UNIVARIATE again, selecting only the output object that shows Quantiles.
6. Run the same PROC UNIVARIATE as in Problem 5. Issue two ODS statements: one to select the MOMENTS output object and the other to send this output to a SAS data set. Run PROC PRINT to see a listing of this data set.

Chapter 20: Creating Charts and Graphs

20.1 Introduction

For the first edition of this book, this chapter was devoted to SAS/GRAPH procedures such a PROC GCHART and PROC GPLOT. Although SAS/GRAPH is still available, you can produce charts and graphs more easily by using PROC SGPLOT (the SG stands for "Statistical Graphics"—although this author thinks it should stand for SAS Graphics instead). Unlike SAS/GRAPH, which requires a separate license, PROC SGPLOT is included in Base SAS.

Three other SG procedures, SGSCATTER, SGPANEL, and SGRENDER, are also included in Base SAS but PROC SGPLOT may be all you need.

20.2 Creating Bar Charts

Suppose you want to see the frequencies of the four blood types in the Blood data set. If you are familiar with either the older procedure called PROC CHART or the SAS/GRAPH procedure PROC GCHART, you will see some similarities to PROC SGPLOT. Below is a program to create a blood type vertical bar chart using PROC SGPLOT:

Program 20.1: Creating a Vertical Bar Chart

```
title "Vertical Bar Chart Example";
proc sgplot data=Learn.Blood;
   vbar BloodType;
run;
```

You use the VBAR statement (stands for "vertical bar") followed by the category variable. Below is the bar chart:

Figure 20.1: Output from Program 20.1

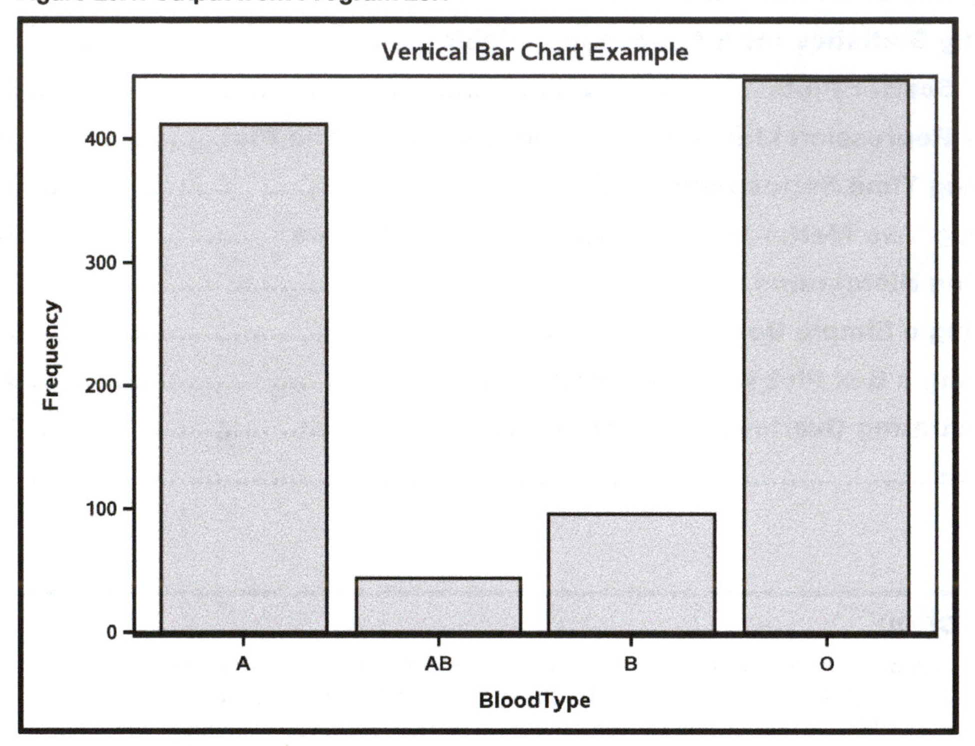

In this bar chart, the height of the bars represents frequency counts. To create a horizontal bar chart, substitute the HBAR statement in place of VBAR as demonstrated in the next program:

Program 20.2: Creating a Horizontal Bar Chart

```
title "Horizontal Bar Chart Example";
proc sgplot data=Learn.Blood;
   hbar BloodType / nofill barwidth=.25;
run;
```

Besides changing the orientation of the bars, this program also include two options that work for both vertical and horizontal bar charts. The first, NOFILL, generates a bar outline. If you plan to print your charts using either a laser or ink jet printer, this will save on toner (or ink). The second option used in this program, BARWIDTH=, allows you to control the width of the bars. The value you choose is the proportion of the maximum default bar width. Here is the output:

Figure 20.2: Output from Program 20.2

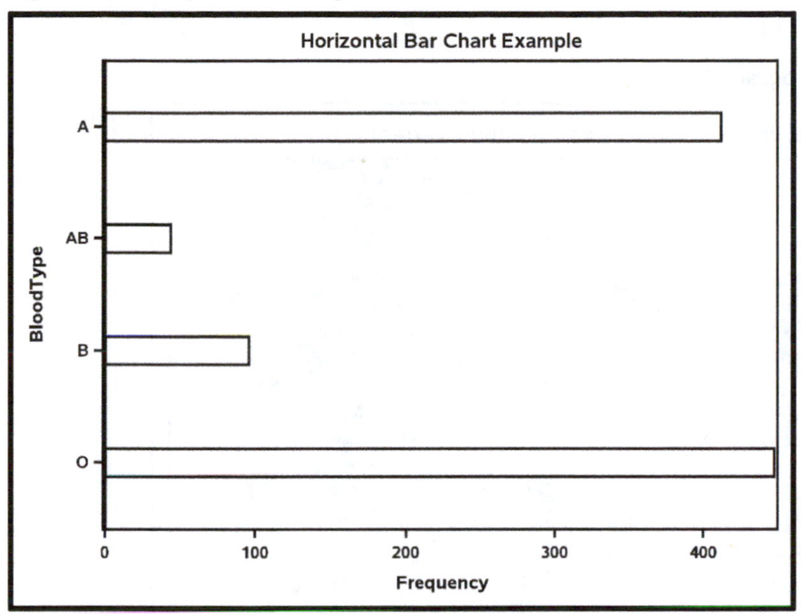

You can create more complex bar charts by displaying the distribution of one variable at each level of a second variable. This is accomplished by adding a GROUP= option to either the VBAR or HBAR statement. Suppose you want to look at frequencies of blood types for males and females. The program below does just that:

Program 20.3: Vertical Bar Chart Example (Two Variables)

```
title "Vertical Bar Chart Example (two variables)";
proc sgplot data=Learn.Blood;
   vbar Gender / group=BloodType;
run;
```

Because GROUP= is an option for the VBAR statement, it is entered following a slash (/). The output is shown below:

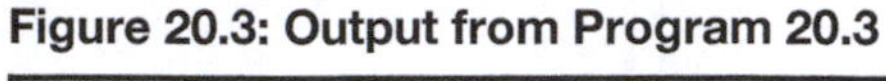

Figure 20.3: Output from Program 20.3

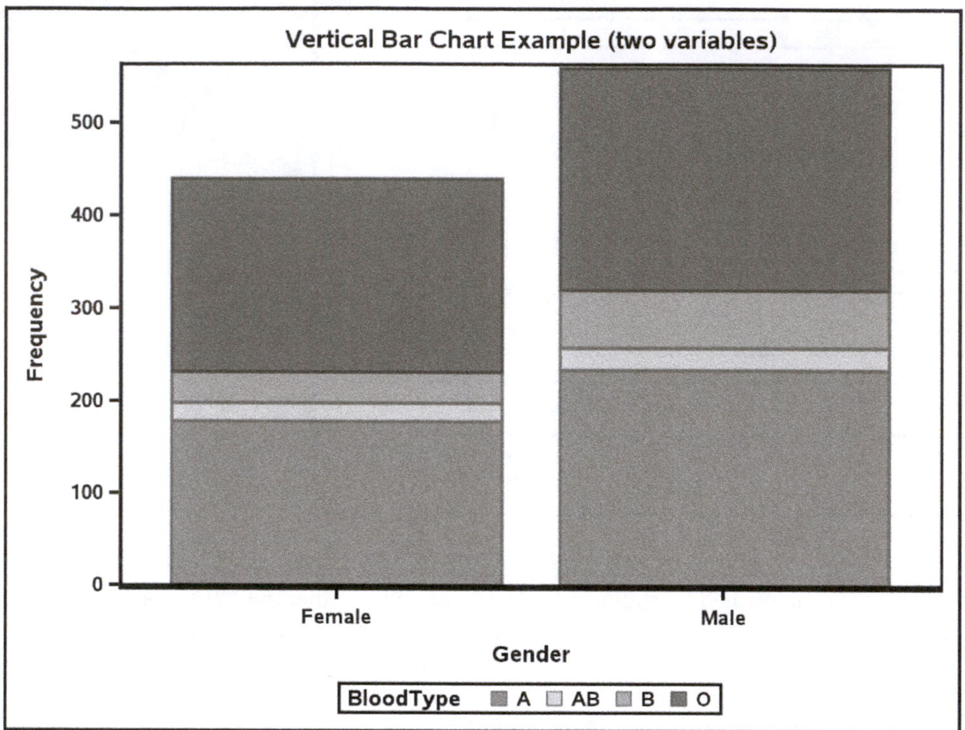

It appears the distribution of blood types is similar in females and males. If you would like to see the groups displayed side by side, add the option GROUPDISPLAY=CLUSTER right after GROUP=Bloodtype.

20.3 Displaying Statistics for a Response Variable

The same plot statements, VBAR and HBAR, can be used to display means or sums for each level of a categorical variable. Instead of the height of each bar representing frequencies, it can represent a mean or sum of a response variable. You accomplish this by including the option RESPONSE= in the VBAR or HBAR statement. In the example below, you want to see the mean cholesterol levels for each blood type.

Program 20.4: Vertical Bar Chart Displaying a Response Variable

```
title "Vertical Bar Chart Displayig a Response Variable";
proc sgplot data=Learn.Blood;
   vbar BloodType / response=Chol stat=mean barwidth=.5 nofill;
run;
```

Here you are requesting that the height of each bar represents the mean cholesteral level for each of the four blood types. You need to include the option, STAT=mean, so that the height of the bars represents means. Without this option, the default statistic of SUM will be used. Here is the output:

Figure 20.4: Output from Program 20.4

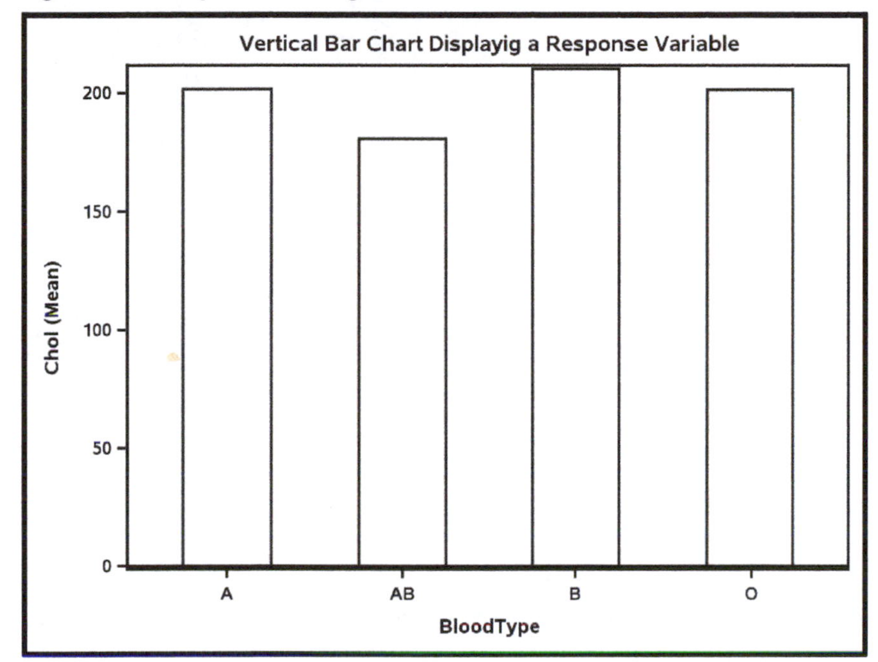

This plot shows the mean cholesterol level for each of the four blood types.

Note: Please keep in mind that this data set is made up and does not represent actual values.

20.4 Creating Scatter Plots

Scatter plots allow you to visually see relationships between two variables. This example uses a data set that is included with SAS software in a library called SASHELP. If you use PROC CONTENTS with option DATA=SASHELP._ALL_, you can see a list of the over 200 data sets in this library. You can also use the SAS HELP icon on the taskbar to learn more about the data sets in this library.

The data set used in this example contains data on various measurements of iris petals. You use the SCATTER statement to produce a scatter plot along with the options X= and Y= to specify which variables go on the x axis and y axis. The program below plots the petal width on the x axis and the petal length on the y axis:

Program 20.5: Simple Scatter Plot

```
title "Simple Scatter Plot";
proc sgplot data=SASHelp.Iris;
   scatter x=PetalWidth y=PetalLength;
run;
```

Figure 20.5: Output from Program 20.5

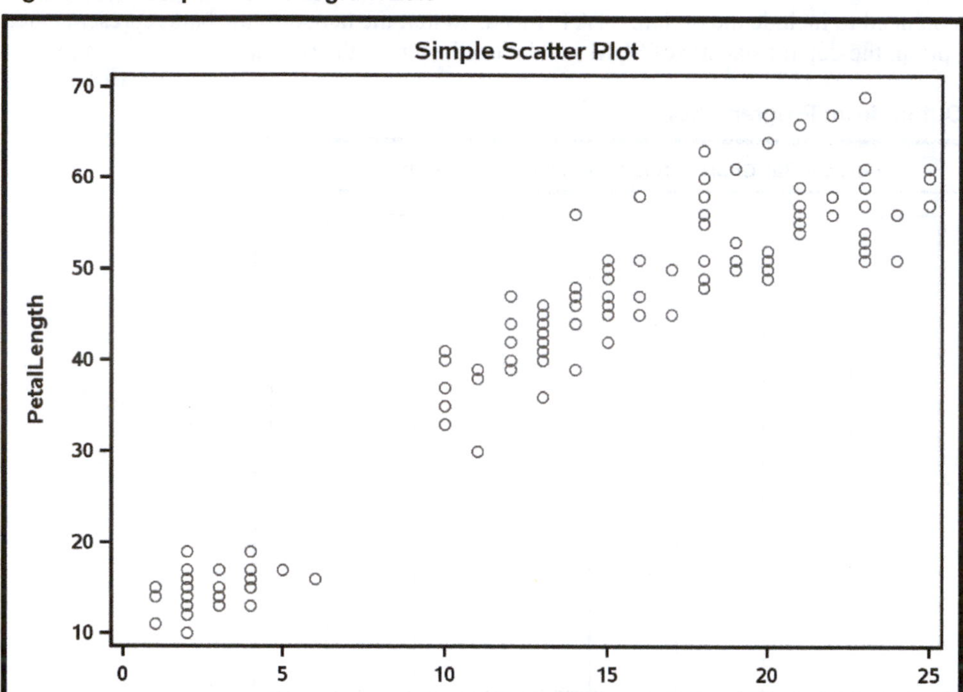

Inspection of this plot would lead you to conclude that there is a strong relationship between petal width and petal length. You could verify this by using statistical techniques such as correlation and regression.

20.5 Adding a Regression Line and Confidence Limits to the Plot

By substituting the REG statement for the SCATTER statement in Program 20.5, you can include a regression line on the plot. The two options, CLM and CLI, add two types of confidence limits to the plot. CLM (confidence limit for the mean) shows you the 95% confidence limits for the mean of Y for any particular value of X. CLI (confidence limits for individual points) shows you the limits where you are 95% confident that an individual data point will be between these limits.

Program 20.6: Scatter Plot with a Regression Line and Confidence Intervals

```
title "Scatter Plot with a Regression Line and Confidence Intervals";
proc sgplot data=SASHelp.Iris;
   reg x=PetalWidth y=PetalLength / CLM CLI;
run;
```

Here is the plot:

Figure 20.6: Output from Program 20.6

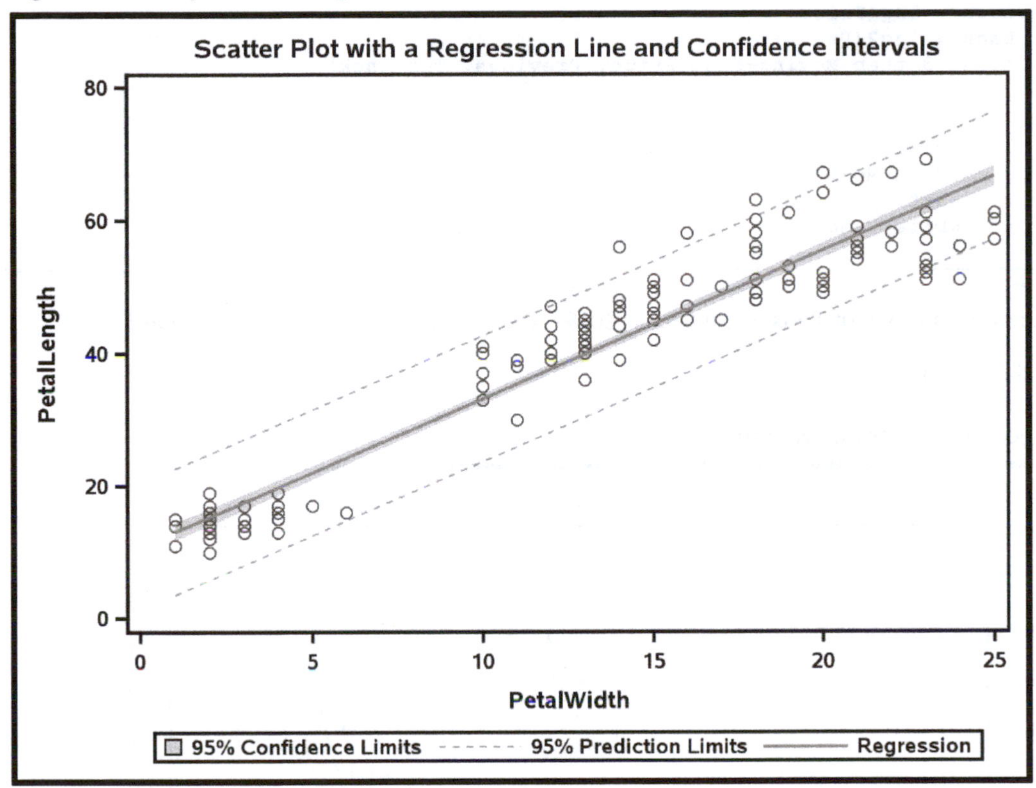

You can see the individual data points, the regression line. and the two confidence limits.

20.6 Generating Time Series Plots

This example starts with a short DATA step to create a moving average. The two functions, LAG and LAG2, return the stock price from the previous day and one day before that and then averages the three values (the current price and the price on the two previous days). You use a SERIES statement along with X= and Y= options to specify the x and y variables for the plot. Each plot produced by PROC SGPLOT is additive—that is, each plot appears in the same graph. Later in this chapter, you will see that you can set a value for transparency if one plot covers a previous plot. In this case, overlapping is not a problem.

Program 20.7: Time Series Plot

```
data Moving_Average;
   set Learn.Stocks;
   Previous = lag(Price);
   Two_Back = lag2(Price);
   if _n_ ge 3 then Moving=mean(Price, Previous, Two_Back);
run;

title "Series Plot";
proc sgplot data=Moving_Average;
   series x=Date y=Price;
   series x=Date y=Moving;
run;
```

The Price versus Day plot starts on January 1, 2017—the moving average plot starts on day3.

Here is the output:

Figure 20.7: Output from Program 20.7

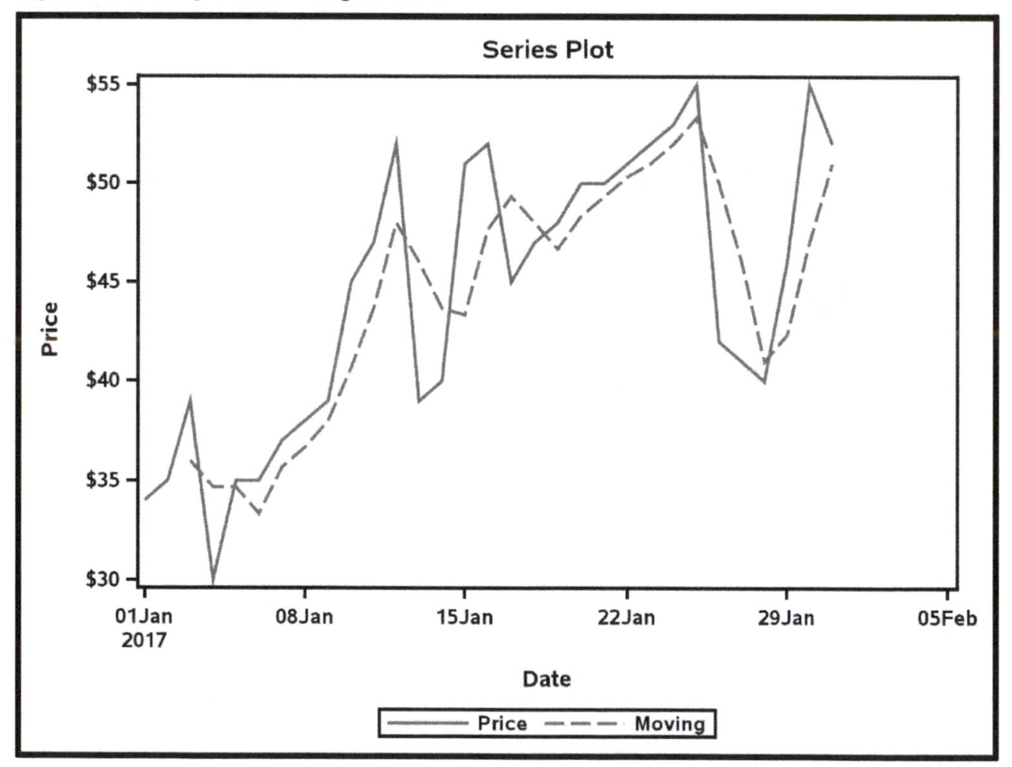

You can see that using a moving average smooths out some of the day-to-day variations.

20.7 Describing Two Methods of Generating Smooth Curves

In addition to providing the capability of connecting data points by straight lines, PROC SGPLOT also provides two non-linear methods of curve fitting. Both of these methods use local regression models to generate smooth curves.

The first method in this section uses local points to fit portions of the curve. To control how closely the smooth curve follows the data points, specify the SMOOTH option=*numeric value* in the PBSPLINE statement.

Program 20.8: Smooth Curves - Splines

```
title "Smooth Curve - Splines";
proc sgplot data=SASHelp.Iris;
   pbspline x=PetalWidth y=PetalLength;
run;
```

You use the PBSPLINE statement and specify the variables on the two axes with the required options X= and Y=. This program uses the iris petal data (see section 20.4) and adds a smooth line to the scatter plot as shown below:

Figure 20.8: Output from Program 20.8

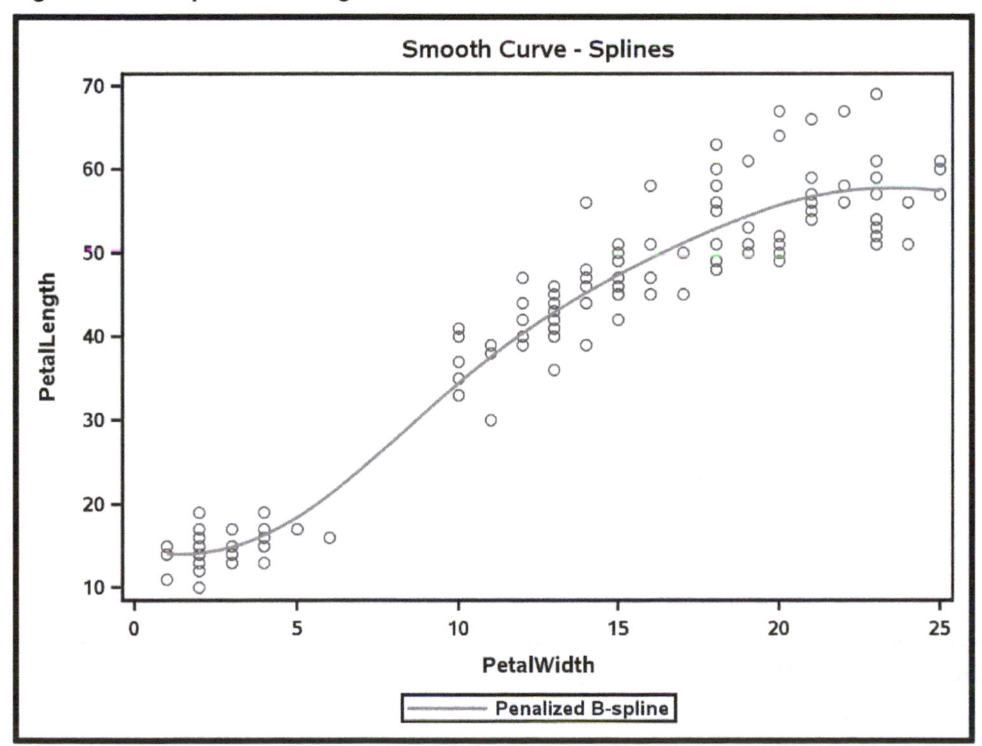

As previously mentioned, you can control the amount of "wiggle" by options in the PBSPLINE statement.

The next smoothing method is called the LOESS method. Although this is not an exact acronym, most documentation for this method says it stands for "LOcal regrESSion." The next program shown uses the same iris data as Program 20.8, but uses the LOESS method instead of PBSPLINE.

Program 20.9: Smooth Curve - LOESS Method

```
title "Smooth Curve - LOESS Method";
proc sgplot data=SASHelp.Iris;
   loess x=PetalWidth y=PetalLength;
run;
```

This program produced the plot below:

Figure 20.9: Output from Program 20.9

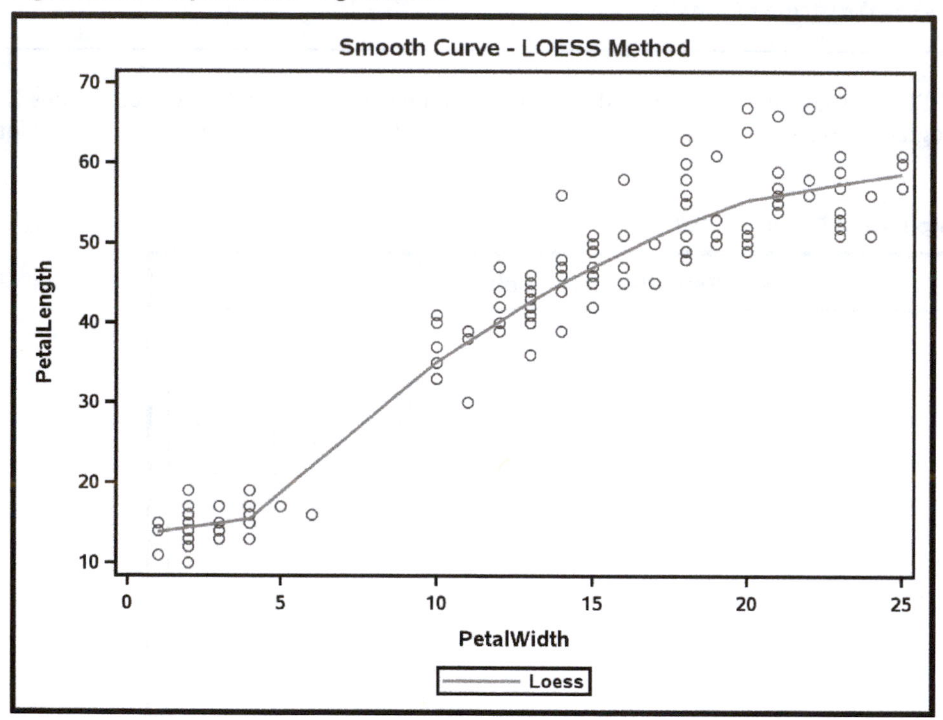

As with PBSPLINE, you can add options to the LOESS statement to control how closely the line follows the points.

20.8 Generating Histograms

If you would like to see a histogram for a variable, use the HISTOGRAM statement followed by the response variable. The program that follows generates a histogram for the variable RBC (red blood cell count) in the Learn.Blood data set. In addition, the DENSITY statement with the response variable RBC overlays a normal curve on the histogram. An alternative to the default normal density curve is the KERNEL density. You can produce the kernel density curve with the option TYPE=KERNEL. Here is the program with the default normal density curve:

Program 20.10: Histogram with a Normal Curve Overlaid

```
title "Histogram with a Normal Curve Overlaid";
proc sgplot data=Learn.Blood;
   histogram RBC;
   density RBC;
run;
```

The output shows the histogram and the overlaid normal density curve. You can include options to control the number or size of the bins in the plot. You use the option BINWITH= to select a bin width—the option NBINS= allows you to specify the number of bins in the histogram.

Figure 20.10: Output from Program 20.10

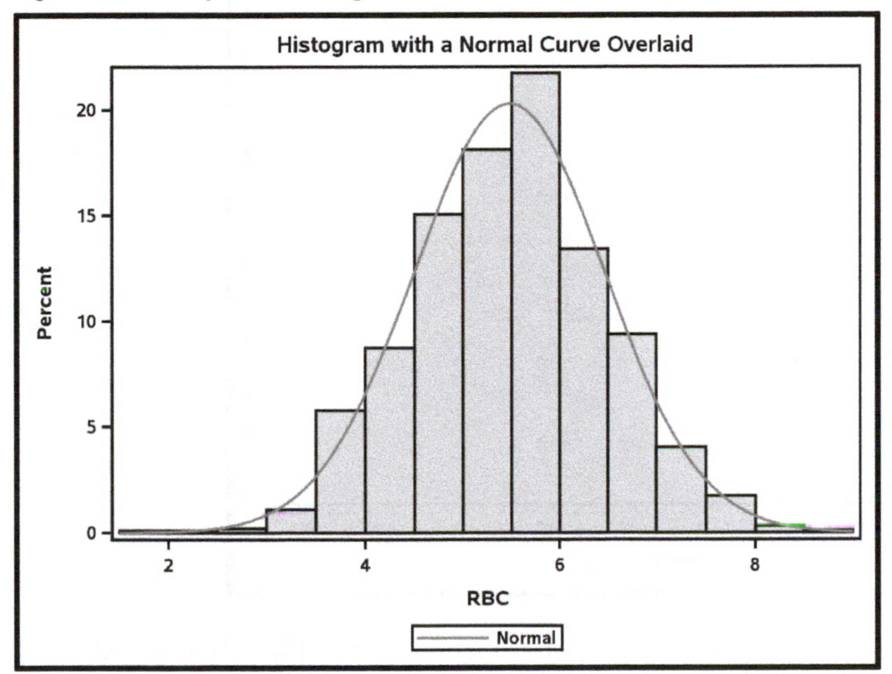

This plot was produced without any options. There are also options for the appearance of the x and y axes.

20.9 Generating a Simple Box Plot

A box plot (also known as a box-and-whisker plot) is a technique used in a field of statistics known as exploratory data analysis (EDA). The plot shows important values such as the median, the first and third quartiles, and data points known as outliers.

To demonstrate a simple box plot, the program below produces a box plot for the variable RBC in the Blood data set. Use the HBOX statement if you want a horizontal plot and the VBOX statement if you prefer a vertical plot.

Program 20.11: Simple Box Plot

```
title "Simple Box Plot";
proc sgplot data=Learn.Blood;
   hbox RBC;
run;
```

Here is the output:

Figure 20.11: Output from Program 20.11

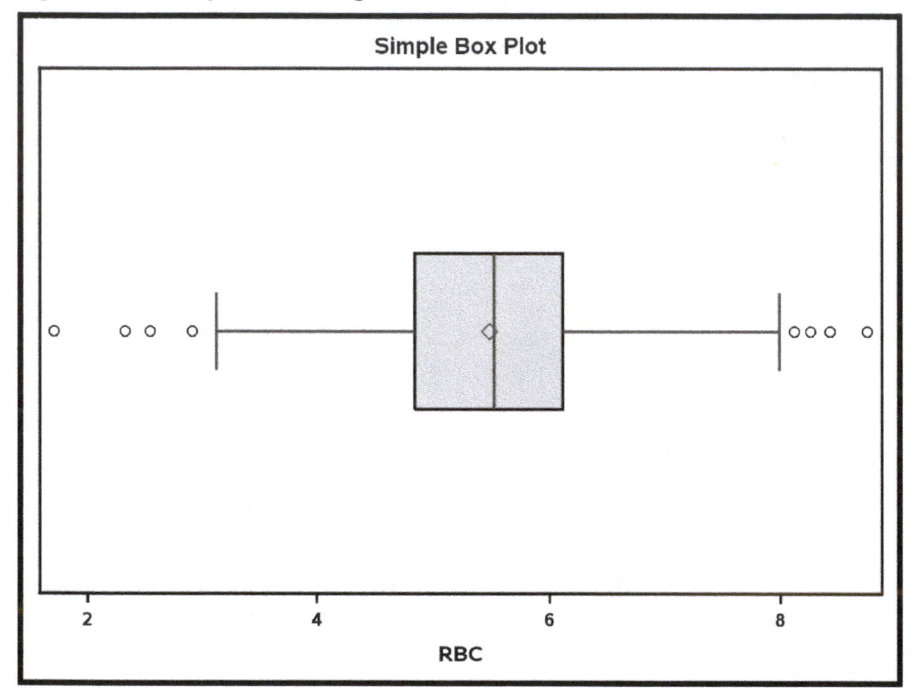

The vertical line inside the box represents the median, and the left and right sides of the box represent the first and third quartiles, respectively. The lines extending from both sides of the box represent a distance of 1.5 interquartile ranges (the distance between the first and third quartiles) on both sides of the box. The diamond inside the box represents the mean. The circles are data points that fall outside 1.5 interquartile ranges (referred to as outliers).

20.10 Producing a Box Plot with a Grouping Variable

To generate a box plot for each value of a grouping variable, add a GROUP= option in the HBOX or VBOX statement. The program below generates a box plot of RBC for each blood type:

Program 20.12: Box Plot with a Grouping Variable

```
title "Box Plot with a Grouping Variable";
proc sgplot data=Learn.Blood;
   hbox RBC / group=BloodType;
run;
```

This program produced the family of box plots shown here:

Figure 20.12: Output from Program 20.12

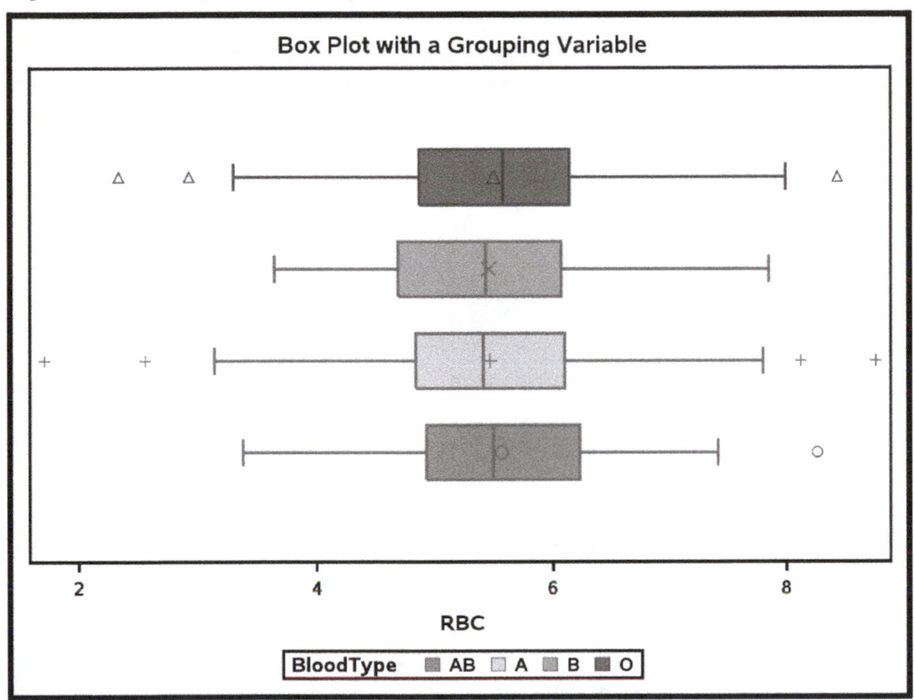

Plots like this provide an excellent visual display that allows you to see the distribution of one variable for each category of another variable.

20.11 Demonstrating Overlays and Transparency

As mentioned earlier, each plot produced by PROC SGPLOT is overlaid on the previous plots. In the case of the series plot or the histogram with a normal density plot, this is not a problem. However, in displays such as bar charts one bar may be hidden by an overlaid plot.

You can set the value of transparency for any plot by using the keyword TRANSPARENCY=. The program below produces two vertical bar charts. The second bar chart sets the value of transparency to .2. In addition, the width of the bars in the second chart is set to .3.

Program 20.13: Demonstrating Overlays and Transparency

```
title "Demonstrating Overlays and Transparency";
proc sgplot data=SASHelp.Iris;
   vbar Species / Response=PetalWidth stat=mean barwidth=.8;
   vbar Species / Response=PetalLength barwidth=.3
                     transparency=.2 stat=mean;
run;
```

Here are the two overlaid charts:

Figure 20.13: Output from Program 20.13

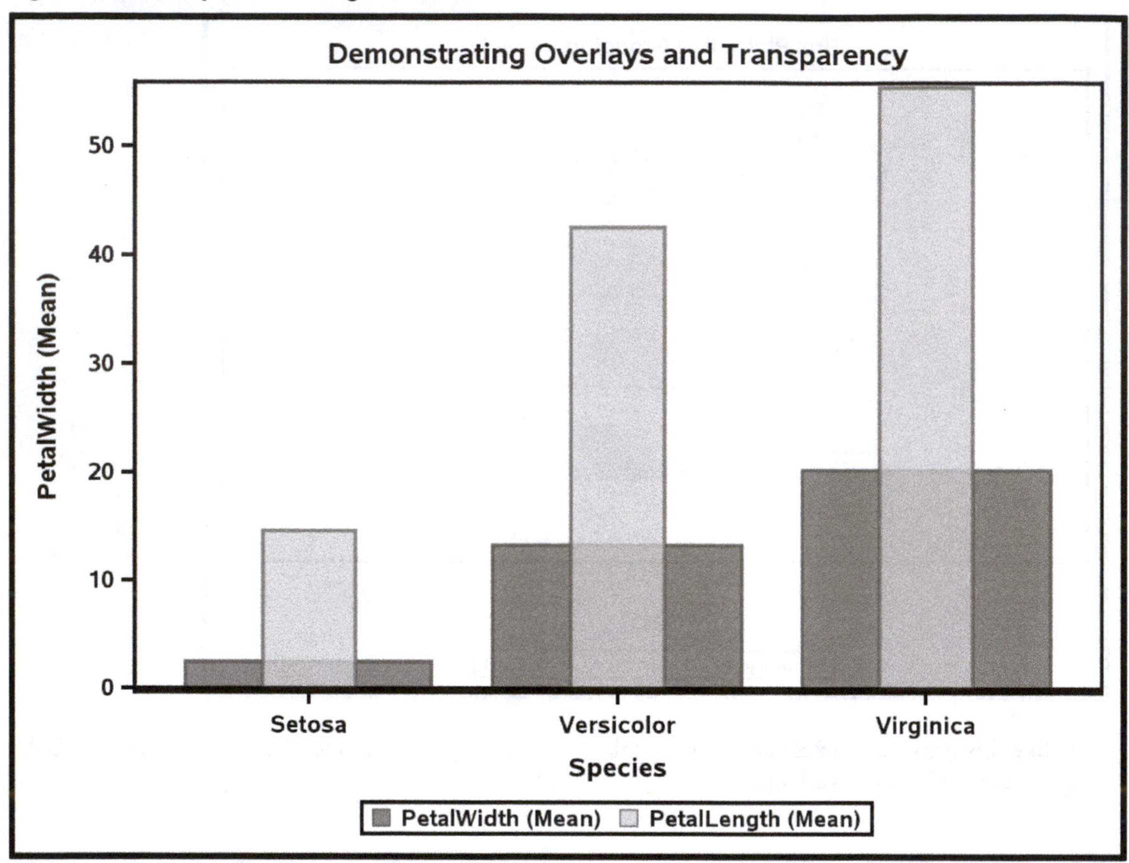

This chapter touched only the basics of PROC SGPLOT. There are several SAS Press books and extensive documentation in SAS Help Center that discuss this procedure. For those of you who have used SAS/GRAPH to produce charts and plots, you will be delighted at how much easier it is to use PROC SGPLOT.

20.12 Problems

Solutions to odd-numbered problems are located at the back of this book. Solutions to all problems are available to professors. If you are a professor, visit the book's companion website at support.sas.com/cody for information about how to obtain the solutions to all of the problems.

1. Using the SASHelp data set, Heart, generate a bar chart showing the frequencies for the variable, Status. It should look like this:

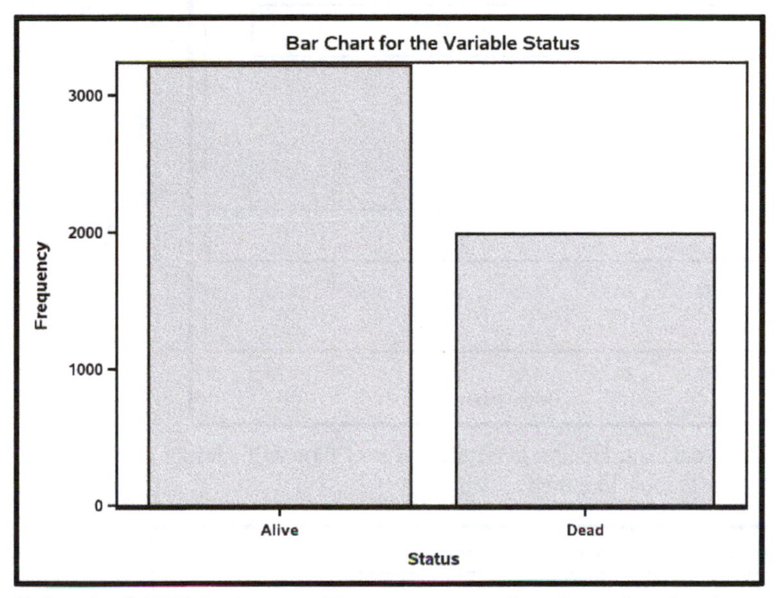

2. Using the SASHelp data set, Heart, generate a bar chart showing the mean Cholesterol value for men and women. The variable names are Cholesterol and Sex. It should look like this:

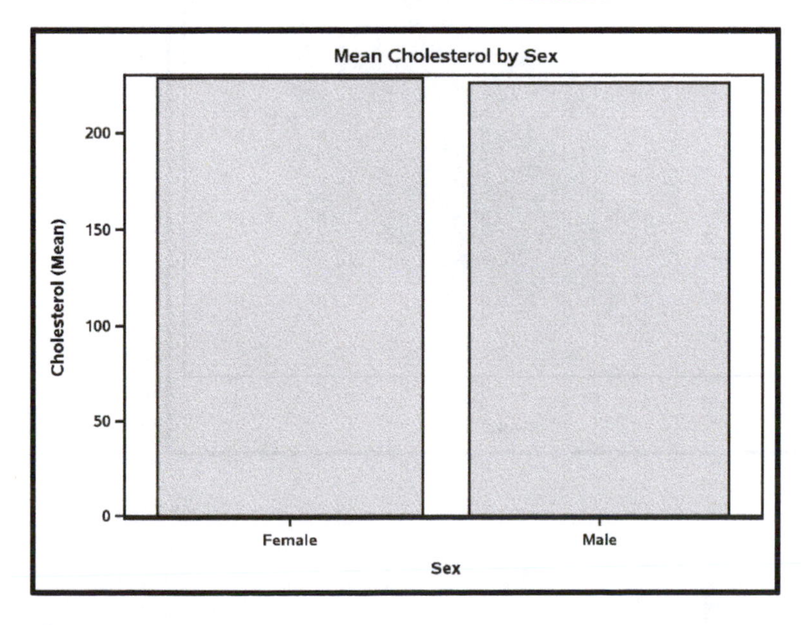

3. Using the SASHelp data set, Heart, generate a horizontal bar chart showing the mean Height for men and women. Use options to make the bars not filled in (outline only) and to make the bars 25% the maximum size. It should look like this:

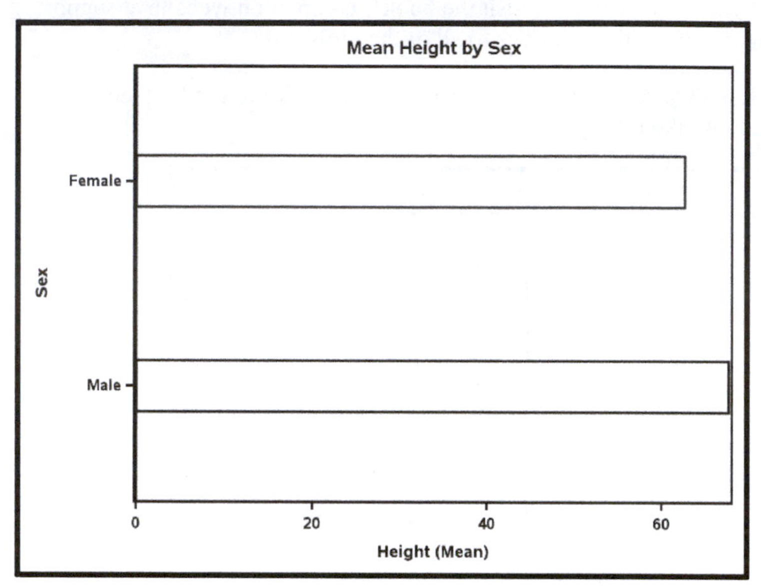

4. Using the SASHelp data set, Health, generate a scatter plot with Height on the x axis and Weight on the y axis. It should look like this:

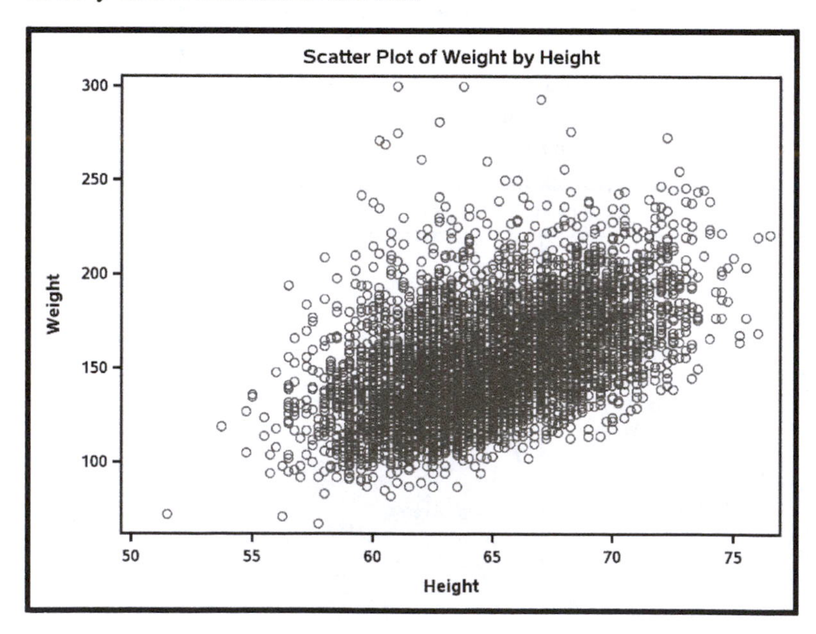

5. Using the SASHelp data set, Heart, create a scatter plot with Height on the x axis and Weight on the y axis. Include a regression line and the two types of confidence limits.

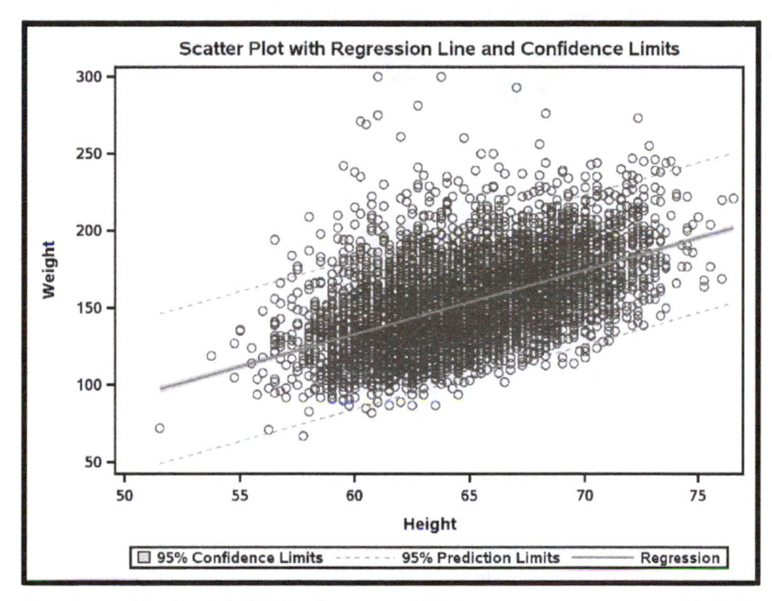

6. Starting with the SASHelp data set, Sales, create a temporary SAS data set (call it Sales) that contains a variable (call it Date) that is a SAS date by using the three variables Month, Day, and Year (remember the MDY function). Be aware that the original SASHelp data set also contains a variable called Date, but it is not a SAS date. You can drop the original Date variable with a DROP= data set option on the SET statement when you bring in data from the SASHelp data set. Next, generate a time series plot with Date on the x axis and Sales on the y axis. It should look like this:

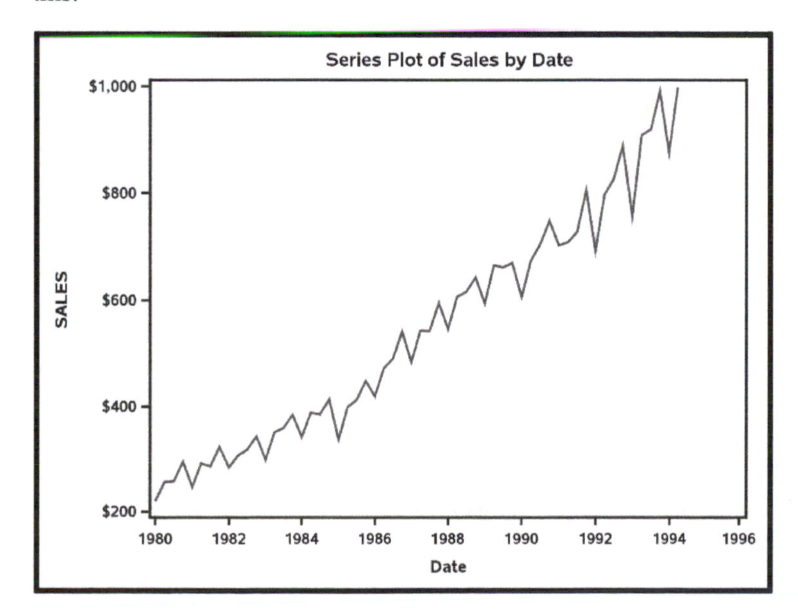

7. Using the first 100 observations from the SASHelp data set, Health (remember the data set option OBS=), use the PBSPLINE technique to plot Height on the x axis and Weight on the y axis, with a smooth curve included on the plot. It should look like this:

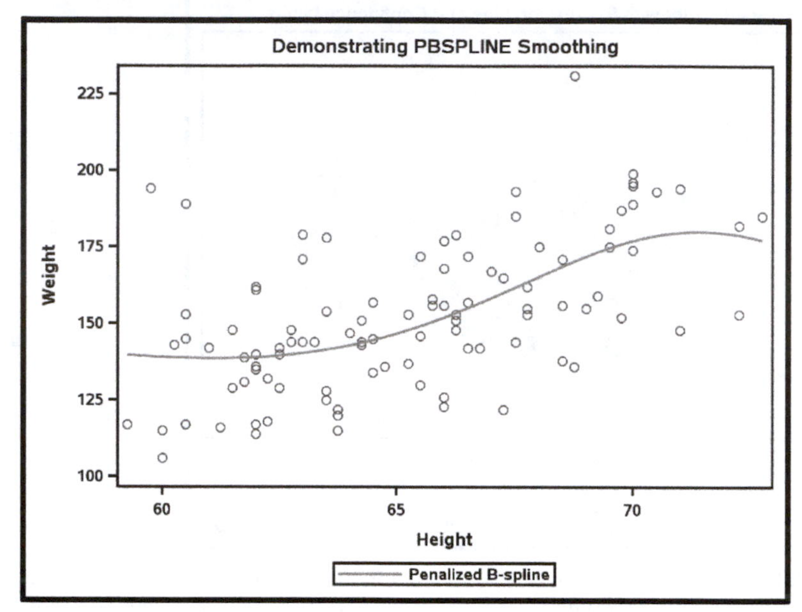

8. Repeat Problem 7, except use the LOESS method of generating a smooth curve. Your output should look like this:

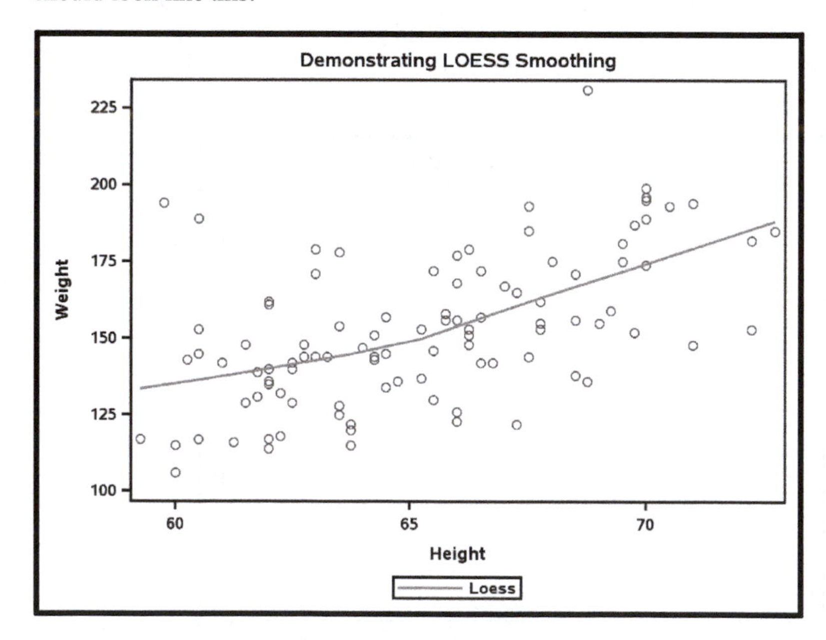

9. Using the SASHelp data set, Heart, generate a histogram for the variable, Cholesterol. It should look like this:

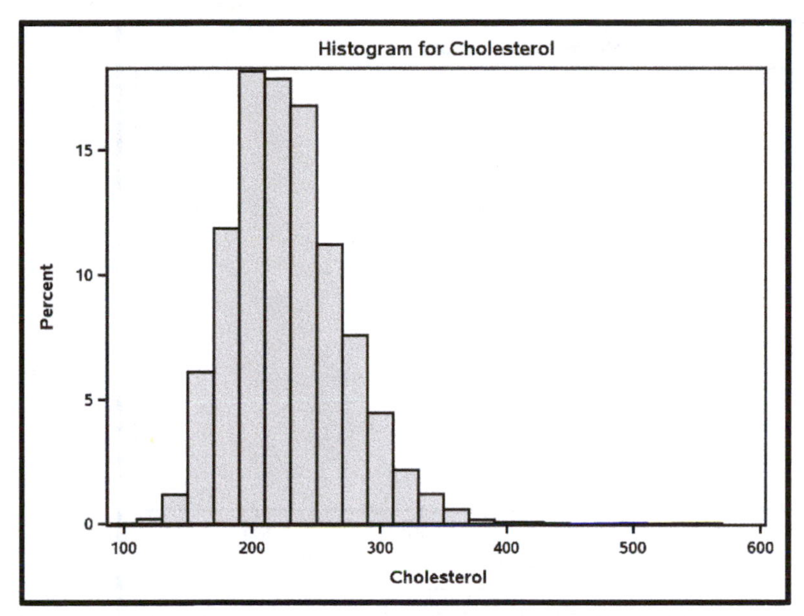

10. Repeat Problem 9, except add a normal curve along with the histogram. It should look like this:

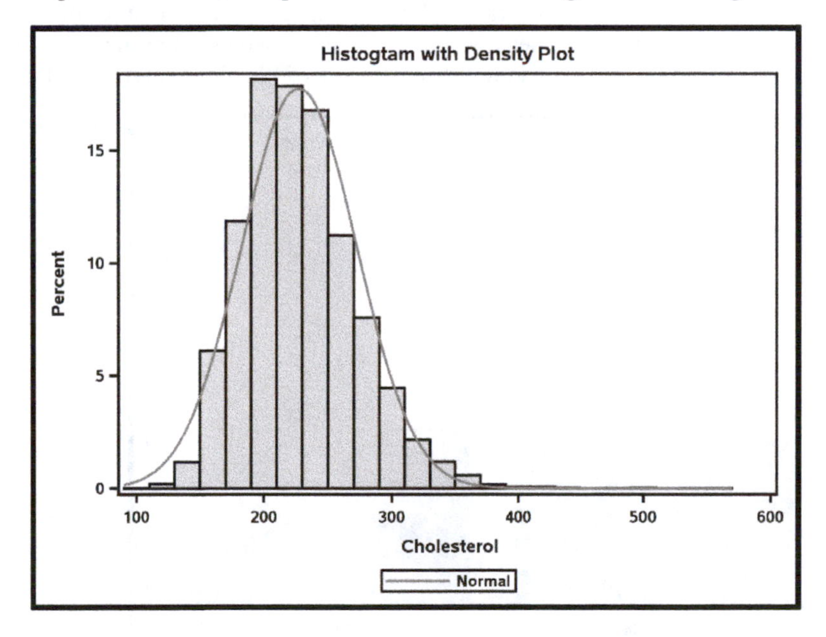

11. Using the SASHelp data set, Heart, generate two horizontal box plots that show the distribution of Cholesterol for men and women (variable Sex). Your plot should look like this:

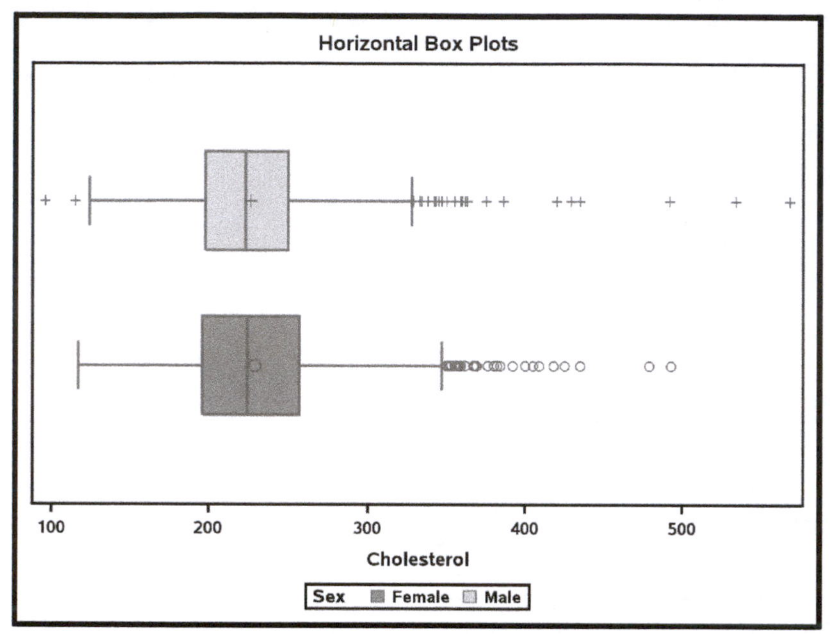

12. Using the SASHelp data set, Heart, generate a plot showing the mean height for men and women. On the same plot, show the mean weight for men and women. For the latter display, set transparency to .2, and make the bar width 25% of the full width. Your chart should look like this:

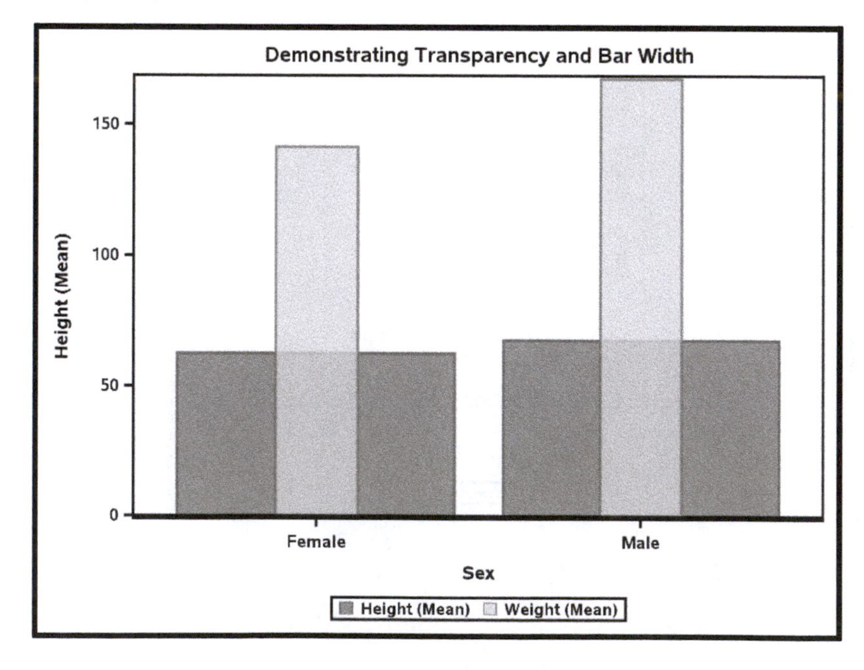

Part 4: Advanced Topics

Chapter 21: Using Advanced INPUT Techniques

21.1 Introduction

Chapter 3 covered the basics of reading raw data. This chapter discusses more advanced input topics. For an excellent reference discussing the INFILE options MISSOVER, PAD, and TRUNCOVER, see Randall Cates' paper on the subject. (Randall Cates, MPH, "Missover, Truncover, and PAD, Oh My!! Or Making Sense of the INFILE and INPUT Statements," Paper 9-26, which is available at www2.sas.com/proceedings/sugi26/p009-26.pdf.

21.2 Handling Missing Values at the End of a Line

Suppose you have a raw data file with three numbers on most of the lines (representing x, y, and z). Some of the lines contain fewer than three numbers. As an example, look at the raw data file **`Missing.txt`**:

```
1 2 3
4 5
6 7 8
9 10 11
```

What happens if you run Program 21.1?

Program 21.1: Missing Values at the End of a Line with List Input

```
data Missing;
   infile 'C:\books\Learning\Missing.txt';
   input x y z;
run;
```

Here is the output:

Figure 21.1: Listing of Data Set Missing

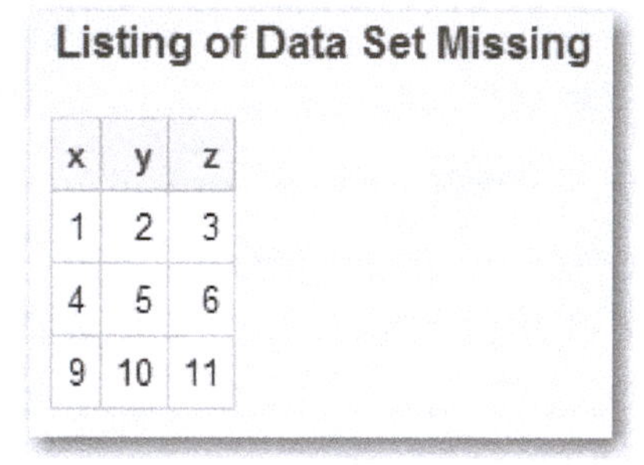

Listing of Data Set Missing

x	y	z
1	2	3
4	5	6
9	10	11

You should notice immediately that something is wrong: There were four lines of raw data but only three observations in the data set. Not only that, SAS read the value of **`6`** on the third line of data as the value for z in the second observation. The remaining two values on the third data line (**`7`** and **`8`**) disappeared. What happened?

The first observation is straightforward (**`x=1`**, **`y=2`**, and **`z=3`**). What happens when SAS reaches Line 2? There are values for x and y but none for z. So, SAS goes to the next line and sets z equal to **`6`**. You are now at the bottom of the DATA step. Because there are more lines of data to be read, control returns back to the top of the DATA step and SAS moves the pointer to the next line of data, ignoring the two other values (**`7`** and **`8`**) on that line, and reads the last three values, **`9`**, **`10`**, and **`11`**, for x, y, and z, respectively. Before you see how to fix this problem, take a look at the SAS log:

Figure 21.2: The Log Window After Running Program 21.1

```
14      data Missing;
15          infile 'c:\books\Learning\Missing.txt';
16          input x y z;
17       run;

NOTE: The infile 'c:\books\Learning\Missing.txt' is:
      Filename=c:\books\Learning\Missing.txt,
      RECFM=V,LRECL=32767,File Size (bytes)=26,
      Last Modified=13Feb2007:14:17:14,
      Create Time=28Nov2017:14:58:38

NOTE: 4 records were read from the infile
      'c:\books\Learning\Missing.txt'.
      The minimum record length was 3.
      The maximum record length was 7.
NOTE: SAS went to a new line when INPUT statement reached past the end
      of a line.
```

The note that SAS went to a new line is important! It tells you exactly what happened and that you need to modify your program.

When you have list input and you have some lines of data with fewer values than you need, you can use the INFILE option MISSOVER to fix things. MISSOVER tells SAS to set all the remaining variables to missing if you have more variables than there are data values in one line of raw data. Here is the program with the MISSOVER option added:

Program 21.2: Using the MISSOVER Option

```
data Missing;
   infile 'C:\books\Learning\Missing.txt' missover;
   input x y z;
run;
```

The resulting data set appears next:

Figure 21.3: Listing of Data Set Missing (with INFILE Option MISSOVER)

Listing of Data Set Missing

x	y	z
1	2	3
4	5	.
6	7	8
9	10	11

Note: This example shows how important it is to look at the SAS log, even if your program runs and produces output.

21.3 Reading Short Data Lines

Take a look at the following file **Short.txt**:

```
         1         2         3
123456789012345678901234567890
------------------------------
001Josuha Tyson    100 97 95
002Helen Ames       87 85
003ShouEn Lu        98 98 92
004Pam Mann        100100 99
```

These lines of data represent a subject number (Columns 1–3), a name (Columns 4–19), and three quiz scores (20–22, 23–25, and 26–28). It's important to know that Line 2 contains a short record and it is not padded with blanks (that is, the person who entered the data pressed the carriage return after typing the 85). Let's see what happens if you read this file with either column or formatted input on a Windows or UNIX platform (flat files like this are padded in mainframe files). Here is a program to do that:

Program 21.3: Reading a Raw Data file with Short Records

```
data Short;
   infile 'C:\books\Learning\Short.txt';
   input Subject  $ 1-3
         Name     $ 4-19
         Quiz1      20-22
         Quiz2      23-25
         Quiz3      26-28;
run;
```

Now look at a listing from PROC PRINT:

Figure 21.4: Listing of Data Set Short

Listing of Data Set Short

Subject	Name	Quiz1	Quiz2	Quiz3
001	Josuha Tyson	100	97	95
002	Helen Ames	87	85	3
004	Pam Mann	100	100	99

Again, you see that something went wrong—very wrong. Four records were read, but the data set only has three observations. Worse yet, Subject 002 has a Quiz3 score of 3.

Even though the INPUT statement instructed SAS to read a value of Quiz3 in Columns 26–28, it went to the next line when it was unable to read those columns in the second record. It then read the subject number in Line 3 as the quiz score. SAS then moved the pointer to the next record and read data for subject 4 correctly. One way to prevent this from happening is to use the PAD option in the INFILE statement. This option pads each of the input records with blanks, out to the end of the logical record. On Windows and UNIX platforms, the default logical record (abbreviated LRECL) is 32,767. You will see later how to change this length.

Program 21.4: Demonstrating the INFILE PAD Option

```
data Short;
   infile 'C:\books\Learning\Short.txt' pad;
   input Subject   $ 1-3
         Name      $ 4-19
         Quiz1       20-22
         Quiz2       23-25
         Quiz3       26-28;
run;
```

After you run this program, you can see by looking at the listing of the data set below that the short record no longer causes a problem:

Figure 21.5: Listing of Data Set Short (with INFILE Option PAD)

Listing of Data Set Short

Subject	Name	Quiz1	Quiz2	Quiz3
001	Josuha Tyson	100	97	95
002	Helen Ames	87	85	.
003	ShouEn Lu	98	98	92
004	Pam Mann	100	100	99

You now have the correct values in the data set. It is a good idea to use the PAD option whenever you are reading files using column or formatted input.

The option TRUNCOVER may be used in place of PAD. It has the effect of the PAD and MISSOVER options combined and is a good overall choice for the INFILE option when you are reading variable length records.

21.4 Reading External Files with Lines Longer Than 32,767 Characters

Because the default logical record length on Windows and UNIX systems is 32,767 bytes, you need to use the INFILE option LRECL to specify record lengths greater than this.

Suppose you are given a raw data file (**`Long.txt`**) where each line contains 50,000 bytes. In order to read this file, you need to specify the record length, like this:

```
infile 'Long.txt' lrecl=50000;
```

If you use the PAD option in combination with the LRECL= option (stands for logical record length), all the records are padded to whatever you specify for the LRECL.

21.5 Detecting the End of the File

There are times when you want to read records from a file and perform certain calculations when you have reached the end of the file. One way to do this is with the END= option in the INFILE statement. Here is an example:

You have a file, **Month.txt**, with monthly sales totals. Here is a listing of the file:

```
Jan 2000
Feb 3000
Mar 2500
Apr 2600
May 1200
Jun 2300
Jul 1000
Aug 2300
Sep 1500
Oct 1900
Sep 2600
Oct 3400
Nov 4000
Dec 1200
```

You want to read this file and print out the total sales for the year. Here's how:

Program 21.5: Demonstrating the END= Option in the INFILE Statement

```
title "Demonstrating the INFILE Option END=";
data _null_;
   file print;
   infile 'C:\books\Learning\Month.txt' end=Last;
   input @1 Month $3.
         @5 MonthTotal 4.;
   YearTotal + MonthTotal;
   if last then
      put "Total for the year is" YearTotal dollar8.;
run;
```

If you need a review of DATA _NULL_, please refer to Chapter 4, Section 10.

The variable name Last in the END= option is a temporary variable that has a value of **false** (0) until the last record is being read from the external file—it then has a value of **true** (1). You can use this logical variable to control the action of your program. In Program 21.5, you are reading monthly values and totaling them with the SUM statement. When you have read the last record in the external file, the variable Last becomes true and the PUT statement executes, writing out the message "Total for the year is" followed by the yearly total. (Because of the FILE PRINT statement, this message is printed to your output device.) Here is a screen shot of the Output window:

Figure 21.6: Output from Program 21.5

Demonstrating the INFILE Option END=

```
Total for the year is $31,500
```

Note: END= is also a SET option and it is used to determine when you are reading the last observation in a SAS data set.

21.6 Reading a Portion of a Raw Data File

Two INFILE options, OBS= and FIRSTOBS=, allow you to read a portion of a raw data file. If you specify OBS=*n* as an INFILE option, SAS stops reading from the file after it has read the *nth* line of data. If you include a FIRSTOBS=*m* INFILE option, SAS starts reading at Line *m*. When used together, you can read any number of contiguous lines of data.

As an example, suppose you only wanted to read the first three months of data from the **Month.txt** file mentioned in Section 21.5. You could use the OBS= INFILE option like this:

Program 21.6: Demonstrating the OBS= INFILE Option to Read the First Three Lines of Data

```
data ReadThree;
   infile 'C:\books\Learning\Month.txt' obs=3;
   input @1 Month $3.
         @5 MonthTotal 4.;
run;
```

Data set ReadThree consists of observations for January, February, and March only.

Note: Using the OBS= INFILE option is particularly useful if you have a very large raw data file and you want to test your program by reading only a few lines of data.

If you don't want to start reading from the first line of a file, you can use the FIRSTOBS= INFILE option to tell SAS where to begin reading data. This is useful if you have one or more header records in a file that you want to skip or if you want to read data from the middle or end of the file. As an example, if you want to read Lines 5 through 7 from the **Month.txt** file, you could use Program 21.7:

Program 21.7: Using the OBS= and FIRSTOBS= INFILE Options Together

```
data Read_5_to_7;
   infile 'C:\books\Learning\Month.txt' firstobs=5 obs=7;
   input @1 Month $3.
         @5 MonthTotal 4.;
run;
```

This program reads Lines 5, 6, and 7 from the file **Month.txt** and the resulting SAS data set contains observations for months May, June, and July. Remember that the OBS= option does not specify how many observations to read but rather the last observation to be read. Think of OBS= as LASTOBS= (although there is no such option as LASTOBS=).

21.7 Reading Data from Multiple Files

SAS has several ways to read from multiple raw data files. If the names are similar, you are free to use the wildcard characters question mark (?) (which substitutes for any single character) or the asterisk (*) (0 or more characters) in your INFILE or FILENAME statements.

Suppose you have four files called **`Quarter1.txt`**, **`Quarter2.txt`**, **`Quarter3.txt`**, and **`Quarter4.txt`**. You want to read data from all four files, so you write the following INFILE statement:

```
infile 'C:\books\Learning\Quarter?.txt';
```

Another way to read multiple files is to execute the INFILE statement conditionally. You can use the END= option in the INFILE statement to detect when you are at the end of one file and use that information to tell SAS to go to another file. To help make this clear, suppose you have two files: **`Alpha.txt`** and **`Beta.txt`**. You want to read all the records from **`Alpha.txt`** followed by all the records from **`Beta.txt`**. One way to do this is by executing Program 21.8:

Program 21.8: Using the END= Option to Read Data from Multiple Files

```
data Combined;
   if Finished = 0 then infile 'Alpha.txt' end=Finished;
   else infile 'Beta.txt';
   input . . .;
   . . .
run;
```

21.8 Reading Data from Multiple Files Using a FILENAME Statement

Another way to read raw data from multiple files is to enter the filenames in a FILENAME statement. To demonstrate this, here is an alternative to Program 21.8:

Program 21.9: Alternative to Program 21.8

```
filename Bilbo ('Alpha.txt' 'Beta.txt');

data Combined;
   infile Bilbo;
   input . . .;
   . . .
run;
```

You may enter as many filenames as necessary when using this method of reading data from multiple files.

21.9 Reading External Filenames from a Data File

For applications where it would be tedious to list all the external filenames in a FILENAME statement, you can have SAS read the names of the external files from a file containing the filenames. Here is an example:

Program 21.10: Reading External Filenames from an External File

```
data ReadMany;
   infile 'C:\books\Learning\Filenames.txt';
   input ExternalNames $ 40.;
   infile dummy filevar=ExternalNames end=Last;
   do until (Last);
      input . . .;
      output;
   end;
run;
```

Here, the data file **`Filenames.txt`** contains the names of the files you want to read. The FILEVAR option in the INFILE statement uses the file whose name is stored in the variable ExternalNames. You need to include a dummy fileref (you can call it anything—SAS ignores it) to preserve proper syntax of the INFILE statement. The DO UNTIL loop reads data from the first file until the last record is read (and the variable Last becomes true). Then the program returns to the top of the DATA step, a new filename is read from the **`Filenames.txt`** file, and processing continues.

A variation on Program 21.10 is to place the names of the external files following a DATALINES statement, as shown in the following program:

Program 21.11: Reading External Filenames Using a DATALINES Statement

```
data ReadMany;
   input ExternalNames $ 40.;
   infile dummy filevar=ExternalNames end=Last;
   do until (Last);
      input . . .;
      output;
   end;
datalines;
C:\books\Learning\Data1.txt
C:\books\Learning\MoreData.txt
C:\books\Learning\Fred.txt
;
```

Placing the filenames in the program may be more convenient than placing them in an external file.

21.10 Reading Multiple Lines of Data to Create One Observation

Back in the "old days" when many flat files had a maximum length of 80 bytes, it was necessary to place data for one subject (observation) on multiple lines. You may even face this problem today.

As an example, you have a raw data file (**`Health.txt`**) with two lines (records) per subject and the following data layout:

Variable	Description	Line #	Starting Column	Length	Type
Subj	Subject Number	1	1	3	Char
DOB	Date of Birth	1	4	10	mm/dd/yyyy
Weight	Weight in Lbs.	1	14	3	Num
HR	Heart Rate	2	4	3	Num
SBP	Systolic Blood Pressure	2	7	3	Num
DBP	Diastolic Blood Pressure	2	10	3	Num

```
00112/25/1944210
    80160100
00205/11/1966102
    88122 76
00308/03/2000 66
    90102 62
```

Here is a program to read this data file and create a temporary SAS data set called Health:

Program 21.12: Reading Multiple Lines of Data to Create One Observation

```
data Health;
   infile 'C:\books\Learning\Health.txt' pad;
   input #1 @1  Subj      $3.
            @4  DOB mmddyy10.
            @14 Weight     3.
         #2 @4  HR         3.
            @7  SBP        3.
            @10 DBP        3.;
   format DOB mmddyy10.;
run;
```

The salient feature of this program is the line pointer, a number sign—or as young people might call it, a hashtag (#)—in the INPUT statement. The # pointer tells SAS which line to read when you have multiple lines of data per observation.

A listing of the SAS data from PROC PRINT is as follows:

Figure 21.7: Listing of Data Set Health

Listing of Data Set Health

Subj	DOB	Weight	HR	SBP	DBP
001	12/25/1944	210	80	160	100
002	05/11/1966	102	88	122	76
003	08/03/2000	66	90	102	62

An alternative way to read this data file is to use a relative line pointer, a forward slash (/), to tell SAS to skip to the next line of input. Rewriting Program 21.12 using this method is shown next:

Program 21.13: Using an Alternate Method of Reading Multiple Lines of Data to Create One SAS Observation

```
data health;
   infile 'C:\books\Learning\Health.txt' pad;
   input @1  Subj      $3.
         @4  DOB mmddyy10.
         @14 Weight     3. /
         @4  HR         3.
         @7  SBP        3.
         @10 DBP        3.;
   format DOB mmddyy10.;
run;
```

Between these two methods, this author prefers the # line pointer over the slash notation because it makes programs easier to read and is less prone to errors.

21.11 Reading Data Conditionally (the Single Trailing @ Sign)

Some data files are more complex than the ones you have seen so far. For example, you may have a mixture of record types in a single file. In order to read values from a file of this type, you need to first read part of the line to determine how to read the remainder of the line. SAS provides you with the single trailing at sign (@) to solve this problem. Here is an example:

You are given a file from a survey taken over two years. Unfortunately, in the second year, a question was added in the middle of the survey. So, the data layout for Year 1 and Year 2 is different. Worse yet, data from both years are completely mixed up in a single file. First, look at the data layout for each of the two years:

Data Layout for Year 2005 Survey

Variable	Description	Starting Column	Length	Type
Number	Survey Number	1	3	Num
Q1	Question 1	4	1	Char
Q2	Question 2	5	1	Char
Q3	Question 3	6	1	Char
Q4	Question 4	7	1	Char
Year	Survey Year	9	4	Char

Data Layout for Year 2006 Survey

Variable	Description	Starting Column	Length	Type
Number	Survey Number	1	3	Num
Q1	Question 1	4	1	Char
Q2	Question 2	5	1	Char
Q2B	Question 2B	6	1	Char
Q3	Question 3	7	1	Char
Q4	Question 4	8	1	Char
Year	Survey Year	9	4	Char

Some sample lines of data from **`Survey56.txt`** follow:

```
001ABED 2005
002AABCD2006
005AADD 2005
007BBCDE2006
009ABABA2006
010DEEB 2005
```

Here is a program that does **NOT** work:

Program 21.14: Incorrect Attempt to Read a File of Mixed Record Types

```
data Survey;
   infile 'C:\books\Learning\Survey56.txt' pad;
   input @9 year $4.; ①
   if year = '2005' then
      input @1 Number
            @4 Q1
            @5 Q2
            @6 Q3
            @7 Q4;
   else if year = '2006' then
      input @1 Number
            @4 Q1
            @5 Q2
            @6 Q2B
            @7 Q3
            @8 Q4;
run;
```

Here is the problem: after SAS processes the first INPUT statement ①, the line pointer moves to the next line in the file so that subsequent data values are read from the next line. To solve this problem, use a trailing @ at the end of the first INPUT statement. This is an instruction to "hold the line" for another INPUT statement in the same DATA step. By "holding the line," we mean to leave the pointer at the present position and not to advance to the next record. The single trailing @ holds the line until another INPUT statement, (without a trailing @) is encountered further down in the DATA step, or the end of the DATA step is reached.

Here is the program, rewritten correctly:

Program 21.15: Using a Trailing @ to Read a File with Mixed Record Types

```
data Survey;
   infile 'C:\books\Learning\Survey56.txt' pad;
   input @9 Year $4. @;
   if Year = '2005' then
      input @1 Number $3.
            @4 Q1 $1.
            @5 Q2 $1.
            @6 Q3 $1.
            @7 Q4 $1.;
   else if Year = '2006' then
      input @1 Number $3.
            @4 Q1 $1.
            @5 Q2 $1.
            @6 Q2B $1.
            @7 Q3 $1.
            @8 Q4 $1.;
run;
```

That little @ sign makes all the difference! Now, after SAS reads a value for Year, it executes the appropriate INPUT statement and the remaining data values are read from the same line of data.

21.12 More Examples of the Single Trailing @ Sign

It might be useful to show some more examples of the single trailing @. Suppose you have a raw data file containing data on males and females. Suppose further that you only want to read data on females. One way would be to read all the variables of interest, including Gender, and test if Gender is equal to **F** (Female). If so, keep the observation; if not, delete it.

This is very inefficient. It is better to read the Gender value, check it, and then only read the remaining values if Gender is equal to **F**.

Let's use the file **Bank.txt** (see the following description) for an example of how this works. Here is the file description:

Variable	Description	Starting Column	Ending Column	Data Type
Subj	Subject Number	1	3	Character
DOB	Date of Birth	4	13	Character
Gender	Gender	14	14	Character
Balance	Bank Account Balance	15	21	Numeric

An efficient program to read data from this file, keeping only data on females, would be as follows:

Program 21.16: Another Example of a Trailing @ Sign

```
data Females;
   infile 'C:\books\Learning\Bank.txt' pad;
   input @14 Gender $1. @;
   if Gender ne 'F' then delete;
   input @1  Subj         $3.
         @4  DOB     mmddyy10.
         @15 Balance        7.;
run;
```

The first INPUT statement reads a single column to determine the value of Gender. The trailing @ prevents SAS from going to a new line. Next, if the value is not **F**, the DELETE statement instructs SAS not to output an observation to the SAS data set and to return control to the top of the DATA step.

21.13 Creating Multiple Observations from One Line of Input

There are times when you want to place data for several subjects on a single line (or someone gave you a file like this). For example, suppose you have some X,Y pairs and want to create a SAS data set. One way is to place one pair of points on each input line, like this:

Program 21.17: Creating One Observation from One Line of Data

```
data Pairs;
   input X Y;
datalines;
1 2
3 4
5 7
8 9
11 14
13 18
21 27
30 40
;
```

This is fine, but it could also be written like this:

Program 21.18: Creating Several Observations from One Line of Data

```
data Pairs;
   input X Y @@;
datalines;
1 2  3 4  5 7  8 9  11 14  13 18  21 27
30 40
;
```

The double @ signs prevent SAS from moving to a new line, even when you reach the bottom of the DATA step (unless there is another INPUT statement in the DATA step without a trailing @@). In other words, the double trailing @ signs allow SAS to read a stream of data, eating through all the values on a line and then going to the next line. **Use this with great care**. If you are missing a data value in the file, everything gets out of whack.

Without the double trailing @ signs in Program 21.18, there would be only two observations (X=1, Y=2 and X=30, Y=40).

21.14 Using Variable and Informat Lists

You can supply a single informat to a list of variables and save some typing. To do this, place a list of variables in parentheses and follow this list with a list of one or more informats, also in parentheses. For example, you could write the following line to read three numeric values (each with an informat of 2.) and five character values (each with an informat of $1.):

```
input @1 (X Y Z Char1-Char5)(3*2. 5*$1.);
```

Notice that each of the informats in this statement is preceded by a multiplier, written as a number followed by an asterisk. In this example, there are as many informats as there are variables. If there are more variables than there are informats, SAS cycles back to the beginning of the informat list to obtain informats for the remaining variables.

You can take advantage of this feature to read a long list of variables that all share the same informat. For example, to read 50 one-digit character values, starting in Column 10, you could use the following code:

```
input @10 (Ques1-Ques50) ($1.);
```

This saves quite a bit of typing.

You can also use an informat list with list input, like this:

```
input (Ques1-Ques50) (: $1.);
```

Don't forget the colon before the informat. The colon says to read a value according to the appropriate informat but to stop reading when a delimiter is encountered.

21.15 Using Relative Column Pointers to Read a Complex Data Structure Efficiently

Besides the absolute column pointer (@), SAS also has a relative column pointer, a plus sign (+). You can use this to move the current pointer left or right of its current position. Let's look at a few examples:

You have seven readings of the variables Time and Temperature arranged in pairs—Time values are in seconds and occupy 2 columns; Temperature values are in degrees Celsius and occupy 4 columns. The data pairs start in Column 5. Here are a few sample lines of data:

```
         1         2         3         4         5
12345678901234567890123456789012345678901234567890
--------------------------------------------------
    1020.11120.21321.61525.01728.72031.42533.8
    1021.71322.81528.01728.92129.92430.42833.7
```

The tedious way to read this file is like this:

```
input @5  Time1 2. @7  T1 4. @11 Time2 2. @13 T2 4.
      @17 Time3 2. @19 T3 4. @23 Time4 2. @25 T4 4.
      @29 Time5 2. @31 T5 4. @35 Time6 2. @37 T6 4.
      @41 Time7 2. @43 T7 4.;
```

If you use a relative column pointer, along with a variable and informat list, this simplifies to the following:

```
input @5 (Time1-Time7)(2. + 4)
      @7 (T1-T7)(4. + 2);
```

This INPUT statement moves the column pointer to Column 5. Then SAS reads a value for Time using a 2. informat and skips four spaces. Because there are still variables in the variable list, this informat (read two spaces and skip four) repeats for the remaining values of Time. Next, the pointer moves back to Column 7. Here you read a value of Temperature using a 4. informat and skip two spaces. This informat is repeated until all the values of Temperature have been read.

If you use either of these INPUT statements in a DATA step, the resulting data sets are identical (except for the order of the variables).

As you saw in this chapter and in Chapter 3, the INPUT statement is one of the most powerful statements in SAS and allows you to read almost any data file.

21.16 Problems

Solutions to odd-numbered problems are located at the back of this book. Solutions to all problems are available to professors. If you are a professor, visit the book's companion website at support.sas.com/cody for information about how to obtain the solutions to all problems.

1. The raw data file `Scores.txt` contains space-delimited data with up to three scores per line. Write a DATA step to read values Score1–Score3 as numeric variables from this file. Be careful because not all lines of data contain three scores. Your resulting data set should have three observations.
2. Repeat Problem 1, except read the text file `Scores_Comma.csv`. This file is similar to `Scores.txt`, except that it is a comma-separated values (CSV) file. Assume that two commas in a row indicate that there is a missing value.
3. The raw data file `Scores_Columns.txt` has three scores per line. Values start in Column 1 and each score occupies two columns. Thus, Score1 is in Columns 1–2, Score2 in Columns 3–4, and Score3 in Columns 5–6. Using column input, read this data file and create a SAS data set.
4. Repeat Problem 3 using formatted input.
5. You want to read observations from the SAS data set Bicycles. Using a DATA _NULL_ step, sum the value of units sold (Units) and the sum of Sales (TotalSales) for all the observations. Call the former TotalUnits and the latter Sum_of_Sales. Use a SET option to test when the last observation from Bicycles is being read and, using a PUT statement, list the values of the two variables TotalUnits and Sum_of_Sales. Your report should look like this:

```
Summary Report from BICYCLES Data Set
---------------------------------------
Total Units Sold is     48,055
Sales Total is    $87,088
```

6. Write a SAS DATA step to read Lines 2 through 5 (inclusive) from **the** `Month.txt` raw data file. This file contains a three-character month value starting in Column 1 and a month total (four digits) starting in Column 5. Use PROC PRINT to verify that you have observations for February through May.
7. You want to read raw data from two files, `File_A.txt` and `File_B.txt`. You want to skip the first line of each file (it contains header information). The remaining lines contain space-delimited values for three variables X, Y, and Z. Use the END= INFILE option to test when you are finished reading from the first file and switch input to the second file. Use PROC PRINT to list the contents of the resulting SAS data set.
8. Two files, `Xyz1.txt` and `Xyz2.txt`, each contain three values on each line, separated by spaces. Write a SAS DATA step to read values for X, Y, and Z from both of these files. Use an INFILE statement using a wildcard character.
9. Repeat Problem 8 using a FILENAME statement and listing the two files by name.

10. The raw data file **Mixed_Recs.txt** contains two types of records. Records with a 1 in Column 16 contain sales records with Date (in mmddyy10. form) starting in Column 1 and Amount in Columns 11–15 (in standard numeric form). Records with a 2 in Column 16 are inventory records and they contain two values, a part number (character, 5 bytes) starting in column 1 and a quantity. These two values are separated by a space. A listing of this file is shown below:

```
10/21/2005   1001
11/15/2005   2001
A13688 250      2
B11112 300      2
01/03/2005 50001
A88778 19       2
```

Write a DATA step to read this file and create two SAS data sets, one called Sales and the other Inventory. The two data sets should look like this:

```
Listing of SALES                    Listing of INVENTORY

Obs        Date     Amount                  Part
                                     Obs    Number    Quantity
 1    10/21/2005       100
 2    11/15/2005       200            1     A13688        250
 3    01/03/2005      5000            2     B11112        300
                                      3     A88778         19
```

11. Each record in the text file **Three_Per_Line.txt** contains three sets of vital signs: heart rate (HR), systolic blood pressure (SBP), and diastolic blood pressure (DBP). Values start in Column 1 and each value takes three columns. Write an INPUT statement that uses a format list and relative column pointers so that you can read the three HRs, the three SBPs, and the three DBPs together. Call the three HR values HR1, HR2, and HR3; the three SPB values SBP1, SBP2, and SBP3; and the three DBP values DBP1, DBP2, and DBP3. Some sample data lines are shown here:

```
068120 80 72130 80 69122 78
072180110 76178102 70178100
054118 70 56118 72 50114 78
```

Chapter 22: Using Advanced Features of User-Defined Formats and Informats

22.1 Introduction

User-defined formats can do much more than make output from SAS procedures more readable. You will see how to use formats to create new variables and to perform table lookups. You can even create your own informats to alter data values as they are being read into a SAS data set. Finally, certain SAS procedures support multi-label formats—that is, the ability to have a single value correspond to more than one format range.

22.2 Using Formats to Recode Variables

Many SAS procedures use formatted values of variables in their processing. For example, PROC FREQ uses formatted values when it computes frequencies; PROC MEANS uses formatted values of CLASS variables in its calculations.

As an example, suppose you want to compute frequencies on age groups (say, 20-year intervals), but your SAS data set contains the variable Age. Rather than create a new variable representing age groups, you can simply write a format and use PROC FREQ, like this:

Program 22.1: Using a Format to Recode a Variable

```
proc format;
   value Agefmt  0 - <20  = '< 20'
                20 - <40  = '20 to 39'
                40 - <60  = '40 to 59'
                60 - high = '60+';
run;

title "Using a Format to Recode a Variable";
proc freq data=Learn.Survey;
   tables Age / nocum nopercent;
   format Age agefmt.;
run;
```

When you run this program, the frequencies are computed for the age groups, not Age, as you can see in this listing:

Figure 22.1: Output from Program 22.1

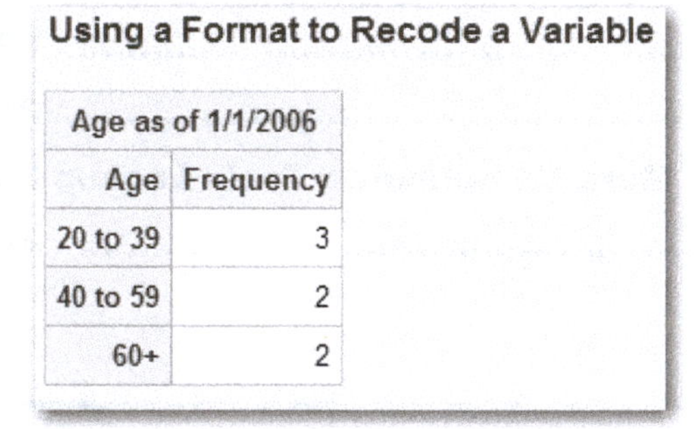

Using a Format to Recode a Variable

Age as of 1/1/2006	
Age	Frequency
20 to 39	3
40 to 59	2
60+	2

This same technique works for CLASS variables in PROC MEANS, PROC SUMMARY, PROC TABULATE, PROC ANOVA, PROC GLM, and several other procedures.

22.3 Using Formats with a PUT Function to Create New Variables

There are times when you want to create a new variable consisting of formatted values. Program 22.2 demonstrates how this is done.

Program 22.2: Using a Format and a PUT Function to Create a New Variable

```
proc format;
   value Agefmt  0 - <20   = '< 20'
                20 - <40   = '20 to 39'
                40 - <60   = '40 to 59'
                60 - high  = '60+';
run;

data Survey;
   set Learn.Survey;
   AgeGroup = put(Age,Agefmt.);
run;

title "Using a Format to Create a Character Variable";
proc print data=Survey;
   id ID;
   var Age AgeGroup;
run;
```

The key to this program is the PUT function. This function takes the value of its first argument, formats this value with the format listed as the second argument, and returns the formatted value. Remember that the PUT function always returns a character value. In this example, AgeGroup is a character variable with a length of 8 bytes (the length of the longest formatted value). Here is a listing of data set Survey:

Figure 22.2: Output from Program 22.2

Using a Format to Create a Character Variable

ID	Age	AgeGroup
001	23	20 to 39
002	55	40 to 59
003	38	20 to 39
004	67	60+
005	22	20 to 39
006	63	60+
007	45	40 to 59

22.4 Creating User-Defined Informats

Although you are probably familiar with using PROC FORMAT to create user-written formats, you may not have used this procedure to create your own informats. You can use user-written informats to alter values as they are read from a raw data file or with the INPUT function to perform a table lookup. Let's look at an example.

Your raw data consists of IDs and letter grades (A+, A, A–, and so on). You want to convert these letter grades into numbers according to a predefined table. Program 22.3 creates an informat called CONVERT and uses that informat to read the letter grades.

Program 22.3: Demonstrating a User-Written Informat

```
proc format;
   invalue Convert 'A+' = 100
                   'A'  = 96
                   'A-' = 92
                   'B+' = 88
                   'B'  = 84
                   'B-' = 80
                   'C+' = 76
                   'C'  = 72
                   'F'  = 65;
run;

data Grades;
   input @1 ID $3.
         @4 Grade Convert2.;
datalines;
001A-
002B+
003F
004C+
005A
;
```

You use an INVALUE statement to create an informat. The rules are similar to the ones you use in creating formats. One tiny difference is that informat names can only be 31 characters in length (including the dollar sign ($) if it will be used to read character data). For those with curious minds, SAS uses 1 byte in the format catalog (the @ sign) to differentiate between formats and informats. (If you look at the entries in your catalog with the FMTLIB option or PROC CATALOG, you will see the @ signs added to the informat names.)

Following the keyword INVALUE, you type the name of the informat you want to create. Just as with formats, you must start the name with a dollar sign if it is to be used to create character data. In the previous program, even though you are reading character values (for example, A+, A, A–), the resulting variable (Grade) is a numeric variable. If you used the name $CONVERT instead of CONVERT in this program, the variable Grade would be character.

Next, you specify ranges and equal sign and labels. Again, you use the same rules as formats.

You use your informat just as you would a built-in SAS informat. Notice that you can add a width to it to specify how many columns of data to read. Here is a listing of data set Grades:

Figure 22.3: Listing of Data Set Grades

Listing of Data Set Grades

ID	Grade
001	92
002	88
003	65
004	76
005	96

It is important to remember that the original character data values are not part of this data set. If you wanted both the original letter grade and its corresponding numeric value, you could read the letter grade as a character variable and use an INPUT function (with the CONVERT informat) to create the numeric grade.

There are some useful options that you can use when you create your informats. UPCASE and JUST are two such options. UPCASE, as the name implies, converts the data values to uppercase before checking on the informat ranges. JUST left-aligns character values. Let's use these two options to enhance Program 22.3.

Program 22.4: Demonstrating Informat Options UPCASE and JUST

```
proc format;
   invalue Convert(upcase just)
            'A+' = 100
            'A'  = 96
            'A-' = 92
            'B+' = 88
            'B'  = 84
            'B-' = 80
            'C+' = 76
            'C'  = 72
            'F'  = 65
         other  =  .;
run;

data Grades;
   input @1 ID $3.
         @4 Grade convert2.;
datalines;
001A-
002b+
003F
004c+
005 A
006X
;

title "Listing of Grades";
proc print data=Grades noobs;
run;
```

Notice that the raw data values being read contain upper- and lowercase values and the value for Student 005 contains a leading blank. We also added a student with an invalid grade (X). To prevent error messages in the SAS log, the informat category OTHER was added so that invalid values are converted to numeric missing values. The following listing shows that all the data values were read correctly:

Figure 22.4: Output from Program 22.4

Listing of Data Set Grades

ID	Grade
001	92
002	88
003	65
004	76
005	96
006	.

22.5 Reading Character and Numeric Data in One Step

There is a little-known feature in user-defined informats—the ability to create an informat that reads both character and numeric data and creates a numeric variable. The examples that follow show ways that you can take advantage of this feature.

For this first example, you have temperature readings on hospital patients. Because many of the patients have a normal temperature (98.6 degrees Fahrenheit), it is convenient to record normal temperatures by entering an **N** as a data value. Some sample lines of data are as follows:

```
101 N 97.3 n N 104.5
```

Notice that some of the Ns are in uppercase and others in lowercase. Before we look at the elegant enhanced informat solution, let's look at a traditional approach:

Program 22.5: A Traditional Approach to Reading a Combination of Character and Numeric Data

```
data Temperatures;
   input Dummy $ @@;
   if upcase(Dummy) = 'N' then Temp = 98.6;
   else Temp = input(Dummy,8.);
   drop Dummy;
datalines;
101 N 97.3 n N 104.5
;
```

Each data value is read as character data. A check is made to see whether this value is equal to an upper- or lowercase **N**. If so, Temp is set to 98.6—if not, the INPUT function performs the character-to-numeric conversion. This works fine, but compulsive programmers (who, me?) are always looking for a more elegant solution. Program 22.6 uses an enhanced numeric informat:

Program 22.6: Using an Enhanced Numeric Informat to Read a Combination of Character and Numeric Data

```
proc format;
   invalue Readtemp(upcase)
                    96 - 106 = _same_
                    'N'      = 98.6
                    other    = .;
run;

data Temperatures;
   input Temp : Readtemp5. @@;
datalines;
101 N 97.3 n N 67 104.5
;
```

The UPCASE option converts any character to uppercase. The keyword _SAME_ in this informat leaves any numeric values in the range 96 to 106 unchanged. Values of **N** are converted to the numeric value of 98.6 and any values that are not in the range of 96 to 106 or equal to **N** are set to a numeric missing value. The technique of using the keyword _SAME_ for valid values and setting other values to missing is one way to screen unwanted extreme values as they are being read. A listing of data set Temperatures is shown next:

Figure 22.5: Listing of Data Set Temperatures

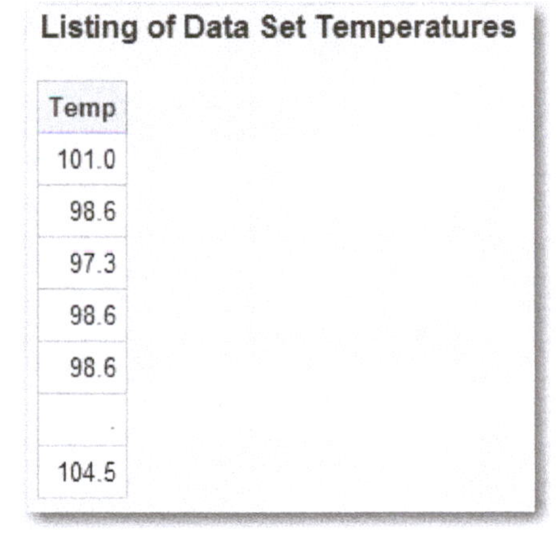
Listing of Data Set Temperatures

Temp
101.0
98.6
97.3
98.6
98.6
.
104.5

The next program reads a combination of letter and number grades. The grades A–F are converted to number grades as follows: **A**=95, **B=85**, **C=75**, and **F=65**. Number grades are not changed. Here is the program:

Program 22.7: Another Example of an Enhanced Numeric Informat

```
proc format;
   invalue ReadGrade(upcase)
      'A' = 95
      'B' = 85
      'C' = 75
      'F' = 65
      other = _same_;
run;

data School;
   input Grade : ReadGrade3. @@;
datalines;
97 99 A C 72 f b
;
```

Here the values **A** through **F** are converted to the appropriate numeric value and all other values are left unchanged. Here is a listing of the data set:

Figure 22.6: Listing of Data Set School

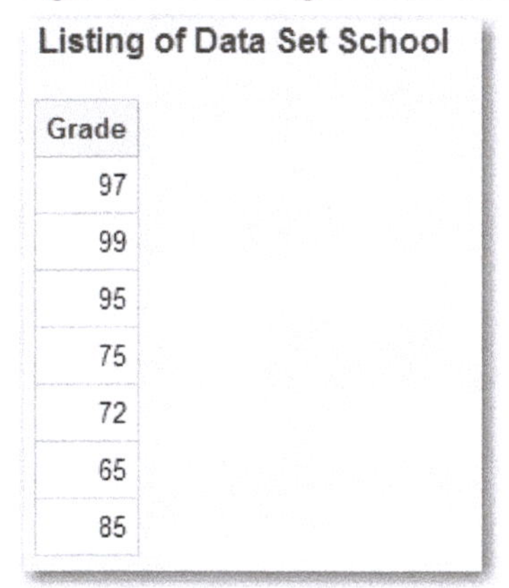

Listing of Data Set School

Grade
97
99
95
75
72
65
85

22.6 Using Formats (and Informats) to Perform Table Lookup

Table lookup is a process where one or more data values are used to retrieve a value for another variable. For example, given an item number from a catalog, you might want to retrieve a product name and a price.

SAS provides you with a variety of ways to perform a table lookup, such as a data set merge, a key value from an index, an array, a format, a hash table, or an informat. Because SAS formats are stored in memory, they are fast and efficient. Program 22.8 uses an item number (**ItemNumber**) to look up a product name and its price. Here is the program:

Program 22.8: Using Formats and Informats to Perform a Table Lookup

```
proc format;
   value NameLookup
     122 = 'Salt'
     188 = 'Sugar'
     101 = 'Cereal'
     755 = 'Eggs'
   other = ' ';
   invalue PriceLookup
     'Salt'   = 3.76
     'Sugar'  = 4.99
     'Cereal' = 5.97
     'Eggs'   = 2.65
      other   = .;
run;

data Grocery;
   input ItemNumber @@;
   Name = put(ItemNumber,NameLookup.);
   Price = input(Name,PriceLookup.);
datalines;
101 755 122 188 999 755
;
```

A user-defined format is used, along with a PUT function, to retrieve a product name from the item number. An informat, relating product names and prices, is used with an INPUT function to retrieve the price. An informat was used for the price lookup because the INPUT function can return a numeric value. Remember that the PUT function always returns a character value. (If you used a format, you would have to perform a character-to-numeric conversion to obtain a numeric value for the price.)

Both the format and informat use the keyword OTHER to assign a missing value to an invalid item code. A listing of data set Grocery follows:

Figure 22.7: Listing of Data Set Grocery

Listing of Data Set Grocery

ItemNumber	Name	Price
101	Cereal	5.97
755	Eggs	2.65
122	Salt	3.76
188	Sugar	4.99
999		.
755	Eggs	2.65

22.7 Using a SAS Data Set to Create a Format

The example in the last section used only a small number of item codes. If you have a large number of values, writing a format statement by hand is tedious. Luckily, SAS allows you to use a specially constructed SAS data set to create a SAS format. The PROC FORMAT option CNTLIN= (control in) allows you to name the SAS data set (or data view) that you want to use.

Control input data sets require a minimum of three variables, named FMTNAME, START, and LABEL. FMTNAME is a character variable that specifies the format or informat name; START is either a single value to be formatted or, if END is also used, the beginning of a range. You should include a $ as the first character of FMTNAME if you are creating a character format. Optionally, you can include a variable called TYPE that is either a `C` for character formats or an `N` for numeric formats.

> **Note:** There are other values of TYPE for informats and picture formats: I specifies a numeric informat; J specifies a character informat; P specifies a picture format.

If you have data values and format labels already stored in a SAS data set, you can use that data set as input to a DATA step (with variables renamed as needed) to create your control data set.

This sounds complicated, but an example should make it clear. You have a permanent SAS data set (Codes) that contains ICD-10 code values (International Classification of Diseases, Version 10) and descriptions. You want to create a format that provides labels for each of the ICD-10 codes. So that you can try this yourself, the program here creates a permanent SAS data set called Codes.

Program 22.9: Creating a Test Data Set That Will be Used to Make a CNTLIN Data Set

```
data Learn.Codes;
   input ICD10 : $5. Description & $21.;
datalines;
020 Plague
022 Anthrax
390 Rheumatic fever
410 Myocardial infarction
493 Asthma
540 Appendicitis
;
```

You want to use this data set to create a control input data set, as follows:

Program 22.10: Creating a CNTLIN Data Set from an Existing SAS Data Set

```
data Control;
   set Learn.Codes(rename=
                   (ICD10 = Start
                    Description = Label));
   retain Fmtname '$ICDFMT'
          Type 'C';
run;

title "Demonstrating an Input Control Data Set";
proc format cntlin=Control fmtlib;
run;
```

The RENAME= data set option renames ICD10 to `Start` and Description to `Label`. A RETAIN statement sets FMTNAME to `$ICDFMT` and TYPE equal to `C`. Using a RETAIN statement is more efficient than an assignment statement, since these values are set at compile time—an assignment statement executes for each iteration of the DATA step. The CNTLIN= option used with PROC FORMAT, names this data set and the FMTLIB option generates a table showing the ranges and format labels. Here are both a listing of the Control data set and the output from PROC FORMAT:

Figure 22.8: Listing of Data Set Control

Listing of Data Set Control

Start	Label	Fmtname	Type
020	Plague	$ICDFMT	C
022	Anthrax	$ICDFMT	C
390	Rheumatic fever	$ICDFMT	C
410	Myocardial infarction	$ICDFMT	C
493	Asthma	$ICDFMT	C
540	Appendicitis	$ICDFMT	C

This data set contains the variables Start, Label, Fmtname, and Type. The FMTLIB option included in the PROC FORMAT statement lists all the formats in the format library. In this example, because you did not explicitly name a library, WORK is used. The screen shot below is the result of the FMTLIB option:

Figure 22.9: Output Produced by the FMTLIB Option

```
      FORMAT NAME: $ICDFMT  LENGTH:   21   NUMBER OF VALUES:    6
   MIN LENGTH:   1  MAX LENGTH:  40  DEFAULT LENGTH:  21  FUZZ:        0
```

START	END	LABEL (VER. V7|V8 03DEC2017:10:10:52)
020	020	Plague
022	022	Anthrax
390	390	Rheumatic fever
410	410	Myocardial infarction
493	493	Asthma
540	540	Appendicitis

Program 22.11 shows how you might use this format:

Program 22.11: Using the CNTLIN= Created Data Set

```
data Disease;
   input ICD10 : $5. @@;
datalines;
020 410 500 493
;

title "Listing of Disease";
proc print data=Disease noobs;
   format ICD10 $ICDFMT.;
run;
```

Here is the output:

Figure 22.10: Output from Program 22.11

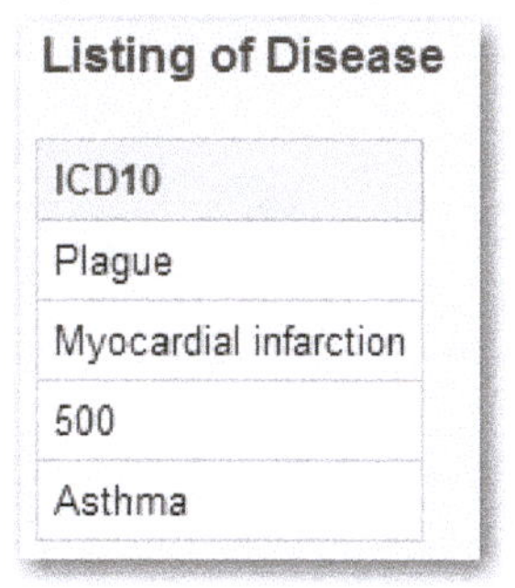

ICD10
Plague
Myocardial infarction
500
Asthma

Notice that there is no format for the ICD-10 code of 500. If you were writing your own PROC FORMAT statements, you could use the keyword OTHER to provide a label for non-matching codes. If you want to accomplish this using a CNTLIN data set, you need a way to assign a value to OTHER. CNTLIN data sets use the variable HLO (stands for *High, Low, Other*) to indicate that you want to use one of these keywords and not an explicit range value.

If you want to assign the label **Not Found** for all ICD-10 codes not in the list, you proceed as follows:

Program 22.12: Adding an OTHER Category to Your Format

```
data Control;
   set Learn.Codes(rename=
                    (ICD10 = Start
                    Description = Label))
                    End = Last;
   retain Fmtname '$ICDFMT'
          Type 'C';
   output;

   if last then do;
      Start = ' ';
      Hlo = 'o';
      Label = 'Not Found';
      output;
   end;
run;
```

You want to add one observation at the bottom of your Control data set with the value of HLO equal to `o` (OTHER) and LABEL equal to `Not Found`. You can do this by using the END= data set option to test for the end of the input data set. The variable Last is true when the last observation in data set Codes has been read. HLO and LABEL are both assigned a value, and a single observation is added to the end of the data set. Here is a listing of data set Control:

Figure 22.11: Listing of Data Set Control

Listing of Data Set Control

Start	Label	Fmtname	Type	Hlo
020	Plague	$ICDFMT	C	
022	Anthrax	$ICDFMT	C	
390	Rheumatic fever	$ICDFMT	C	
410	Myocardial infarction	$ICDFMT	C	
493	Asthma	$ICDFMT	C	
540	Appendicitis	$ICDFMT	C	
	Not Found	$ICDFMT	C	o

The last observation in this data set contains the instructions for setting OTHER equal to `Not Found`. Although we decided to set Start to a missing value in this program, it was not actually necessary because the keyword OTHER, not an actual range, is being used.

When this new format is used in Program 22.11, the ICD-10 code of 500 is now labeled as `Not Found`. Here are both the output from PROC FORMAT showing that the OTHER category was added to the format and a listing of the Disease data set:

Figure 22.12: Format Created by the FMTLIB Option

```
        FORMAT NAME: $ICDFMT   LENGTH:    21    NUMBER OF VALUES:    7
    MIN LENGTH:    1   MAX LENGTH:   40   DEFAULT LENGTH:   21   FUZZ:         0
```

START	END	LABEL (VER. V7|V8 03DEC2017:10:34:16)
020	020	Plague
022	022	Anthrax
390	390	Rheumatic fever
410	410	Myocardial infarction
493	493	Asthma
540	540	Appendicitis
OTHER	**OTHER**	Not Found

Notice the category OTHER is now labeled as `Not Found`.

Here is the listing of data set Control with the updated $ICDFMT format:

Figure 22.13: Listing of Data Set Control (with OTHER Added)

Listing of Disease

Unformatted_ICD	ICD10
020	Plague
410	Myocardial infarction
500	Not Found
493	Asthma

Both the unformatted and formatted values of the ICD data are shown so that it is clear that the code of 500 now shows **Not Found**.

22.8 Updating and Maintaining Your Formats

Suppose you want to add some ICD-10 codes to your existing $ICDFMT format. The easiest way to accomplish this is to first use the CNTLOUT= option of PROC FORMAT to create a data set containing all the formatting information. You can then use this data set in a DATA step to add new codes or modify existing ones.

As an example, suppose you want to add two new ICD-10 codes to your $ICDFMT format, 427.5 (bronchitis) and 466 (cardiac arrest). The following program does the trick:

Program 22.13: Updating an Existing Format Using the CNTLOUT= Data Set Option

```
proc format cntlout=Control_Out;
   select $ICDFMT;
run;

data New_Control;
   length Label $ 21;
   set Control_Out end=Last;
   output;

   if Last then do;
      Hlo = ' ';
      Start = '427.5';
      End = Start;
      Label = 'Cardiac Arrest';
      output;
      Start = '466';
      End = Start;
      Label = 'Bronchitis';
      output;
   end;

run;

proc format cntlin=New_Control;
   select $ICDFMT;
run;
```

You first run PROC FORMAT with the CNTLOUT= option, creating an output data set containing all of the information necessary to re-create the format. A SELECT statement allows you to choose which format you want to use to create this data set. Here is a listing of this Control_Out data set:

Figure 22.14: Listing of Data Set Control_Out

Listing of Data Set Control_Out (selected variables)

FMTNAME	START	END	LABEL	MIN	MAX	DEFAULT	LENGTH
ICDFMT	020	020	Plague	1	40	21	21
ICDFMT	022	022	Anthrax	1	40	21	21
ICDFMT	390	390	Rheumatic fever	1	40	21	21
ICDFMT	410	410	Myocardial infarction	1	40	21	21
ICDFMT	427.5	427.5	Cardiac Arrest	1	40	21	21
ICDFMT	466	466	Bronchitis	1	40	21	21
ICDFMT	493	493	Asthma	1	40	21	21
ICDFMT	540	540	Appendicitis	1	40	21	21
ICDFMT	**OTHER**	**OTHER**	Not Found	1	40	21	21

Notice that there are additional variables that PROC FORMAT uses when creating a format. There are actually even more variables in this data set that were omitted from the listing. You do not need to concern yourself with these variables—SAS uses them as needed. The next step is to add observations to the end of this data set that contain the new codes and their labels. You can do this in a DATA step, by first reading all of the existing format information and then adding the new ones. (Alternatively, you could create a data set of the new codes and use PROC APPEND to add these observations to the end of the existing Control data set.) A LENGTH statement ensures that the storage length for Label is sufficient for any new labels you want to create.

Program 22.13 uses the data set option END= to determine when you have read the last observation from data set Control_Out. At this point, the statements in the DO group execute, outputting two observations to the end of the new data set. It is important to set the variable HLO (High, Low, Other) to missing and to set the variable End equal to the Start values. Because these variables exist in the Control_Out data set and are read with a SET statement, they are automatically retained. Therefore, if you do not assign values to these variables, they retain the values they had in the last observation in data set Control_Out.

Finally, you use the CNTLIN= option to re-create the $ICDFMT format. Instead of the FMTLIB option, a SELECT statement is used to identify which format you want to list. When you use a SELECT statement with PROC FORMAT, it is not necessary to include the FMTLIB option as well. Here is the output from PROC FORMAT, showing that the format now contains the two additional codes:

Figure 22.15: Format Listing Due to the SELECT Statement

Listing of the New $ICDFMT Format

FORMAT NAME: $ICDFMT LENGTH: 21 NUMBER OF VALUES: 9
MIN LENGTH: 1 MAX LENGTH: 40 DEFAULT LENGTH: 21 FUZZ: 0

START	END	LABEL (VER. V7|V8 03DEC2017:12:53:38)
020	020	Plague
022	022	Anthrax
390	390	Rheumatic fever
410	410	Myocardial infarction
427.5	427.5	Cardiac Arrest
466	466	Bronchitis
493	493	Asthma
540	540	Appendicitis
OTHER	**OTHER**	Not Found

Notice that the two new formats for 427.5 and 466 are included.

22.9 Using Formats within Formats

When you define format or informat labels, you can also include the name of a SAS or user-written format or informat, rather than a text string in place of a label. Here is an example.

You want to read dates from January 1, 2018, to December 31, 2018, using the MMDDYY10. informat. Dates before January 1, 2018, should be formatted as `Not Open` and dates after December 31, 2018, should be formatted as `Too Late`. You can use nested formats as follows to accomplish this task:

Program 22.14: Demonstrating Nested Formats

```
proc format;
   value Registration low - <'01Jan2018'd   = 'Not Open'
                '01Jan2018'd - '31Dec2018'd = [mmddyy10.]
                '01Jan2019'd - high         = 'Too Late';
run;
```

The format MMDDYY10. is placed in square brackets where you normally place a format label. Program 22.15 uses this format to process registration dates and produce a list of dates and names.

Program 22.15: Using the Nested Format in a DATA Step

```
data Conference;
   input @1  Name $10.
         @11 Date mmddyy10.;
   format Date Registration.;
datalines;
Smith     10/21/2018
Jones     11/13/2017
Harris    01/03/2018
Arnold    02/12/2019
;
```

A listing of data set Conference is shown here:

Figure 22.16: Listing of Data Set Conference

Listing of Data Set Conference

Name	Date
Smith	10/21/2018
Jones	Not Open
Harris	01/03/2018
Arnold	Too Late

This is a very powerful feature of PROC FORMAT. It saves having to write an IF-THEN-ELSE statement in a DATA step.

The next example uses nested user-written informats. In this example, you are given data on benzene exposure. Some of the values are actual benzene levels in parts per million, while other values are the year a worker was exposed. For each of the years from 1946 to 1952, benzene levels were tabulated. For years outside this range, the actual benzene levels were reported. Some sample data from **`Benzene.txt`**, consisting of either a year or a benzene level, is shown next:

```
001 90
002 1950
003 217
004 1952
005 177
```

All of the benzene levels are much less than any of the year values, so there is no confusion whether to read a value as a year or a benzene level. Creating a data set of IDs and benzene levels is greatly simplified by nesting informats, like this:

Program 22.16: Creating a Data Set of Benzene Levels

```
proc format;
   invalue YearExp 1946 = 250
                   1947 = 244
                   1948 = 240
                   1949 = 200
                   1950 = 188
                   1951 = 150
                   1952 = 100;
   invalue Exp Low - <1946  = [7.1]
               1946 - 1952  = [Yearexp.]
               1952< - high = [7.1];
run;

data Benzene;
   infile 'C:\books\Learning\Benzene.txt';
   input ID Exposure : Exp4.;
run;
```

The first informat, YEAREXP, substitutes a benzene level for each of the years from 1946 to 1952. The informat EXP uses a SAS 7.1 informat for values less than 1946 or greater than 1952. Otherwise, the YEAREXP informat is used.

Here is a listing of the data set:

Figure 22.17: Listing of Data Set Benzene

Listing of Data Set Benzene

ID	Exposure
1	9.0
2	188.0
3	21.7
4	100.0
5	17.7

This problem could have also been solved using formats instead of informats. To use formats, you would need to read a raw data value and use a PUT function to obtain a formatted value. Because the result of a PUT function is a character value, you would also need to use an INPUT function to obtain a benzene level as a numeric value. Creating informats makes the process much simpler.

22.10 Multilabel Formats

Under normal circumstances, you get an error message if any of your format ranges overlap. However, you can create a format with overlapping ranges if you use the MULTILABEL option in the VALUE statement. Certain procedures that use multilabels can then use the MULTILABEL format (MLF) to produce tables showing all of the format ranges. Here is an example.

You want to see the variable Age (from the Survey data set) broken down two ways: one, in 20-year intervals, and the other, by a split at 50 years old. You first create a MULTILABEL format like this:

Program 22.17: Creating a MULTILABEL Format

```
proc format;
   value AgeGroup (multilabel)
    0 - <20    = '0 to <20'
    20 - <40   = '20 to <40'
    40 - <60   = '40 to <60'
    60 - <80   = '60 to <80'
    80 - high = '80 +'

    0 - <50    = 'Less than 50'
    50 - high = '> or = to 50';
run;
```

Without the MULTILABEL option, PROC FORMAT issues an error message about overlapping ranges and fails to create the format.

You can use this format in PROC MEANS, PROC SUMMARY, and PROC TABULATE. Here is a PROC MEANS example:

Program 22.18: Using a MULTILABEL Format with PROC MEANS

```
title "Demonstrating a Multilabel Format";
title2 "PROC MEANS Example";
proc means data=Learn.Survey;
   class Age / MLF;
   var Salary;
   format Age AgeGroup.;
run;
```

It is important to use the MLF option in the CLASS statement if you want to use this feature of the format. Here is the output:

Figure 22.18: Output from Program 22.18

Demonstrating a Multilabel Format
PROC MEANS Example

Analysis Variable : Salary Yearly Salary						
Age as of 1/1/2006	N Obs	N	Mean	Std Dev	Minimum	Maximum
20 to <40	3	3	29186.67	6798.13	23060.00	36500.00
40 to <60	2	2	76111.50	16.2634560	76100.00	76123.00
60 to <80	2	2	109000.00	26870.06	90000.00	128000.00
> or = to 50	3	3	98041.00	26857.01	76123.00	128000.00
Less than 50	4	4	40915.00	24104.46	23060.00	76100.00

Inspection of this output shows the average Salary broken down by Age in two ways: in 20-year intervals, and by a split at 50 years old.

You can use MULTILABEL formats with PROC TABULATE to create interesting tables. Here is a program to compute frequencies in a table of Age by Gender:

Program 22.19: Demonstrating a Multilabel Format

```
title "Demonstrating a Multilabel Format";
title2 "PROC TABULATE Example";
proc tabulate data=Learn.Survey;
   class Age Gender / MLF;
   table Age ,
         Gender;
   format Age AgeGroup.;
run;
```

This produces the following output:

Figure 22.19: Output from Program 22.19

Demonstrating a Multilabel Format
PROC TABULATE Example

	Gender	
	Female	Male
	N	N
Age as of 1/1/2006		
20 to <40	.	3
40 to <60	2	.
60 to <80	1	1
> or = to 50	2	1
Less than 50	1	3

Notice that certain age groups are not shown in the table. For example, there were no subjects in the 0 to <20 age group. To see all of the possible format ranges in your table, you can use the PRELOADFMT option in the CLASS statement. Preloaded formats allow PROC TABULATE to list categories for which there are no data values.

Here is the PROC TABULATE code with the PRELOADFMT option and two additional options to control the printing of missing values. (They are explained following the program.)

Program 22.20: Using the PRELOADFMT, PRINTMISS, and MISSTEXT Options with PROC TABULATE

```
title "Demonstrating a Multilabel Format";
title2 "PROC TABULATE Example";
proc tabulate data=Learn.Survey;
   class Age Gender / MLF preloadfmt;
   table Age,
         Gender / printmiss  misstext=' ';
   format Age AgeGroup.;
run;
```

The PRELOADFMT option forces the procedure to list all of the format ranges. This option has no effect without the PRINTMISS option in the TABLE statement. Remember that PRINTMISS is an instruction to include categories that contain one or more missing values of a CLASS variable. Finally, the MISSTEXT= option tells the procedure what character (or characters) you would like printed that represents missing values (the default is a period). Here you choose a blank to represent missing values. Finally, the output:

Figure 22.20: Output from Program 22.20

Demonstrating a Multilabel Format
PROC TABULATE Example

	Gender	
	Female	Male
	N	N
Age as of 1/1/2006		
0 to <20		
20 to <40		3
40 to <60	2	
60 to <80	1	1
80 +		
> or = to 50	2	1
Less than 50	1	3

Now all the Age ranges appear in the table and the missing values print as blanks.

22.11 Using the INPUTN Function to Perform a More Complicated Table Lookup

This final example of the chapter performs a two-way table lookup using a function that allows you to compute an informat name in a DATA step. This section is a bit complicated so you may either want to skip it or refer to SASHelp Center at http://go.documentation.sas.com.

The problem: you have a table of years (1944 to 1949) and job codes (A through E). Each combination of year and job code has a benzene exposure (in parts per million) associated with it. The following table lists these values:

	Job Code				
Year	**A**	**B**	**C**	**D**	**E**
1944	220	180	210	110	90
1945	202	170	208	100	85
1946	150	110	150	60	50
1947	105	56	88	40	30
1948	60	30	40	20	10
1949	45	22	22	10	8

One way to compute an exposure, given a year and job code, is to create an informat for each year, assigning each of the job codes a label containing the exposure value. An informat is used instead of a format because you can use an INPUT (or, as you will see, an INPUTN) function with the user-defined informats to obtain a numeric value directly. If you did this "by hand," you would proceed like this:

Program 22.21: Partial Program Showing How to Create Several Informats

```
proc format;
   invalue Exp1944fmt (upcase)
     'A' = 220
     'B' = 180
     'C' = 210
     'D' = 110
     'E' = 90
   other = .;
   invalue Exp1945fmt (upcase)
     'A' = 202
     'B' = 170
     'C' = 208
     'D' = 100
     'E' = 85
   other = .;
   invalue Exp1946fmt (upcase)
     'A' = 150
      . . .
run;
```

It takes far less work to create an input control data set (CNTLIN) to create the required informats. You proceed as follows:

Program 22.22: Creating Several Informats with a Single CNTLIN Data Set

```
data Exposure;
   retain Type 'I' Hlo 'U';
   length Fmtname $ 10;
   do Year = 1944 to 1949;
      Fmtname = cats('Exp',Year,'fmt');
      do Start = 'A','B','C','D','E';
         End = Start;
         input Label : $3. @;
         output;
      end;
   end;
   drop Year;
datalines;
220   180   210   110   90
202   170   208   100   85
150   110   150    60   50
105    56    88    40   30
 60    30    40    20   10
 45    22    22    10    8
;

title "Creating the Exposure Format";
proc format cntlin=Exposure fmtlib;
run;
```

`Type=I` tells PROC FORMAT that you want to create a numeric informat. The value of `U` for the HLO variable tells the procedure to use the UPCASE option in the INVALUE statement. You want informat names of the following form:

Exp***yyyy***fmt

Here, ***yyyy*** is a year value from 1944 to 1949. The CATS function concatenates (joins) each of the arguments after first stripping off any leading or trailing blanks. This function also allows numeric arguments and performs the numeric-to-character conversion as well. Here are the first few observations in the Exposure data set:

Figure 22.21: Partial Listing of Data Set Exposure

Listing of Data Set Exposure

Type	Hlo	Fmtname	Start	End	Label
I	U	Exp1944fmt	A	A	220
I	U	Exp1944fmt	B	B	180
I	U	Exp1944fmt	C	C	210
I	U	Exp1944fmt	D	D	110
I	U	Exp1944fmt	E	E	90
I	U	Exp1945fmt	A	A	202
I	U	Exp1945fmt	B	B	170
I	U	Exp1945fmt	C	C	208
I	U	Exp1945fmt	D	D	100

By the way, if you want to see selected informats using the SELECT statement of PROC FORMAT, you need to include an @ sign as the first character in the informat name. For example, to list the contents of EXP1944FMT and EXP1945FMT, you would use Program 22.23:

Program 22.23: Using a SELECT Statement to Display the Contents of Two Informats

```
proc format;
   select @Exp1944fmt @Exp1945fmt;
run;
```

When you use a SELECT statement with PROC FORMAT, it is not necessary to include the FMTLIB option as well.

Let's continue with our program. Now that you have an informat for each of the six years, all you have to do is use the year and job code to create an informat name and use that name with an INPUTN function to retrieve the exposure. Here is the program:

Program 22.24: Using User-Defined Informats to Perform a Table Lookup Using the INPUTN Function

```
data Read_Exp;
   input Worker $ Year JobCode $;
   Exposure = inputn(JobCode,cats('Exp',Year,'fmt8.'));
datalines;
001 1944 B
002 1948 E
003 1947 C
005 1945 A
006 1948 d
;
```

You cannot use the INPUT function in this program because with that function the informat name must be a constant. The INPUTN function allows you to create the informat name in the DATA step. By the way, the 'N' in the name INPUTN stands for *numeric*. You need to tell SAS ahead of time if the informat you are going to create is going to be character or numeric (it can't tell if the name starts with a $ until it is created). There are two corresponding functions, PUTN and PUTC, that allow you to create character or numeric format names in the DATA step.

When this program runs, the informat names are created with the CATS function. Each year value creates a different informat name to be used in the INPUTN function. Here is a listing of the resulting data set:

Figure 22.22: Listing of Data Set Read_Exp

Listing of Data Set Read_Exp

Worker	Year	JobCode	Exposure
001	1944	B	180
002	1948	E	10
003	1947	C	88
005	1945	A	202
006	1948	d	20

The last few sections of this chapter may seem, at first glance, to be a "bit off the deep end." However, if you try experimenting and writing a few programs of your own using these ideas, you will see how powerful these tools can be.

22.12 Problems

Solutions to odd-numbered problems are located at the back of this book. Solutions to all problems are available to professors. If you are a professor, visit the book's companion website at support.sas.com/cody for information about how to obtain the solutions to all problems.

1. You are given a SAS data set (BloodPressure) with variables SBP and DBP (systolic blood pressure and diastolic blood pressure). You want to see frequencies for SBP and DBP grouped as follows: SBP pressures below 140 are to be called **`Normal`** and pressures of 140 and above are to be called **`High SBP`**. For DBP, pressures below 90 are to be called **`Normal`** and pressures of 90 and above are to be called **`High DBP`**. Create an appropriate format and use PROC FREQ to produce these frequencies. Do not use a DATA step. Use the appropriate options to omit the cumulative frequencies and percentages from the PROC FREQ output.
2. Given the SAS data set Grades, produce frequencies of Grade (a numeric variable) grouped as follows:

   ```
   Less than 65 = F
   65 to less than 75 = C
   75 to less than 85 = B
   85 to 100 = A
   ```

   Create a format and use PROC FREQ to list these frequencies. Do not use a DATA step.
3. Using the same definitions for high and low SBP and DBP from Problem 1, create a new, temporary SAS data set (BloodPressure) with two new variables (SBPGroup and DBPGroup).
4. Using the same definitions for letter grades from Problem 2, create a new, temporary SAS data set (Grades) with a new variable (LetterGrade) created by using a PUT function with a user-defined format.
5. Using the same definitions for letter grades from Problem 2, create a new, temporary SAS data set (Grades) with a new variable LetterGrade (with values of **`A`**, **`B`**, **`C`**, and **`F`**). Do this by creating an informat to read the raw scores from the text file **`NumGrades.txt`**.
6. You want to read the following line of data:

   ```
   A 6.7 X b c a 10.9 11.6 C
   ```

   The numbers represent blood lead levels and the letters A, B, and C represent default values from three labs (A=4, B=5, and C=3.5). An X represents a missing value. Use an enhanced numeric informat to read this line of data with the letters converted to the appropriate values (notice that the letters are in mixed case).
7. Use the SAS data set DxCodes to create a control data set to be used to create a character format (DXCODES). This data set contains variables Dx (character diagnosis codes) and Description. The Description variable contains the labels you want to use for each of the Dx codes. Use a SELECT statement to display this format.

8. Repeat Problem 7, except add the necessary statements in your DATA step to assign a label of **`Not Found`** to any value that does not match a format range (equivalent to **`OTHER=Not Found`**).
9. Starting with the permanent SAS data set Gym, create a listing using PROC PRINT that labels all dates from January 1, 1990 to December 31, 2004 as **`Too Early`** and dates from after January 1, 2007 as **`Too Late`**. Leave all other dates unchanged (and listed using the MMDDYY10. format). **Hint:** Create a format that uses the embedded format MMDDYY10. for dates in the appropriate range and labels the early and late dates as defined.

Chapter 23: Restructuring SAS Data Sets

23.1 Introduction

The term restructuring, also called transposing, means to take a data set with one observation per subject and convert it to a data set with many observations per subject, or vice versa. Why would you want or need to do this? There are some operations that are most easily performed when all the information per subject (or other unit of analysis) is contained in a single observation. Other operations are more convenient when there are several observations per subject.

Several of the statistical procedures require data to be stored one way or the other, depending on what type of analysis is required.

This chapter demonstrates how to restructure data sets using DATA step approaches and PROC TRANSPOSE.

23.2 Converting a Data Set with One Observation per Subject to a Data Set with Several Observations per Subject: Using a DATA Step

This first example uses a small data set (OnePer) where each subject has from one to three diagnosis codes (Dx1–Dx3). Here is a listing of this data set:

Figure 23.1: Listing of Data Set OnePer

Listing of Data Set Learn.OnePer

Subj	Dx1	Dx2	Dx3
001	450	430	410
002	250	240	.
003	410	250	500
004	240	.	.

Notice that some subjects have three diagnosis codes, some two, and one (Subject 004) only one. How do you obtain a frequency distribution for diagnosis codes? For example, Subject 001 had code 410 as the third diagnosis and Subject 003 has this same code listed as Dx1. It would be easier to compute frequencies on the diagnosis codes if the data set were structured like this:

Figure 23.2: Listing of Data Set ManyPer

Listing of Data Set Learn.ManyPer

Subj	Visit	Diagnosis
001	1	450
001	2	430
001	3	410
002	1	250
002	2	240
003	1	410
003	2	250
003	3	500
004	1	240

Data set ManyPer has from one to three observations per subject. Here is a program to make this conversion:

Program 23.1: Creating a Data Set with Several Observations per Subject from a Data Set with One Observation per Subject

```
data Learn.ManyPer;
   set Learn.OnePer;
   array Dx{3};

   do Visit = 1 to 3;
      if missing(Dx{Visit}) then delete;
      Diagnosis = Dx{Visit};
      output;
   end;

   keep Subj Diagnosis Visit;
run;
```

Although you don't have to use arrays to solve this problem, it does make the program more compact. The DX array has three elements: Dx1, Dx2, and Dx3 (remember that if you leave off the variable list, the variable names default to the array name with the digits 1 to *n* appended to the end).

Let's take the time to describe in detail how this program works (feel free to skip this section if this program seems intuitively clear to you). The DO loop starts with Visit set equal to `1`. The MISSING function tests if the value Dx{1} (equal to Dx1) is missing. For Subject 001, none of the diagnosis codes are missing, so the IF statement is never true for this subject. A new variable, Diagnosis, is set equal to the value of the array element Dx{1}, which is the same as the variable Dx1, which is equal to `450`. At this point, the program data vector (PDV) contains the following:

Subj	Dx1 <drop>	Dx2 <drop>	Dx3 <drop>	Visit	Diagnosis
001	450	430	410	1	450

Because of the KEEP statement, only the variables Subj, Visit, and Diagnosis are written out to data set ManyPer at the bottom of the DO loop.

During the next iteration of the DO loop, Visit is equal to `2`, Diagnosis is equal to `430`, and these values are written out to the second observation in data set ManyPer. Finally, Visit is set to `3`, Diagnosis is set to `410`, and the third observation is written to the output data set.

The DATA step has now reached the bottom. Because the end of the file on the OnePer data set has not been reached, a new observation from data set OnePer is brought into the PDV and the process continues. Because the third diagnosis (Dx3) is missing for Subject 002, the MISSING function returns a value of **`true`** and the DELETE statement executes. A DELETE statement returns control back to the top of the data set and an observation is not written out to data set ManyPer. (It is also assumed that if there are any missing Dx codes, they come after the nonmissing codes.)

Note: If you replaced DELETE with CONTINUE, this program would work even if the missing Dx codes were stored in any of the Dx variables. See Chapter 8 for a description of the CONTINUE statement.

23.3 Converting a Data Set with Several Observations per Subject to a Data Set with One Observation per Subject: Using a DATA Step

What if you want to go the other way, creating a data set with one observation per subject from one with several observations per subject? This process is a bit more complicated, and you need to take a few special precautions.

As an example, suppose you started with data set ManyPer and wanted to create a data set that looked like OnePer. Here is one way to do it:

Program 23.2: Creating a Data Set with One Observation per Subject from a Data Set with Several Observations per Subject

```
proc sort data=Learn.ManyPer out=ManyPer;
   by Subj Visit;
run;

data OnePer;
   set ManyPer;
   by Subj Visit;
   array Dx{3};
   retain Dx1-Dx3;
   if first.Subj then call missing(of Dx1-Dx3);
   Dx{Visit} = Diagnosis;
   if last.Subj then output;
   keep Subj Dx1-Dx3;
run;
```

You first sort the input data set by Subj. (The original data set was already in Subj order, but the sort was included to make the program more general and, although not needed for this example, it was also sorted by visit date.) Next, you set up an array to hold the three Dx values and retain these three variables. You need to retain these three variables because they do not come from a SAS data set and are, by default, set equal to a missing value for each iteration of the DATA step. The RETAIN statement prevents this from happening.

Next, when you start processing the first visit for each subject, you set the three values of Dx to missing. If you don't do this, a subject with fewer than three visits may wind up with a diagnosis from the previous subject. The CALL MISSING routine can set any number of numeric and/or character values to missing at one time. As with many of the SAS functions and CALL routines, if you use a variable list in the form Var1–Var*n*, you need to precede the variable list with the keyword OF.

If you need to review First. and Last. variables, please refer to a discussion of these variables in the next chapter.

Next, you assign the value of Diagnosis to the appropriate Dx variable (Dx1 if **`Visit=1`**, Dx2 if **`Visit=2`**, and Dx3 if **`Visit=3`**).

Finally, if you are processing the last visit for a patient, you output a single observation keeping the variables Subj and Dx1–Dx3.

A listing of data set OnePer is identical to the one shown earlier in this chapter.

23.4 Converting a Data Set with One Observation per Subject to a Data Set with Several Observations per Subject: Using PROC TRANSPOSE

PROC TRANSPOSE can also be used to restructure SAS data sets. Sometimes, PROC TRANSPOSE provides a quick and simple solution—sometimes the PROC TRANSPOSE solution can be quite complicated. In general, a DATA step solution gives you more control over the restructuring process. As you will see in this first PROC TRANSPOSE example, you need to add some data set options (see Program 23.4) to achieve the same results as the original DATA step solution.

The following program attempts to solve the same problem as in Section 23.2. You start out with a program that restructures the OnePer data set, but it does not completely solve the problem. Here is the code:

Program 23.3: Using PROC TRANSPOSE to Convert a Data Set with One Observation per Subject into a Data Set with Several Observations per Subject (First Attempt)

```
***Note: data set already in Subject order;
proc transpose data=Learn.OnePer
               out=ManyPer;
   by Subj;
   var Dx1-Dx3;
run;
```

PROC TRANSPOSE takes an input data set and outputs a data set where the original rows become columns and the original columns become rows. This program includes a BY SUBJECT statement that performs the operation for each value of Subject. The result is the data set listed next.

Figure 23.3: Listing of ManyPer (Transpose Method)

Listing of Data Set ManyPer

Subj	_NAME_	COL1
001	Dx1	450
001	Dx2	430
001	Dx3	410
002	Dx1	250
002	Dx2	240
002	Dx3	.
003	Dx1	410
003	Dx2	250
003	Dx3	500
004	Dx1	240
004	Dx2	.
004	Dx3	.

This is almost what you want. All that is needed is to rename the variable COL1 to Diagnosis, eliminate the _NAME_ variable, and remove the observations with missing Dx values. All these goals are accomplished by using some data set options in the output data set, as follows:

Program 23.4: Using PROC TRANSPOSE to Convert a Data Set with One Observation per Subject into a Data Set with Several Observations per Subject

```
proc transpose data=Learn.OnePer
               out=t_ManyPer(rename=(col1=Diagnosis)
                             drop=_Name_
                             where=(Diagnosis is not null));
   by Subj;
   var Dx1-Dx3;
run;
```

The RENAME= option renames COL1 to Diagnosis, while the DROP= option eliminates the _NAME_ variable from the data set. Finally, a WHERE= data set option removes observations where Diagnosis is missing. The result is identical to the output from Program 23.1.

23.5 Converting a Data Set with Several Observations per Subject to a Data Set with One Observation per Subject: Using PROC TRANSPOSE

The last example in this chapter shows how to convert a data set with several observations per subject into a data set with one observation per subject using PROC TRANSPOSE.

Program 23.5: Using PROC TRANSPOSE to Convert a SAS Data Set with Several Observations per Subject into One with One Observation per Subject

```
proc transpose data=Learn.ManyPer prefix=Dx
               out=OnePer(drop=_NAME_);
   by Subj;
   id Visit;
   var Diagnosis;
run;
```

The PREFIX= option to the procedure and an ID statement create an output data set identical to the one produced by Program 23.2.

Figure 23.4: Listing of OnePer (Transpose Method)

Listing of Data Set OnePer

Subj	Dx1	Dx2	Dx3
001	450	430	410
002	250	240	.
003	410	250	500
004	240	.	.

The PREFIX= option combines the prefix value (`Dx`) with the values of Visit (`1`, `2`, and `3`) to create the three variables Dx1, Dx2, and Dx3. PROC TRANSPOSE knew to use the values of Visit to create these variable names because it was identified as an ID variable in the procedure.

23.6 Problems

Solutions to odd-numbered problems are located at the back of this book. Solutions to all problems are available to professors. If you are a professor, visit the book's companion website at support.sas.com/cody for information about how to obtain the solutions to all problems.

1. A listing of data set Wide, containing the variables Subj, X1–X5, and Y1–Y5, is shown here:

Listing of Data Set Wide

Subj	X1	X2	X3	X4	X5	Y1	Y2	Y3	Y4	Y5
001	8	5	6	5	4	10	20	30	40	50
002	7	5	6	4	5	11	33	29	34	56
003	2	2	4	5	6	22	38	21	20	34

Using a DATA step, create a temporary SAS data set (Long) using data set Wide as input. This data set should contain Subj, Time, X, and Y, with five observations per subject. A partial listing of data set Long looks like this:

Listing of Long

Subj	Time	X	Y
001	1	8	10
001	2	5	20
001	3	6	30
001	4	5	40
001	5	4	50
002	1	7	11
002	2	5	33
002	3	6	29
	4	4	34

2. Using the SAS data set Narrow (shown here), create a new, temporary SAS data set (Stretch) where the five scores for each subject are contained in a single observation, with the variable names S1–S5. S1 is the Score at Time 1, S2 is the Score at Time 2, etc. Do this using a DATA step.

Partial Listing of Data Set Narrow

Subj	Time	Score
001	1	7
001	2	6
001	3	5
001	4	5
001	5	4
002	1	8
002	2	7
002	3	6
002	4	6
002	5	6

Data set Stretch should look like this:

Listing of Streach

Obs	Subj	S1	S2	S3	S4	S5
1	001	7	6	5	5	4
2	002	8	7	6	6	6
3	003	8	7	6	6	5

3. Repeat Problem 1 using PROC TRANSPOSE. Do this only for the variables X1–X5. Your resulting data set should look like this:

Listing of Data Set Long

Subj	Time	X	Y
001	1	8	10
001	2	5	20
001	3	6	30
001	4	5	40
001	5	4	50
002	1	7	11
002	2	5	33
002	3	6	29
002	4	4	34
002	5	5	56
003	1	2	22
003	2	2	38
003	3	4	21
003	4	5	20
003	5	6	34

4. Repeat Problem 2 using PROC TRANSPOSE.

Chapter 24: Working with Multiple Observations per Subject

24.1 Introduction

You might encounter data sets with multiple observations per subject (or other groupings such as transactions on a single day or observations grouped in other ways). These data structures are sometimes referred to as *longitudinal data.*

Because SAS processes one observation at a time, you need special techniques to perform calculations across observations. For example, if you have a data set representing patient visits to a clinic, you might want to compare patient values from one visit to the next or from the first visit to the last visit. You may want to retrieve the data from the first or last visit for each patient.

The good news is that SAS has all the tools you need to accomplish these tasks. This chapter shows you how to use these tools.

24.2 Identifying the First or Last Observation in a Group

For most of the examples in this chapter, you will use a data set (Clinic) consisting of medical data on patients who visit a clinic. Here is a listing of this data set:

Figure 24.1: Listing of Data Set Clinic

Listing of Data Set Learn.Clinic

ID	VisitDate	Dx	HR	SBP	DBP
101	10/21/2005	GI Problems	68	120	80
101	02/25/2006	Cold	68	122	84
255	09/01/2005	Routine Visit	76	188	100
255	12/18/2005	Routine Visit	74	180	95
255	02/01/2006	Heart Problems	79	210	110
255	04/01/2006	Heart Problems	72	180	88
303	10/10/2006	Routine Visit	72	138	84
409	09/01/2005	Injury	88	142	92
409	10/02/2005	Routine Visit	72	136	90
409	12/15/2006	Routine Visit	68	130	84
712	04/06/2006	Infection	58	118	70
712	04/15/2006	Infection	56	118	72

Variables in this data set are patient ID (ID), visit date (VisitDate), diagnosis (Dx), heart rate (HR), systolic blood pressure (SBP), and diastolic blood pressure (DBP). Notice that some patients have several visits, but one patient, Number 303, has a single visit.

One of the important tasks when dealing with data sets like this is to identify when you are processing the first or the last observation for each patient (or the variable that you want to use to identify a group). Program 24.1 shows how to do this:

Program 24.1: Creating FIRST. and LAST. Variables

```
proc sort data=Learn.Clinic out=Clinic;
   by ID VisitDate;
run;

data Last;
   set clinic;
   by ID;
   put ID= VisitDate= First.ID= Last.ID=;
   if last.
run;
```

The key to this program is the BY statement following the SET statement. In this example, the data set was also sorted by VisitDate within each ID so that the first observation for each ID would correspond to the first visit date and the last observation for each ID would correspond to the last visit date.

This BY statement following a SET statement creates two temporary SAS variables, First.ID and Last.ID. The PUT statement in the program was added so that you can see the value of these two temporary variables.

Here is the section of the SAS log showing these data values.

Note: The shading was added to help identify multiple visits for each patient.

Log File Showing FIRST.ID and LAST.ID Values

```
ID=101 VisitDate=10/21/2005 FIRST.ID=1 LAST.ID=0
ID=101 VisitDate=02/25/2006 FIRST.ID=0 LAST.ID=1
ID=255 VisitDate=09/01/2005 FIRST.ID=1 LAST.ID=0
ID=255 VisitDate=12/18/2005 FIRST.ID=0 LAST.ID=0
ID=255 VisitDate=02/01/2006 FIRST.ID=0 LAST.ID=0
ID=255 VisitDate=04/01/2006 FIRST.ID=0 LAST.ID=1
ID=303 VisitDate=10/10/2006 FIRST.ID=1 LAST.ID=1
ID=409 VisitDate=09/01/2005 FIRST.ID=1 LAST.ID=0
ID=409 VisitDate=10/02/2005 FIRST.ID=0 LAST.ID=0
ID=409 VisitDate=12/15/2006 FIRST.ID=0 LAST.ID=1
ID=712 VisitDate=04/06/2006 FIRST.ID=1 LAST.ID=0
ID=712 VisitDate=04/15/2006 FIRST.ID=0 LAST.ID=1
```

It should be pretty clear what these two variables represent: First.ID is `true` (`1`) for the first visit for each patient and `false` (`0`) otherwise. Last.ID is `true` for the last visit for each patient and `false` otherwise. For Patient Number 303, both First.ID and Last.ID are `true` (it is both the first and the last visit for this patient).

To create a data set consisting of the last visit for each patient, the program uses a subsetting IF statement to check when the value of Last.ID is equal to `1`. Here is a listing of data set Last:

Figure 24.2: Listing of Data Set Last

Listing of Data Set Last

ID	VisitDate	Dx	HR	SBP	DBP
101	02/25/2006	Cold	68	122	84
255	04/01/2006	Heart Problems	72	180	88
303	10/10/2006	Routine Visit	72	138	84
409	12/15/2006	Routine Visit	68	130	84
712	04/15/2006	Infection	56	118	72

A common data processing requirement is to perform such operations as setting counters equal to 0,when you are processing the first observation in a BY group and outputting counts when you are processing the last observation in a BY group.

Program 24.2 uses a DATA step to count the number of visits for each patient. (We will demonstrate later how to do this using PROC FREQ.)

Program 24.2: Counting the Number of Visits per Patient Using the DATA Step

```
data Count;
   set Clinic;
   by ID;
   *Initialize counter at first visit;
   if first.ID then N_visits = 0;
   *Increment the visit counter;
   N_visits + 1;
   *Output an observation at the last visit;
   if last.ID then output;
run;
```

Each time you encounter a new ID, you set the variable N_visits equal to `0`. You then increment this variable by one for each iteration of the DATA step, using a sum statement. Finally, an observation is written out to data set Count when you are processing the last visit for each patient. Here is a listing of data set Count:

Figure 24.3: Listing of Data Set Count

Listing of Data Set Count

ID	VisitDate	Dx	HR	SBP	DBP	N_visits
101	02/25/2006	Cold	68	122	84	2
255	04/01/2006	Heart Problems	72	180	88	4
303	10/10/2006	Routine Visit	72	138	84	1
409	12/15/2006	Routine Visit	68	130	84	3
712	04/15/2006	Infection	56	118	72	2

Notice that the values for VisitDate, Dx, HR, SBP, and DBP are values from the last visit for each patient.

24.3 Counting the Number of Visits Using PROC FREQ

Instead of using a DATA step, you can use PROC FREQ to compute the number of visits per patient. Here is the solution:

Program 24.3: Using PROC FREQ to Count the Number of Observations in a BY Group

```
proc freq data=Learn.Clinic noprint;
   tables ID / out=Counts;
run;
```

The NOPRINT option is included because you want PROC FREQ to create a data set and you do not need printed output. In this program, you name the data set Counts. PROC FREQ uses the variable names Count and Percent for the variables representing the frequencies and percentages for each value of the variable listed in the TABLES statement. This is clear if you look at the listing of data set Counts created by Program 24.3.

Figure 24.4: Listing of Data Set Counts

Listing of Data Set Counts

ID	COUNT	PERCENT
101	2	16.6667
255	4	33.3333
303	1	8.3333
409	3	25.0000
712	2	16.6667

You can use the RENAME= and DROP= or KEEP= data set options to control what variables are in this data set and what they are called. For example, the next program renames Count to N_Visits and drops the Percent variable. In addition, a MERGE statement adds the variable N_Visits to each observation in the original Clinic data set. Here is the program:

Program 24.4: Using the RENAME= and DROP= Data Set Options to Control the Output Data Set

```
proc freq data=Clinic noprint;
   tables ID / out=Counts (rename=(count = N_Visits)
                           drop=percent);
run;

data Clinic;
   merge Learn.Clinic Counts;
   by ID;
run;
```

The RENAME= data set option renames the variable Count (the name chosen by SAS) to N_Visits. The DROP= data set option tells SAS that you don't want the variable Percent in the new data set. You don't need to sort data set Counts because it will be in ID order. The default ordering for PROC FREQ is to order values by their internal (unformatted) values. Data set Clinic is already in ID order. Here is the result:

Figure 24.5: Listing of Data Set Clinic

Listing of Data Set Clinic

ID	VisitDate	Dx	HR	SBP	DBP	N_Visits
101	10/21/2005	GI Problems	68	120	80	2
101	02/25/2006	Cold	68	122	84	2
255	09/01/2005	Routine Visit	76	188	100	4
255	12/18/2005	Routine Visit	74	180	95	4
255	02/01/2006	Heart Problems	79	210	110	4
255	04/01/2006	Heart Problems	72	180	88	4
303	10/10/2006	Routine Visit	72	138	84	1
409	09/01/2005	Injury	88	142	92	3
409	10/02/2005	Routine Visit	72	136	90	3
409	12/15/2006	Routine Visit	68	130	84	3
712	04/06/2006	Infection	58	118	70	2
712	04/15/2006	Infection	56	118	72	2

You now have the same result produced by Program 24.2. You may prefer the DATA step approach or PROC FREQ to do the counting for you.

24.4 Computing Differences between Observations

Suppose you want to see changes in heart rate (HR), systolic blood pressure (SBP), and diastolic blood pressure (DBP) from visit to visit. The LAG function provides an easy way to compare values from the present observation to ones in a previous observation. Here is a program to output changes from one visit to the next:

Program 24.5: Computing Differences between Observations

```
data Difference;
   set Clinic;
   by ID;
   *Delete patients with only one visit;
   if first.ID and last.ID then delete;
   Diff_HR = HR - lag(HR);
   Diff_SBP = SBP - lag(SBP);
   Diff_DBP = DBP - lag(DBP);
   if not first.ID then output;
run;
```

There are a few important points to notice about this program. First, you delete patients with only one visit (when both First.ID and Last.ID are true). Next, you use the LAG function to obtain values of HR, SBP, and DBP from the previous visit. Let's take a moment and follow the DATA step logic.

The first patient (`ID=101`) has more than one visit, so the first IF statement is not true. The values of the three Diff variables are all missing during the first iteration of the DATA step. First.ID is true, so NOT First.ID is false, and an observation is not written out to the data set.

During the second iteration of the DATA step, the three Diff variables represent the difference between the current values and the values from the last visit. The last IF statement is true, and an observation is written out to data set Difference.

The next observation is the first visit for Patient Number 255. Again, this patient has more than one visit, so the first IF statement is not true. The three Diff variables represent the difference between the three values for Patient Number 255 and the last visit for Patient Number 101! This seems wrong, but hold on, this will work out OK since you are not going to output an observation this time (the last IF statement is not true). You can think of this iteration of the DATA step as "priming the pump" so that the LAG function returns the correct values on the next iteration of the DATA step. You might be tempted to execute the three DIFF statements conditionally. **Do not do this**. You need to execute the LAG function for every iteration of the DATA step in this program so that you get the correct difference values.

The next observation from the Clinic data set is the second visit for Patient Number 255. Now the three DIFF statements correctly compute the difference between the current values for Patient Number 255 and the values from the previous visit. Because the last IF statement is now true, this observation is written out to the Difference data set.

You may think this explanation is too long and tedious—and you may be right. However, it is really important that you understand the logic of this program if you are to use the LAG function correctly.

By the way, for you compulsive programmers (not me!), you can use the DIF function to compute the three differences directly. Refer to Chapter 11 for details on this function.

Here is a listing of data set Difference:

Figure 24.6: Listing of Data Set Difference

Listing of Data Set Difference

ID	VisitDate	Dx	HR	SBP	DBP	N_Visits	Diff_HR	Diff_SBP	Diff_DBP
101	02/25/2006	Cold	68	122	84	2	0	2	4
255	12/18/2005	Routine Visit	74	180	95	4	-2	-8	-5
255	02/01/2006	Heart Problems	79	210	110	4	5	30	15
255	04/01/2006	Heart Problems	72	180	88	4	-7	-30	-22
409	10/02/2005	Routine Visit	72	136	90	3	-16	-6	-2
409	12/15/2006	Routine Visit	68	130	84	3	-4	-6	-6
712	04/15/2006	Infection	56	118	72	2	-2	0	2

The value of HR, SBP, and DBP represent the values for the current visit. The three Diff values represent these values minus the values from the previous visit.

24.5 Computing Differences between the First and Last Observation in a BY Group Using the LAG Function

This section demonstrates how to compute the difference between the first and last visit for each patient. Remember how you were cautioned never to execute the LAG function conditionally? The program that follows does just that—and it works:

Program 24.6: Computing Differences between the First and Last Observation in a BY Group

```
data First_Last;
   set Clinic;
   by ID;

   *Delete patients with only one visit;
   if first.ID and last.ID then delete;

   if first.ID or last.ID then do;
      Diff_HR = HR - lag(HR);
      Diff_SBP = SBP - lag(SBP);
      Diff_DBP = DBP - lag(DBP);
   end;

   if last.ID then output;
run;
```

The first time the LAG function executes is when the first visit for each patient is being processed. The next time the LAG function executes is during the last visit. Therefore, the differences are the values in the last visit minus the values from the first visit (**the last time the LAG function executed**). Notice also that an observation is only written out during the last visit for each patient. Here is a listing of this data set:

Figure 24.7: Listing of Data Set First_Last

Listing of Data Set First_Last

ID	VisitDate	Dx	HR	SBP	DBP	N_Visits	Diff_HR	Diff_SBP	Diff_DBP
101	02/25/2006	Cold	68	122	84	2	0	2	4
255	04/01/2006	Heart Problems	72	180	88	4	-4	-8	-12
409	12/15/2006	Routine Visit	68	130	84	3	-20	-12	-8
712	04/15/2006	Infection	56	118	72	2	-2	0	2

The DIFF variables in this listing represent the difference between values recorded at the first visit minus values recorded at the last visit.

24.6 Computing Differences between the First and Last Observation in a BY Group Using a RETAIN Statement

Because SAS has so many useful tools, there are often several ways to solve the same problem. This section describes how to accomplish the same task as the previous section, except you use retained variables instead of the LAG function to accomplish the task.

Using retained variables is one of the best ways to "remember" values from previous observations. By default, remember that variables that do not come from SAS data sets are set to a missing value during each iteration of the DATA step. A RETAIN statement allows you to tell SAS not to do this. Armed with this knowledge, look at the following program:

Program 24.7: Demonstrating the Use of Retained Variables

```
data First_Last;
   set Clinic;
   by ID;

   *Delete patients with only one visit;
   if first.ID and last.ID then delete;

   retain First_HR First_SBP First_DBP;

   if first.ID then do;
      First_HR = HR;
      First_SBP = SBP;
      First_DBP = DBP;
   end;

   if last.ID then do;
      Diff_HR = HR - First_HR;
      Diff_SBP = SBP - First_SBP;
      Diff_DBP = DBP - First_DBP;
      output;
   end;

   drop First_: ;
run;
```

You start out the same as Program 24.6. Next, you use a RETAIN statement so that the three variables, First_HR, First_SBP, and First_DBP, are not set back to missing when the DATA step iterates. (The RETAIN statement operates at compile time and is not an executable statement.)

Next, when you are processing the first visit for each patient, you set the three retained variables equal to the values of HR, SBP, and DBP, respectively. These values remain in the program data vector (PDV) and are not replaced until the first visit for a new patient is processed. When you reach the last visit for each patient, you compute the three difference values and output them.

One final comment on this program involves the DROP statement. The value `First_:` refers to all variables that begin with the characters `First_`.

The resulting data set is identical to the one created in Program 24.6.

24.7 Using a Retained Variable to "Remember" a Previous Value

This section describes another problem where using a retained variable greatly simplifies the code. For each patient in your Clinic data set, you want to check if the patient had a systolic blood pressure (SBP) over 140 during any of the visits. Here is the program:

Program 24.8: Using a Retained Variable to "Remember" a Previous Value

```
data Hypertension;
   set Learn.Clinic;
   by ID;

   retain HighBP;

   if first.ID then HighBP = 'No ';
   if SBP gt 140 then HighBP = 'Yes';
   if last.ID then output;

run;
```

This program uses the retained variable HighBP to "remember" if a patient ever had a systolic blood pressure over 140. As each new patient is processed, HighBP is set to **No**. Then, if any SBP is greater than 140, HighBP is set to **Yes**. Because this value is retained, it remains equal to **Yes** even if the SBP is less than 140 on all subsequent visits. When SAS reaches the last visit for each patient, an observation is written to the Hypertension data set. Here is a listing of Hypertension:

Figure 24.8: Listing of Data Set Hypertension

Listing of Data Set Hypertension

ID	VisitDate	Dx	HR	SBP	DBP	HighBP
101	02/25/2006	Cold	68	122	84	No
255	04/01/2006	Heart Problems	72	180	88	Yes
303	10/10/2006	Routine Visit	72	138	84	No
409	12/15/2006	Routine Visit	68	130	84	Yes
712	04/15/2006	Infection	56	118	72	No

Even though SAS operates one observation at a time, you can use the techniques in this chapter to perform computations between observations in a SAS data set.

24.8 Problems

Solutions to odd-numbered problems are located at the back of this book. Solutions to all problems are available to professors. If you are a professor, visit the book's companion website at support.sas.com/cody for information about how to obtain the solutions to all problems.

1. Data set DailyPrices contains a stock symbol and from 1 to 5 observations showing the price for each date. Here is a listing of this data set:

Listing of Data Set DailyPrices

Symbol	Date	Price
CSCO	01/01/2007	19.75
CSCO	01/02/2007	20.00
CSCO	01/03/2007	20.50
CSCO	01/04/2007	21.00
IBM	01/01/2007	76.00
IBM	01/02/2007	78.00
IBM	01/03/2007	75.00
IBM	01/04/2007	79.00
IBM	01/05/2007	81.00
LU	01/01/2007	2.55
LU	01/02/2007	2.53
AVID	01/01/2007	41.25
BAC	01/01/2007	51.00
BAC	01/02/2007	51.00
BAC	01/03/2007	51.20
BAC	01/04/2007	49.90
BAC	01/05/2007	52.10

Note: These prices are made up but reflect the approximate prices when the first edition of this book was written in 2007.

Create a listing showing the stock symbol, the date, and the price from the most recent date for each stock.

2. Create a temporary SAS data set containing the average price for each stock and the number of values used to create this average. In addition, this data set should contain the minimum and maximum price for each stock. A listing of this data set should look like this:

Listing of Summary

Symbol	Price_Mean	Price_N	Price_Min	Price_Max
AVID	41.2500	1	41.25	41.25
BAC	51.0400	5	49.90	52.10
CSCO	20.3125	4	19.75	21.00
IBM	77.8000	5	75.00	81.00
LU	2.5400	2	2.53	2.55

3. Create a temporary SAS data set with two variables, N_Days and Symbol, by writing a DATA step and using the SAS data set DailyPrices as input. N_Days represents the number of observations for each stock.

4. Repeat Problem 3 using PROC FREQ to count the number of observations for each unique stock symbol.

5. Using the SAS data set DailyPrices, compute the difference between the price on the last day minus the price on the first day. Omit any stocks that have only one observation. Do this using a DATA step and retained variables.

6. Repeat Problem 5 using the LAG or DIF function.

7. Using the SAS data set DailyPrices, compute day-to-day differences for all stocks that have more than one observation.

Chapter 25: Introducing the SAS Macro Language

25.1 Introduction

Although the SAS macro language is usually thought of as an advanced topic, there are aspects of this language that are useful and easy to use, even to the beginning or intermediate SAS programmer. This chapter gives you the tools to use macro variables and write simple macros. Macros are particularly useful if you want to make your SAS programs more flexible and allow them to be used in different situations without having to rewrite entire programs. Macro variables also provide a useful method for passing information from one DATA step to another. You can even select which procedures within a large program to run, depending on the day of the week or the values in your data.

If you want to learn more about the SAS macro language, I highly recommend two books. The first book, by Michele Burlew, might be more appropriate to beginning programmers (Michele M. Burlew, *SAS Macro Programming Made Easy, Third Edition* (Cary, NC: SAS Institute Inc., 2014). The second book, by Art Carpenter, covers more advanced topics. (Art Carpenter, *Carpenter's Complete Guide to the SAS Macro Programming Language, Third Edition* (Cary, NC: SAS Institute Inc., 2016).

25.2 Macro Variables: What Are They?

You may have seen programs with funny-looking variable names such as &VAR or &SYSDATE in them. Perhaps you have seen statements such as %LET or other statements containing percent signs. When you submit a SAS program for processing, before SAS starts to compile and then execute your program, it first checks for the existence of ampersands (&) or percent signs (%) in the program and calls in the macro processor if it sees one. In a sense, the macro processor works like the find-and-replace feature of Microsoft Word—it looks for all the macro variables (names starting with ampersands) and replaces them with text values. There are several ways to assign values to macro variables and you will see several described in this chapter.

25.3 Some Built-In Macro Variables

There are several built-in macro variables available to you as soon as you begin a SAS session. Two useful ones are &SYSDATE9 and &SYSTIME. As you may guess, the former stores the date you started your SAS session (in DATE9. format) and the latter stores the time the session started. As an example, you could use these automatic macro variables to include a date and time in a TITLE statement, like this:

Program 25.1: Using an Automatic Macro Variable to Include a Date and Time in a Title

```
title "The Date is &Sysdate9 - the Time is &Systime";
proc print data=Learn.Test_Scores noobs;
run;
```

If you started your SAS session at 10:02 on December 10, 2017, the macro processor would first resolve the macros variables (substitute text for the macro variable) before running the program. The listing would look like this:

Figure 25.1: Output from Program 25.1

The Date is 10DEC2017 - the Time is 10:02

ID	Score1	Score2	Score3
1	90	95	98
2	78	77	75
3	88	91	92

This is a good time to tell you that if you place a macro variable inside quotation marks, SAS resolves only macro variables that are inside **double** quotation marks. If you used single quotation marks in Program 25.1, the title would read:

Figure 25.2: Output If You Use Single Quotation Marks in the Title

The Date is &Sysdate9 - the Time is &Systime

ID	Score1	Score2	Score3
1	90	95	98
2	78	77	75
3	88	91	92

25.4 Assigning Values to Macro Variables with a %LET Statement

You can assign a value to a macro variable with a %LET statement. You usually place %LET statements in open code, that is, not inside a DATA or PROC step. Here is a simple example:

Program 25.2: Assigning a Value to a Macro Variable with a %LET Statement

```
%let Var_List = RBC WBC Chol;

title "Using a Macro Variable List";
proc means data=Learn.Blood n mean min max maxdec=1;
   var &Var_List;
run;
```

Every place in your program where you use the macro variable &Var_List, SAS substitutes the text RBC WBC Chol. This is a simple but useful trick when you need to repeat lists of variables.

Here is another example of using a %LET statement. Suppose you want to generate a data set containing random integers from 1 to 100. You want the program to be flexible so that you can choose how many random numbers to generate each time you run the program. Here is the program:

Program 25.3: Another Example of Using a %LET Statement

```
%let n = 3;

data Generate;
   do Subj = 1 to &n;
      x = ceil(100*rand('uniform'));
      output;
   end;
run;

title "Data Set with &n Random Numbers";
proc print data=Generate noobs;
run;
```

Notice that the %LET statement comes before the DATA step. When this program runs, each occurrence of `&n` is replaced with a `3`. While it would be easy to use the SAS Program Editor to change a 3 to some other value each time you run this program, imagine that this value was used many times in a longer program. By simply editing the single %LET statement, you can change the value of `n` everywhere in the program.

25.5 Demonstrating a Simple Macro

So far, you have seen macro variables used in a program. Now it's time to see an actual SAS macro. Macros can be entire SAS programs or just pieces of SAS code. You start a macro with a %MACRO statement and end it with a %MEND (macro end) statement. Here is an example.

You want to make Program 25.3 more general. Not only do you want to change the number of random numbers you generate, you want to change the starting and ending values of these numbers as well. The macro that follows does just that:

Program 25.4: Writing a Simple Macro

```
%macro Gen(n,Start,End);
   data Generate;
      do Subj = 1 to &n;
       x = int(rand('uniform') * (&End - &Start + 1) + &start);
         output;
      end;
   run;

   proc print data=generate noobs;
      title "Randomly Generated Data Set with &n Obs";
      title2 "Values are Integers from &Start to &End";
   run;

%mend Gen;
```

The macro name is GEN and there are three positional arguments: n, Start, and End. To generate four random integers from 1 to 100, you would call the macro like this:

```
%Gen(4,1,100)
```

The macro calling statement consists of a percent sign followed by the macro name. In this calling statement, you specify the values that you want to substitute for n, Start, and End in parentheses following the macro name. If you submit this macro with the option MPRINT turned on (`options mprint;`), you see the actual SAS code that was generated. Here is a listing of the SAS log (with the MPRINT option in effect):

Figure 25.3: Listing of the SAS Log with Option MPRINT Turned On

```
383  %gen(4,1,100)
MPRINT(GEN):   data Generate;
MPRINT(GEN):   do Subj = 1 to 4;
MPRINT(GEN):   x = int(rand('uniform') * (100 - 1 + 1) + 1);
MPRINT(GEN):   output;
MPRINT(GEN):   end;
MPRINT(GEN):   run;

NOTE: The data set WORK.GENERATE has 4 observations and 2 variables.
NOTE: DATA statement used (Total process time):
      real time           0.01 seconds
      cpu time            0.01 seconds


MPRINT(GEN):   proc print data=generate noobs;
MPRINT(GEN):   title "Randomly Generated Data Set with 4 Obs";
MPRINT(GEN):   title2 "Values are Integers from 1 to 100";
MPRINT(GEN):   run;
```

Notice that each macro variable was replaced by the value specified in the calling sequence.

Here is the output:

Figure 25.4: Output from Program 25.4

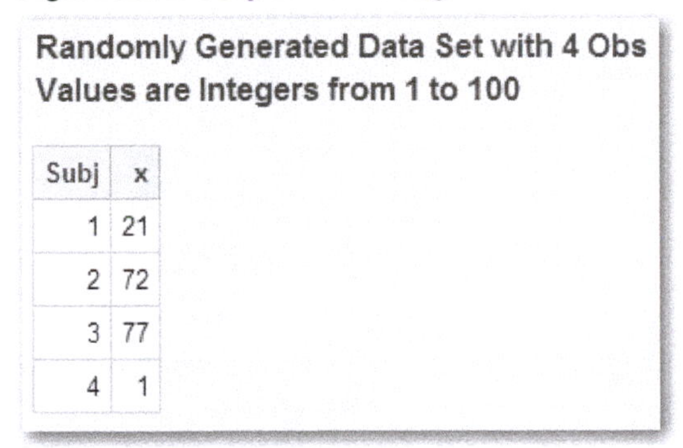

Randomly Generated Data Set with 4 Obs
Values are Integers from 1 to 100

Subj	x
1	21
2	72
3	77
4	1

When you call the macro, there is no semicolon at the end of the statement. If you look at the SAS code that was generated (in the previous SAS log), you can see that each SAS statement that was generated ends in a semicolon. Remember that SAS macros may generate pieces of SAS code, sometimes even parts of other SAS statements. Therefore, following the macro call with a semicolon could result in an error.

25.6 Describing Positional and Keyword Macro Parameters

The macro shown in Program 25.4, used what are called *positional parameters*. That is, each of the macro variables in parentheses following the macro name are separated by commas. When you call this macro, you also list the values that you want to substitute for each macro variable, in order, separated by commas.

There is an alternate (and superior way, at least the belief of this author) to specify the macro parameters. It is called *keyword parameters*. You follow each macro variable in the list of parameters with an equal sign. When you call the macro, you also list the name of the parameter, followed by an equal sign, followed by the value that you want to substitute for the macro variable.

What follows is a repeat of Program 25.4, rewritten to use keyword parameters. A recommended way to document these parameters is also included in the program.

Program 25.5: Program 25.4 Rewritten to Use Keyword Parameters

```
%macro Gen(n=,        /* number of random numbers */
           Start=,    /* Starting value           */
           End=       /* Ending value             */);
   /*********************************************
   Example: To generate 4 random numbers from
   1 to 100 use:
   %Gen(n=4, Start=1, End=100)
   *********************************************/
   data Generate;
      do Subj = 1 to &n;
       x = int(rand('uniform') * (&End - &Start + 1) + &start);
         output;
      end;
   run;
```

```
    proc print data=Generate noobs;
       title "Randomly Generated Data Set with &n Obs";
       title2 "Values are Integers from &Start to &End";
    run;
 %mend Gen;
```

You would call the macro this way:

```
%Gen(n=4, Start=1, End=100)
```

This method looks more tedious than simply listing the parameters with commas. However, here are some of the advantages of keyword parameters:

- It provides better documentation. Someone reading your calling routine can see what values are associated with each macro variable.
- You can list your parameters in any order. When you call positional parameters, they have to be listed in the same order that they were listed in the macro definition.
- You can supply default values when you write your macro. If you call the macro and leave out a parameter that you defined with a default value that value will be used—if you include a value for that parameter, it will override the default value.

As an example, suppose you want to write a macro that prints the first *n* observations of a data set. This macro also prints out a title stating how many observations were printed. Finally, you want to set a default value of 10 observations if you call the macro without specifying the value of *n*.

Such a macro is shown below in Program 25.6:

Program 25.6: Macro Demonstrating Keyword Parameters and Default Values

```
%macro Print(Dsn=, /* Name of the data set to print */
             n=10  /* The number of obs to print, default
                      is 10 */);

title "Listing of Data Set &Dsn";
title2 "First &n Observations";
proc print data=&Dsn(Obs=&n) noobs;
run;
%mend Print;
```

If you call the macro like this:

```
%Print(Dsn=Learn.Blood, n=3)
```

the output looks like this:

Figure 25.5: Output From Calling the Macro

Listing of Data Set Learn.Blood
First 3 Observations

Subject	Gender	BloodType	AgeGroup	WBC	RBC	Chol
1	Female	AB	Young	7710	7.40	258
2	Male	AB	Old	6560	4.70	.
3	Male	A	Young	5690	7.53	184

Notice that setting n = 3 overrode the default of 10. If you call the macro like this (without specifying a value for n):

```
%Print(Dsn=Learn.Blood)
```

You will obtain a listing of the first 10 observations in the Learn.Blood data set. Considering all the advantages of keyword parameters, this author will use this method in the remaining macros in this chapter.

Note: This is a change from the first edition where all the macro examples used positional parameters. Even "old dogs" can learn new tricks!

25.7 A Word about Tokens

The SAS macro processor must figure out where your macro variables start and end. In the previous macro, each time a macro variable was used, it was followed by a character that was not valid in a variable name. Suppose you submitted the following SAS code:

Program 25.7: Demonstrating a Problem with Resolving a Macro Variable

```
%let Prefix = abc;

data &Prefix123;
   x = 3;
run;
```

You are hoping to generate a data set called abc123. However, take a look at the SAS log:

Figure 25.6: SAS Log After Running Program 25.7

```
76
77     %let Prefix = abc;
78
WARNING: Apparent symbolic reference PREFIX123 not resolved.
79        data &Prefix123;
                -
                22
                200
ERROR 22-322: Syntax error, expecting one of the following: a name,
              a quoted string, /, ;, _DATA_, _LAST_, _NULL_.

ERROR 200-322: The symbol is not recognized and will be ignored.

80           x = 3;
81        run;
```

This message tells you that the macro processor cannot resolve (find) a macro variable with the name PREFIX123. SAS assumes that the macro variable name is PREFIX123 since this is a valid macro variable name. To tell SAS that the macro variable name is &PREFIX, you use a period to indicate where the macro variable name ends, like this:

Program 25.8: Program 25.7 Corrected

```
%let Prefix = abc;

data &prefix.123;
   x = 3;
run;
```

The period following the word Prefix tells the macro processor that &PREFIX is the macro variable name, not &PREFIX123. Notice that the SAS log no longer shows an error:

Figure 25.7: SAS Log after Running Program 25.8

```
82
83     %let Prefix = abc;
84
85       data &Prefix.123;
86          x = 3;
87       run;

NOTE: The data set WORK.ABC123 has 1 observations and 1 variables.
NOTE: DATA statement used (Total process time):
      real time           0.00 seconds
      cpu time            0.00 seconds
```

Notice that the value `ABC` was substituted for the macro variable `&Prefix` to generate the data set name `work.ABC123`.

25.8 Another Example of Using a Macro Variable as a Prefix

Suppose you want to use a macro variable to specify a libref (the first part of a two-part data set name). You might be tempted to write a program like this:

Program 25.9: Using a Macro Variable as a Prefix (Incorrect Version)

```
%let libref = Learn;

proc print data=&libref.Survey;
   title "Listing of Survey";
run;
```

Here's what happens when you try to run this program:

Figure 25.8: SAS Log after Running Program 25.9

```
94
95       %let libref = Learn;
96
97       proc print data=&libref.Survey;
ERROR: File WORK.LEARNSURVEY.DATA does not exist.
98          title "Listing of Survey";
99       run;
```

What happened? The period between the &libref and the name Survey was eaten up by the macro processor. It told the macro processor that the name of the macro variable was &libref. With the period removed, SAS was looking to create a data set called Work.LearnSurvey. The solution (as strange as it looks) is to use two periods: one to tell the macro processor where the macro variable name ends and the other to separate the libref from the data set name. The corrected program is as follows:

Program 25.10: Using a Macro Variable as a Prefix (Corrected Version)

```
%let libref = learn;

proc print data=&libref..survey;
   title "Listing of SURVEY";
run;
```

Here is the SAS log after running this program:

Figure 25.9: SAS Log After Running Program 25.10

```
238      %let libref = Learn;
239      proc print data=&libref..Survey;
240         title "Listing of Survey";
241      run;

NOTE: There were 7 observations read from the data set LEARN.SURVEY.
NOTE: PROCEDURE PRINT used (Total process time):
      real time           0.03 seconds
      cpu time            0.01 seconds
```

You can see that SAS resolved the macro variable and the result was a data set called Learn.Survey.

25.9 Using a Macro Variable to Transfer a Value between DATA Steps

Macro variables defined outside of a macro are, by default, global. That is, they maintain their value during a SAS session. This makes them a very useful tool for transferring values between DATA steps.

As an example, suppose you ran PROC MEANS to compute the means of the variables RBC and WBC from the Blood data set and wanted to compare each individual value of RBC and WBC against the mean values. One way to do this is to include a conditional SET statement in your DATA step, like this:

```
if _n_ = 1 then set Means;
```

Here, Means is the one-observation data set produced by PROC MEANS. Another way, which we describe next, uses macro variables to hold the mean values. Here is the program:

Program 25.11: Using Macro Variables to Transfer Values from One DATA Step to Another

```
proc means data=learn.Blood noprint;
   var RBC WBC;
   output out=Means mean= / autoname;
run;

data _null_;
   set Means;
   call symputx('AveRBC',RBC_Mean);
   call symputx('AveWBC',WBC_Mean);
run;

data New;
   set learn.Blood(obs=5 keep=Subject RBC WBC);
   Per_RBC = RBC / &AveRBC;
   Per_WBC = WBC / &AveWBC;
   format Per_RBC Per_WBC percent8.;
run;
```

PROC MEANS creates a data set (called Means in this example) consisting of one observation and two variables (RBC_Mean and WBC_Mean), which represent the mean values of RBC and WBC, respectively.

Note: As a reminder, the AUTONAME option in the OUTPUT statement of PROC MEANS names the variables in the output data set automatically by using the variable names and adding an underscore, followed by the name of the statistic. That is why the two means are called RBC_Mean and WBC_Mean.

The DATA _NULL_ step uses a CALL routine called SYMPUTX. This CALL routine assigns the value of a DATA step variable to a macro variable. The first argument in this CALL routine is the name of a macro variable, and the second argument is the name of a DATA step variable. You cannot use a %LET statement here because you don't know the value of RBC_Mean or WBC_Mean ahead of time.

When this DATA step executes, each of the two macro variables (RBC_Ave and WBC_Ave) will be the two mean values. The values of these two macro variables are not available in the same DATA step, so a final DATA step is needed.

In the final DATA step, the two variables Percent_RBC and Percent_WBC are computed by dividing the value in the current observation by the mean value. You may wonder why this statement doesn't multiply the resulting value by 100 to create a percentage. The SAS format PERCENT not only adds a percent sign, it multiplies by 100 as well.

Here is a listing of data set Compare_to_Mean:

Figure 25.10: Partial Listing of Data Set Compare_to_Mean

Listing of Data Set Compare_to_Mean

Subject	WBC	RBC	Percent_RBC	Percent_WBC
1	7710	7.40	135%	109%
2	6560	4.70	86%	93%
3	5690	7.53	137%	81%
4	6680	6.85	125%	95%
5	.	7.72	141%	.
6	6140	3.69	67%	87%
7	6550	4.78	87%	93%
8	5200	4.96	90%	74%
9	.	5.66	103%	.
[illegible]	7710	5.5[illegible]	10[illegible]%	109%

Hopefully, this very brief introduction to SAS macros has shown you that the SAS macro language is not as scary as you might have thought and has given you the courage to try using them in your own programs.

25.10 Problems

Solutions to odd-numbered problems are located at the back of this book. Solutions to all problems are available to professors. If you are a professor, visit the book's companion website at support.sas.com/cody for information about how to obtain the solutions to all problems.

1. Use PROC PRINT to list the first five observations from data set Stocks. Use the system macro variables &SYSDAY (returns the day of the week), &SYSDATE, and &SYSTIME to customize the title (see the following example, which was run on a Friday):

   Listing Produced on Sunday, 10DEC2017 at 10:02

Date	Price
01/01/2017	$34
01/02/2017	$35
01/03/2017	$39
01/04/2017	$30
01/05/2017	$35

2. Rewrite the following program using %LET to assign the starting and ending values of the DO loop to macro variables. Be sure these values also are printed in the PROC PRINT output.

```
data Sqrt_Table;
   do n = 1 to 5;
      Sqrt_n = sqrt(n);
      output;
   end;
run;

title "Square Root Table from 1 to 5";
proc print data=Sqrt_Table noobs;
run;
```

3. Create a macro (call it PRINT_N) to produce a listing of the first *n* observations from a selected data set. Use the following program as a guide, replacing the data set name and the number of observations to print as calling arguments to the macro.

```
title "Listing of the first 5 Observations from "
      "Data set Stocks";
proc print data=Learn.Stocks(obs=5) noobs;
run;
```

Use the macro to list the first four observations from SAS data set Bicycles. Your listing should look like this:

Listing of the First 4 Observations from Data Set Learn.Bicycles

Country	Model	Manuf	Units	UnitCost	TotalSales
USA	Road Bike	Trek	5000	$2,200	$11,000
USA	Road Bike	Cannondale	2000	$2,100	$4,200
USA	Mountain Bike	Trek	6000	$1,200	$7,200
USA	Mountain Bike	Cannondale	4000	$2,700	$10,800

4. Turn the following program into a macro (call it Stats), making it more general. Use as calling arguments the input data set name (Dsn), the CLASS variables (Class), and variables listed in the VAR statement (Vars):

```
title "Statistics from Data Set Learn.Bicycles";
proc means data=Learn.Bicycles n mean min max maxdec=1;
   class Country;
   var Units TotalSales;
run;
```

Test your macro by calling it like this:

```
%Stats(Dsn=Learn.Bicycles, Class=Country, Vars=Units TotalSales)
```

5. List three variables (TimeMile, RestPulse, and MaxPulse) from the data set Fitness. In addition to these three variables, compute three new variables (call them P_TimeMile, P_RestPulse, and P_MaxPulse) that represent these three values as a percentage of the mean for all subjects in the data set. Do this by first computing the three means using PROC MEANS. Next, in a DATA _NULL_ step, use CALL SYMPUTX to create three macro variables representing the means of these three values. Finally, in a DATA step, use these three macro variables in assignment statements to compute the percentage values.

Chapter 26: Introducing the Structured Query Language

26.1 Introduction

PROC SQL (structured query language—usually pronounced *sequel*) offers an alternative to the DATA step for querying and combining SAS data sets. There are some tasks that PROC SQL can perform much better and easier than the DATA step. Other tasks may be easier or more efficient using a DATA step. You may also be more familiar with SQL versus the DATA step or vice versa. The best advice is to learn both and use the tool that works best for you in each situation.

This chapter only touches on the basics of PROC SQL. There are some excellent books published by SAS that you will probably want to own. Please check SAS Help Center at http://go.documentation.sas.com or books in the SAS Press series for more information on PROC SQL

26.2 Some Basics

Programmers familiar with SAS syntax may have some difficulty getting started with PROC SQL. For example, you use commas, not spaces, to separate variable and data set names in PROC SQL. You may also find yourself putting semicolons where they don't belong. Let's start out with some simple examples of querying a data set and creating a SAS data set by subsetting observations from a larger SAS data set.

Before we show you a program, a few words on terminology are in order. The following table lists SAS terms and the corresponding SQL terms:

SAS Term	SQL Equivalent
Data set	Table
Observation	Row
Variable	Column

For the first few examples, we will be working with a SAS data set called Health. Here is a listing:

Figure 26.1: Listing of Data Set Health

Listing of Data Set Health

Subj	Height	Weight
001	68	155
003	74	250
004	63	110
005	60	95

If you want to print a subset of this data set, selecting all subjects with heights over 65 inches, you could submit the following query:

Program 26.1: Demonstrating a Simple Query from a Single Data Set

```
title "Subjects from Health with Height > 65";
proc sql;
   select Subj,
          Height,
          Weight
   from Learn.Health
   where Height gt 65;
quit;
```

This query starts with a SELECT keyword where you list the variables you want. Notice that the variables in this list are separated by commas (spaces do not work). The keyword FROM names the data set you want to read. Finally, a WHERE clause describes the particular subset you want.

SELECT, FROM, and WHERE form a single query, which you end with a single semicolon. In this example, you are not creating an output data set, so, by default, the result of this query is sent as a listing in the Output window (or whatever output location you have specified). Finally, the query ends with a QUIT statement. You do not need a RUN statement because PROC SQL executes as soon as a complete query has been specified. If you don't include a QUIT statement, PROC SQL remains in memory for another query.

Here is the output that resulted from this program:

Figure 26.2: Output from Program 26.1

Subjects from Health with Height > 65

Subj	Height	Weight
001	68	155
003	74	250

If you want to select all the variables from a data set, you can use an asterisk (*), like this:

Program 26.2: Using an Asterisk to Select all the Variables in a Data Set

```
proc sql;
   select *
   from Learn.Health
   where Height gt 65;
quit;
```

If you want the result of the query to be stored in a SAS data set, include a CREATE TABLE statement, like this:

Program 26.3: Using PROC SQL to Create a SAS Data Set

```
proc sql;
   create table Height65 as
   select *
   from Learn.Health
   where Height gt 65;
quit;
```

26.3 Joining Two Tables (Merge)

Two tables, Health (used in the last example) and Demographic, are used to demonstrate various ways to perform joins using PROC SQL. Here are listings of these two tables:

Figure 26.3: Listing of Tables Health and Demographic

Listing of Data Set Health

Subj	Height	Weight
001	68	155
003	74	250
004	63	110
005	60	95

Listing of Data Set Demographic

Subj	DOB	Gender	Name
001	10/15/1960	M	Friedman
002	08/01/1955	M	Stern
003	12/25/1988	F	McGoldrick
005	05/28/1949	F	Chien

You can select variables from two tables by listing all the variables of interest in the SELECT clause and listing the two data sets in the FROM clause. If a variable has the same name in both data sets, you need a way to distinguish which data set to use. Here's how it's done:

Program 26.4: Joining Two Tables (Cartesian Product)

```
title "Demonstrating a Cartesian Product";
proc sql;
   select Health.Subj,
          Demographic.Subj,
          Height,
          Weight,
          Name,
          Gender
   from Learn.Health,
        Learn.Demographic;
quit;
```

Because the column Subj is in both tables, you prefix the variable name with the table name. You will see in a minute that you can simplify this a bit. The result from this query is called a *Cartesian product* and it represents all possible combinations of rows from the first table with rows from the second table. The listing shows two columns, both with the heading of Subj.

Here is a listing of the result:

Figure 26.4: Output from Program 26.4

Demonstrating a Cartesian Product

Subj	Subj	Height	Weight	Name	Gender
001	001	68	155	Friedman	M
001	002	68	155	Stern	M
001	003	68	155	McGoldrick	F
001	005	68	155	Chien	F
003	001	74	250	Friedman	M
003	002	74	250	Stern	M
003	003	74	250	McGoldrick	F
003	005	74	250	Chien	F
004	001	63	110	Friedman	M
004	002	63	110	Stern	M
004	003	63	110	McGoldrick	F
004	005	63	110	Chien	F
005	001	60	95	Friedman	M
005	002	60	95	Stern	M
005	003	60	95	McGoldrick	F
005	005	60	95	Chien	F

If you use this same query to create a SAS data set, there will only be one variable called Subj. If you would like to keep both values of Subj from each data set, you can rename the columns, like this:

Program 26.5: Renaming the Two Subj Variables

```
title "Renaming the Two Subj Variables";
proc sql;
   select Health.Subj as Health_Subj,
          Demographic.Subj as Demog_Subj,
          Height,
          Weight,
          Name,
          Gender
   from Learn.Health,
        Learn.Demographic;
quit;
```

Running this query results in the following:

Figure 26.5: Partial Output from Program 26.5

Renaming the Two Subj Variables

Health_Subj	Demog_Subj	Height	Weight	Name	Gender
001	001	68	155	Friedman	M
001	002	68	155	Stern	M
001	003	68	155	McGoldrick	F
001	005	68	155	Chien	F
003	001	74	250	Friedman	M
003	002	74	250	Stern	M
003	003	74	250	McGoldrick	F

Notice that the two subject columns are renamed.

A Cartesian product is especially useful when you want to perform matches between names in two tables that are similar (sometimes called a fuzzy merge). The number of rows in this table is the number of rows in the first table times the number of rows in the second table. In practice, you will want to add a WHERE clause to restrict which rows to select. In the program that follows, we add a WHERE clause to select only those rows where the subject number is the same in the two tables.

Besides adding a WHERE clause, the next program also shows how to distinguish between two columns both with a heading of Subj. Finally, this next program uses a simpler method of naming the two Subj variables in the SELECT clause. Here it is:

Program 26.6: Using Aliases to Simplify Naming Variables

```
title "Matching Subj Numbers from Both Tables";
proc sql;
   select H.Subj as Subj_Health,
          D.Subj as Subj_Demog,
          Height,
          Weight,
          Name,
          Gender
   from Learn.Health as H,
        Learn.Demographic as D
   where H.Subj eq D.Subj;
quit;
```

First take a look at the FROM clause. To make it easier to name variables with the same name from different tables, you create aliases for each of the tables, `H` and `D`, in this program. You can use these aliases as a prefix in the SELECT clause (`H.Subj` and `D.Subj`). In this program, a WHERE clause was added, restricting the result to rows where the subject number is the same in both tables. Here is the result:

Figure 26.6: Matching Subj Numbers from Both Tables

Matching Subj Numbers from Both Tables

Subj_Health	Subj_Demog	Height	Weight	Name	Gender
001	001	68	155	Friedman	M
003	003	74	250	McGoldrick	F
005	005	60	95	Chien	F

Only subjects who are in both tables are listed here. In SQL terminology, this is called an *inner join.* It is equivalent to a merge in a DATA step where each of the two data sets contributes to the merge.

Just so this is clear, here is the same (well almost—in PROC SQL, you get two Subj variables) result using a DATA step:

Program 26.7: Performing an Inner Join Using a DATA Step

```
proc sort data=Learn.Health out=Health;
   by Subj;
run;

proc sort data=Learn.Demographic out=Demographic;
   by Subj;
run;

data Inner;
   merge Health(in=in_Health)
         Demographic(in=in_Demographic);
   by Subj;
   if in_Health and in_Demographic;
run;

title "Performing an Inner Join Using a DATA Step";
proc print data=Inner;
   id Subj;
run;
```

Isn't it nice that you don't have to sort the data sets first when you use SQL? (Although PROC SQL does the sorts for you.) Below is a listing of the data set Inner:

Figure 26.7: Output from Program 26.7

Performing an Inner Join Using a DATA Step

Subj	Height	Weight	DOB	Gender	Name
001	68	155	10/15/1960	M	Friedman
003	74	250	12/25/1988	F	McGoldrick
005	60	95	05/28/1949	F	Chien

This is identical to the PROC SQL table except for the fact that there is only one Subj variable.

26.4 Left, Right, and Full Joins

An alternative to Program 26.6 is to separate the two table names with the term INNER JOIN, like this:

Program 26.8: Performing an Inner Join

```
title "Demonstrating an Inner Join (Merge)";
proc sql;
   select H.Subj as Subj_Health,
          D.Subj as Subj_Demog,
          Height,
          Weight,
          Name,
          Gender
   from Learn.Health as H inner join
        Learn.Demographic as D
   on H.Subj eq D.Subj;
quit;
```

One of the rules of SQL is that when you use the keyword JOIN to join two tables, you use an ON clause instead of a WHERE clause. (You may further subset the result with a WHERE clause.)

If you write your inner join this way, it is easy to replace the term INNER JOIN with one of the following: LEFT JOIN, RIGHT JOIN, or FULL JOIN.

A left join includes all the rows from the first (left) table and those rows from the second table where there is a corresponding value in the first table. A right join includes all rows from the second (right) table and only matching rows from the first table. A full join includes all rows from both tables (equivalent to a merge in a DATA step). The following program demonstrates these three joins:

Program 26.9: Demonstrating a Left, Right, and Full Join

```
proc sql;
   title "Left Join";
   select H.Subj as Subj_Health,
          D.Subj as Subj_Demog,
          Height,
          Gender
   from Learn.Health as H left join
        Learn.demographic as D
   on H.Subj eq D.Subj;

   title "Right Join";
   select H.Subj as Subj_Health,
          D.Subj as Subj_Demog,
          Height,
          Gender
   from Learn.Health as H right join
        Learn.demographic as D
   on H.Subj eq D.Subj;

   title "Full Join";
   select H.Subj as Subj_Health,
          D.Subj as Subj_Demog,
          Height,
          Gender
   from Learn.health as H full join
        Learn.demographic as D
   on H.Subj eq D.Subj;
quit;
```

The results are as follows:

Figure 26.8: Output from Program 26.9

Left Join

Subj_Health	Subj_Demog	Height	Gender
001	001	68	M
003	003	74	F
004		63	
005	005	60	F

Right Join

Subj_Health	Subj_Demog	Height	Gender
001	001	68	M
	002	.	M
003	003	74	F
005	005	60	F

Full Join

Subj_Health	Subj_Demog	Height	Gender
001	001	68	M
	002	.	M
003	003	74	F
004		63	
005	005	60	F

Inspection of this output should help make clear the distinctions among the different types of joins.

26.5 Concatenating Data Sets

In a DATA step, you concatenate two data sets by naming them in a single SET statement. In PROC SQL, you use a UNION operator to concatenate selections from two tables. Unlike the DATA step, there are different "flavors" of UNION operators. Here is a summary.

Operator	Description
Union	Matches by **column position** (not column name) and drops duplicate rows
Union All	Matches by **column position** but does not drop duplicate rows.
Union Corresponding	Matches by **column name** and drops duplicate rows.
Union All Corresponding	Matches by **column name** and does not drop duplicate rows
Except	Matches by **column name** and drops rows found in both tables
Intersection	Matches by **column name** and keeps unique rows in both tables

The UNION ALL CORRESPONDING operator is equivalent to naming two data sets in a SET statement in a DATA step.

It is very important to realize that a UNION operator without the keyword CORRESPONDING results in the two data sets being concatenated by column position, not column name. This is illustrated in the examples here.

The table New_Members was created to illustrate various types of unions. Here is the listing:

Figure 26.9: Listing of Data Set New_Members

Listing of Data Set New_Members

Subj	Gender	Name	DOB
010	F	Ostermeier	03/05/1977
013	M	Brown	06/07/1999

For reference, here is the listing of data set Demographic:

Figure 26.10: Listing of Demographic

Listing of Data Set Demographic

Subj	DOB	Gender	Name
001	10/15/1960	M	Friedman
002	08/01/1955	M	Stern
003	12/25/1988	F	McGoldrick
005	05/28/1949	F	Chien

Suppose you want to add these new members to the Demographic data set and call the new data set Complete_List. Here's how to do it using PROC SQL. (Notice that the columns are not in the same order as the Demographic data set.)

Program 26.10: Concatenating Two Tables

```
proc sql;
   create table Complete_List as
   select *
   from Learn.Demographic union all corresponding
   select *
   from Learn.New_Members
quit;
```

The resulting table contains all the rows from Demographic followed by all the rows from New_Members.

Figure 26.11: Output from Program 26.10

Listing of Data Set Complete_List

Subj	DOB	Gender	Name
001	10/15/1960	M	Friedman
002	08/01/1955	M	Stern
003	12/25/1988	F	McGoldrick
005	05/28/1949	F	Chien
010	03/05/1977	F	Ostermeier
013	06/07/1999	M	Brown

If you leave out the keyword CORRESPONDING, here is the result (SAS log):

Figure 26.12: Resulting Log when CORRESPONDING is Omitted

```
90       proc sql;
91          create table Complete_List as
92          select *
93          from Learn.Demographic union all
94          select *
95          from Learn.New_Members
96       quit;
ERROR: Column 2 from the first contributor of UNION ALL is not the same
       type as its counterpart from the second.
ERROR: Column 4 from the first contributor of UNION ALL is not the same
       type as its counterpart from the second.
```

If, by chance, the data types match column by column in the two data sets, SQL will perform the union. To understand this, here is another data set, New_Members_Order, where the order of the columns is changed:

Figure 26.13: Listing of New_Members_Order

Listing of Data Set New_Members_Order

Name	DOB	Gender	Subj
Ostermeier	03/05/1977	F	010
Brown	06/07/1999	M	013

Each of the four columns of New_Members_Order has the same data type (character or numeric) as the four columns of Demographic. So, if you omit the CORRESPONDING keyword when performing a union of these two data sets, you have the following result:

Figure 26.14: Result of Omitting the Keyword CORRESPONDING

Listing of Data Set Complete_List
With the Keyword CORRESPONDING Omitted

Subj	DOB	Gender	Name
001	10/15/1960	M	Friedman
002	08/01/1955	M	Stern
003	12/25/1988	F	McGoldrick
005	05/28/1949	F	Chien
Ostermeier	03/05/1977	F	010
Brown	06/07/1999	M	013

You can now see why you need to choose the correct UNION operator when concatenating two data sets.

26.6 Using Summary Functions

You can use functions such as MEAN and SUM to create new variables that represent means or sums of other variables. You can also create new variables within the query.

The following program shows one of the strengths of PROC SQL, which is the ability to add a summary variable to an existing table.

Suppose you want to express each person's height in the Health data set as a percentage of the mean height of all the subjects. Using a DATA step, you would first use PROC MEANS to create a data set containing the mean height. You would then combine this with the original data set and perform the calculation. (See the section about combining detail and summary data in Chapter 10 for an example.) PROC SQL makes this task much easier. Let's take a look.

Program 26.11: Using a Summary Function in PROC SQL

```
proc sql;
   select Subj,
          Height,
          Weight,
          mean(Height) as Ave_Height,
          100*Height/calculated Ave_Height as
             Percent_Height
   from Learn.Health
quit;
```

The mean height is computed using the MEAN function. This value is also given the variable name Ave_Height. When you use this variable in a calculation, you need to precede it with the keyword CALCULATED, so that PROC SQL doesn't look for the variable in one of the input data sets. Here is the result:

Figure 26.15: Output from Program 26.11

Demonstrating a Summary Function in PROC SQL

Subj	Height	Weight	Ave_Height	Percent_Height
001	68	155	66.25	102.6415
003	74	250	66.25	111.6981
004	63	110	66.25	95.09434
005	60	95	66.25	90.56604

Notice how much easier this is using PROC SQL compared to a DATA step.

26.7 Demonstrating the ORDER Clause

PROC SQL can also sort your table if you use an ORDER clause. For example, if you want the subjects in the Health table in height order, use the following:

Program 26.12: Demonstrating the ORDER Clause

```
title "Listing in Height Order";
proc sql;
   select Subj,
          Height,
          Weight
   from Learn.Health
   order by Height;
quit;
```

The result (not shown) is a listing of the variables Subj, Height, and Weight in order of increasing Height.

26.8 An Example of Fuzzy Matching

One of the strengths of PROC SQL is its ability to create a Cartesian product. As mentioned earlier in this chapter, a Cartesian product is a pairing of every row in one table with every row in another table. Here are two tables: Demographic (used in many of the other examples in this chapter) and Insurance.

Figure 26.16: Listing of Data Sets Demographic and Insurance

Listing of Data Set Demographic

Subj	DOB	Gender	Name
001	10/15/1960	M	Friedman
002	08/01/1955	M	Stern
003	12/25/1988	F	McGoldrick
005	05/28/1949	F	Chien

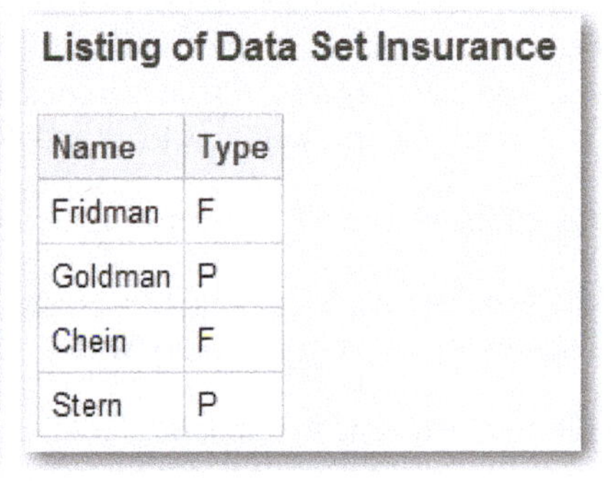

Listing of Data Set Insurance

Name	Type
Fridman	F
Goldman	P
Chein	F
Stern	P

You want to join (merge) these tables by Name, allowing for slight misspellings of the names. Here is an SQL query that does just that:

Program 26.13: Using PROC SQL to Perform a Fuzzy Match

```
title "Example of a Fuzzy Match";
proc sql;
   select Subj,
          Demographic.Name,
          Insurance.Name
   from Learn.Demographic,
        Learn.Insurance
   where spedis(Demographic.Name,Insurance.Name) le 25;
quit;
```

The SPEDIS (spelling distance) function allows for misspellings (see Chapter 12). The WHERE clause operates on every combination of names from the two tables and selects those names that are within a spelling distance of 25. In practice, you would want to compare other variables such as Gender and DOB between two files to increase the likelihood that a valid match is being made. Take a look at the listing that follows to see which names were matched by this program:

Figure 26.17: Output from Program 26.13

Example of a Fuzzy Match

Subj	Name	Name
001	Friedman	Fridman
005	Chien	Chein
002	Stern	Stern

You could easily modify this program so that the output would not include names that match exactly.

We have only touched the surface of what you can do with SQL. Hopefully, this introduction to SQL will encourage you to learn more.

26.9 Problems

Solutions to odd-numbered problems are located at the back of this book. Solutions to all problems are available to professors. If you are a professor, visit the book's companion website at support.sas.com/cody for information about how to obtain the solutions to all problems.

1. Use PROC SQL to list all the observations from data set Inventory where Price is greater than 20.

2. Repeat Problem 1, except use PROC SQL to create a new, temporary SAS data set (Price20) containing all observations from Inventory where the Price is greater than 20.

3. Use PROC SQL to create a new, temporary, SAS data set (N_Sales) containing the observations from Sales where Region has a value of `North`. Include only the variables Name and TotalSales in the new data set.

4. Data sets Inventory and Purchase are shown here:

Listing of Data Set Inventory

Model	Price
M567	$23.50
S888	$12.99
L776	$159.98
X999	$29.95
M123	$4.59
S776	$1.99

Listing of Data Set Purchase

CustNumber	Model	Quantity
101	L776	1
102	M123	10
103	X999	2
103	M567	1

Use PROC SQL to list all purchased items showing the Cust Number, Model, Quantity, Price, and a new variable, Cost, equal to the Price times the Quantity.

5. Data sets Left and Right are shown here. Use PROC SQL to create a new, temporary SAS data set (Both) containing Subj, Height, Weight, and Salary. Do this three ways: first, include only those subjects who are in both data sets, second, include all subjects from both data sets, and third, include only those subjects who are in data set Left.

Listing of Data Set Left

Subj	Height	Weight
001	68	155
002	75	220
003	65	99
005	79	266
006	70	190
009	61	122

Listing of Data Set Right

Subj	Salary
001	$46,000
003	$67,900
004	$28,200
005	$98,202
006	$88,000
007	$57,200

6. Write the necessary PROC SQL statements to accomplish the same goal as the program here:

```
data allproducts;
   set learn.inventory learn.newproducts;
run;
```

7. Write the necessary PROC SQL statements to accomplish the same goal as the program here:

```
data third;
   set learn.first learn.second;
run;
```

Be careful! The order of the variables is not the same in both data sets. Also, some subject numbers are in both data sets.

8. Use PROC SQL to list the values of RBC (red blood cells) and WBC (white blood cells) from the Blood data set. Include two new variables in this list: Percent_RBC and Percent_WBC. These variables are the values of RBC and WBC expressed as a percentage of the mean value for all subjects. The first few observations in this list should look like this:

Listing of Data Set Right

RBC	WBC	Percent_RBC	Percent_WBC
7.4	7710	134.9497	109.4708
4.7	6560	85.71127	93.14248
7.53	5690	137.3204	80.78974
6.85	6680	124.9196	94.8463
7.72	.	140.7853	.
3.69	6140	67.29247	87.17909

9. In a similar manner to Problem 8, use the Blood data set to create a new, temporary SAS data set (Percentages) containing the variables Subject, RBC, WBC, MeanRBC, MeanWBC, Percent_RBC, and Percent_WBC.

10. Run the program here to create two temporary SAS data sets, XXX and YYY.

```
data xxx;
   input NameX : $15. PhoneX : $13.;
datalines;
Friedman (908)848-2323
Chien (212)777-1324
;

data yyy;
   input NameY : $15. PhoneY : $13.;
datalines;
Chen (212)888-1325
Chambliss (830)257-8362
Saffer (740)470-5887
;
```

Then write the PROC SQL statements to perform a fuzzy match between the names in each data set. List the observations where the names are within a spelling distance (the SPEDIS function) of 25. The result should be only one observation, as follows:

Listing of Fuzzy

Obs	NameX	PhoneX	NameY	PhoneY
1	Chien	(212)777-1324	Chen	(212)888-1325

Chapter 27: Introducing Perl Regular Expressions

27.1 Introduction

Regular expressions were first introduced in a language called Perl, developed for UNIX and Linux systems. Perl regular expressions were added to SAS starting with SAS 9. They are used to describe text patterns. For example, you can write an expression that matches any Social Security number (three digits, a dash, two digits, a dash, and four digits). Therefore, you can use a regular expression to test whether a specific pattern is present.

27.2 Describing the Syntax of Regular Expressions

There are entire books devoted to regular expressions. The goal of this chapter is to describe some basic aspects of regular expressions and provide some examples of how they can be used.

A regular expression (called a *regex* by programmers) starts with a delimiter (most often a forward slash), followed by a combination of characters and metacharacters. Metacharacters describe character classes such as all digits or all punctuation marks. The expression ends with the same delimiter that you started with. For example, a regular expression for a Social Security number is:

```
/\d\d\d-\d\d-\d\d\d\d/
```

`\d` is the metacharacter for any digit. The two dashes in the expression are just that—dashes. Any character that is not defined as a special character, such as the dashes in the preceding expression, is a literal character. Even spaces count as characters in a regular expression.

Because writing `\d` four times is tedious, this expression can be rewritten as:

```
/\d\d\d-\d\d-\d{4}/
```

You can probably guess that the {4} following the `\d` says to repeat `\d` four times.

You can create sets of characters using square brackets ([and]). All the uppercase letters are represented by [A-Z]. All uppercase and lowercase letters are represented by [A-Za-z].

Likewise, the digits 0-9 can be represented by [0-9]. If you are using ASCII, you can also use `\d` instead of [0-9].

In SAS, all the regular expression functions begin with the letters PRX, which stands for Perl Regular Expression). Because this chapter is an introduction to the PRX functions, only two of these functions will be described. For a more detailed discussion of the SAS PRX functions, please look at *SAS Functions by Example*, 2nd edition, written by this author and published by SAS Press.

27.3 Testing That Social Security Numbers Are in Standard Form

Before a discussion of metacharacters and other details of regular expressions, let's look at a simple program to verify that a list of Social Security numbers conforms to the standard form. Here is such a program:

Program 27.1: Using a Regex to Test Social Security Values

```
title "Checking Social Security Numbers Using a Regular Expression";

data _null_;
   file print;
   input SS $11.;
   if not prxmatch("/\d\d\d-\d\d-\d\d\d\d/",SS) then
      put "Error for SS Number " SS;
datalines;
123-45-6789
123456789
123-ab-9876
999-888-7777
;
```

You may not be familiar with the DATALINES statement. Instead of writing an external file and using an INFILE statement, you can place your data right after the DATALINES statement. The INPUT statement will read the values as if they were in an external file.

The PRXMATCH function takes two arguments. The first argument is a regular expression, and the second argument is the string that you are examining. If the string contains a pattern described by the regular expression, the function returns the starting position for the pattern. If the pattern described by the regular expression is not found in the string, the function returns a zero. In Program 27.1, the regular expression is describing a pattern of a typical Social Security number.

Here is the output:

Figure 27.1: Output from Program 27.1

Checking Social Security Numbers Using a Regular Expression

```
Error for SS Number 123456789
Error for SS Number 123-ab-9876
Error for SS Number 999-888-777
```

Each of the numbers listed does not conform to the standard Social Security number format.

27.4 Checking for Valid ZIP Codes

You can use a program similar to Program 27.1 to verify that an address contains the valid form of a ZIP code, either a five-digit code or a five-digit code followed by a dash, followed by four more digits (ZIP code +4).

A regular expression to check either the five digit or ZIP+4 code is:

```
/\d{5}(-\d{4})?/
```

The first part of this expression is pretty clear—it says to search for five digits. The expression in parenthesis matches a dash followed by four digits. The question mark following the parenthesis means to search for zero or 1 occurrences of the previous expression. Therefore, this expression matches either of the two valid US ZIP code formats.

You can run the following program to test this expression:

Program 27.2: Testing the Regular Expression for US ZIP Codes

```
title "Testing the Regular Expression for US ZIP Codes";
data _null_;
   file print;
   input Zip $10.;
  if not prxmatch("/\d{5}(-\d{4})?/",Zip) then
      put "Invalid ZIP Code " Zip;
datalines;
12345
78010-5049
12Z44
ABCDE
08822
;
```

This program reads in the ZIP code (allowing for up to 10 characters) and prints out an error message for any code that does not match the regular expression. Here is the output from Program 27.2:

Figure 27.2: Output from Program 27.2

```
Testing the Regular Expression for US ZIP Codes

Invalid ZIP Code 12Z44
Invalid ZIP Code ABCDE
```

Notice that all valid ZIP codes, either the five-digit codes or the ZIP + 4 codes were validated.

27.5 Verifying That Phone Numbers Are in a Standard Form

You want to search a list of phone numbers and identify any numbers that do not conform to the form:

```
(ddd)ddd-dddd
```

Where `d` is any digit. Writing a regular express for this is a bit tricky because both the open and closed parentheses have a special meaning in a regular expression. In the previous example, you may have noticed that the parentheses around the dash and final four digits allowed the question mark (repeat the previous expression 0 or 1 times) to be placed outside the final closed parenthesis. In regular expressions, parentheses indicate a grouping of characters or metacharacters. How do you indicate that you are searching for either the open or closed parenthesis in a standard phone number? The answer is that you place a backward slash (\) before either of these special characters to signify that you mean to treat them as characters and not grouping symbols. The backslash character is sometimes referred to as an *escape character*. Therefore, the regular expression for a phone number in the form discussed here is:

```
/\(\d\d\d\)\d\d\d-\d\d\d\d/
```

Notice the backslash before the open and closed parentheses. Writing the validation program for standard phone numbers is identical to either Program 27.1 or Program 27.2, substituting the expression for a phone number for Social Security numbers or ZIP codes. Here it is:

Program 27.3: Using a Regex to Check for Phone Numbers in Standard Form

```
title "Checking that Phone Numbers are in Standard Form";
data _null_;
   file print;
   input Phone $13.;
  if not prxmatch("/\(\d\d\d\)\d\d\d-\d\d\d\d/",Phone) then
      put "Invalid Phone Number " Phone;
datalines;
(908)432-1234
800.343.1234
8882324444
(888)456-1324
;
```

Output from this program is shown next:

Figure 27.3: Output from Program 27.3

Checking that Phone Numbers are in Standard Form

```
Invalid Phone Number 800.343.1234
Invalid Phone Number 8882324444
```

It is just about impossible for this author to leave this section without showing you how to convert any phone number that includes an area code into a standard number. In Chapter 12 that focuses on character functions, you saw how the COMPRESS function can be used with a k (keep) modifier to keep selected characters and to remove everything else. You can use the k modifier to your advantage in converting nonstandard phone numbers. Program 27.4 reads the data from Program 27.3 and converts all of the phone numbers to the standard form:

Program 27.4: Converting Phone Numbers to Standard Form

```
data Standard;
   length Standard $ 13;
   input Phone $13.;
   Digits = compress(Phone,,'kd');
   Standard = cats('(',substr(Digits,1,3),')',substr(Digits,4,3),
                   '-',substr(Digits,7,4));
   drop Digits;
datalines;
(908)432-1234
800.343.1234
8882324444
(888)456-1324
;
```

The COMPRESS function with the kd (keep the digits) modifiers extracts the digits from the phone numbers. To create a standard phone number, you use the CATS function to concatenate all the necessary pieces (including the parentheses and the dash), and you use the SUBSTR function to extract the appropriate digits from the original phone numbers. Here is a listing of the resulting data set:

Figure 27.4: Listing of Data Set Standard

Listing of Data Set Standard

Standard	Phone
(908)432-1234	(908)432-1234
(800)343-1234	800.343.1234
(888)232-4444	8882324444
(888)456-1324	(888)456-1324

27.6 Describing the PRXPARSE Function

Many of the more advanced PRX functions and call routines require that you first run a function called PRXPARSE to obtain a *return code* that you use in place of a regular expression in these functions and call routines. As a matter of fact, you can use PRXPARSE to create a return code that you can use in place of the regular expression in the PRXMATCH function, which were used in the previous programs in this chapter. To see how this works, here is Program 27.1, which was rewritten to use a combination of the PRXPARSE and PRXMATCH functions.

Program 27.5: Demonstrating a Combination of PRXPARSE and PRXMATCH Functions

```
title "Checking Social Security Numbers Using a Regular Expression";
title2 "Using a Combination of PRXPARSE and PRXMATCH Functions";

data _null_;
   file print;
   input SS $11.;
   Return_Code = prxparse("/\d\d\d-\d\d-\d\d\d\d/");
   if not prxmatch(Return_Code,SS) then
      put "Error for SS Number " SS;
datalines;
123-45-6789
123456789
123-ab-9876
999-888-7777
;
```

The PRXPARSE function "compiles" the regular expression and assigns a sequential number (starting at 1 and increasing for each use of PRXPARSE in a DATA step) to a variable (called Return_Code) in this example. You can use this variable in place of the actual regular expression as the first argument of PRXMATCH.

Although the PRXMATCH function allows you to enter the regular expression as the first argument, many of the other PRX functions and call routines require you to use PRXPARSE first. The output from Program 27.5 is identical to that in Figure 27.3

It is a good practice to execute the PRXPARSE function only once and retain the return code. (Although it is not necessary if the argument to the PRXPARSE function is a constant, as it is here). You can rewrite Program 27.5 like this:

Program 27.6: Rewriting Program 27.5 to Demonstrate a Program Written by a Compulsive Programmer

```
title "Checking Social Security Numbers Using a Regular Expression";
title2 "Using a Combination of PRXPARSE and PRXMATCH Functions";

data _null_;
   file print;
   input SS $11.;

   retain Return_Code;
   if _n_ = 1 then
      Return_Code = prxparse("/\d\d\d-\d\d-\d\d\d\d/");

   if not prxmatch(Return_Code,SS) then
      put "Error for SS Number " SS;
datalines;
123-45-6789
123456789
123-ab-9876
999-888-7777
;
```

Because you execute the PRXPARSE function only once, you need a RETAIN statement so that the variable Return_Code does not get set to Missing at the next iteration of the DATA step.

Hopefully, you will see that regular expressions can be incredibly useful when you need to verify a pattern rather than an exact match of characters. You can use Google to search for regular expressions for almost any term or phrase. Beware because some of the results may be incorrect or overly complicated. One site recommended by this author is www.stackoverflow.com.

27.7 Problems

Solutions to odd-numbered problems are located at the back of this book. Solutions to all problems are available to professors. If you are a professor, visit the book's companion website at support.sas.com/cody for information about how to obtain the solutions to all problems.

1. You have a list of license plate numbers and want to extract any number that does not have the form of three uppercase letters followed by three digits. Use this list of numbers to test your program:

   Note: SASMAN is my Texas license plate and SASJEDI belongs to my friend Mark Jordan.

   ```
   ABC123
   SASMAN
   SASJEDI
   345XYZ
   low987
   WWW999
   ```

2. Use the telephone numbers in Program 27.3 but assume you want the numbers to be in the form `ddd.ddd.dddd` where `d` is any digit. Be aware that a period is a special character in a regular expression. It is a wildcard that stands for any character. Therefore, you must precede the period with a backslash when you write your expression.

3. Repeat Problem 2 using a combination or PRXPARSE and PRXMATCH.

Solutions to Odd-Numbered Exercises

This SAS file contains the solutions to all the odd-numbered problems in the text: Learning SAS by Example: A Programmer's Guide, 2nd edition.

You need to modify any libname and infile statements so that they point to the appropriate folder on your computer. The simplest way to convert all the libname and infile statements in this file is to find the string:

```
c:\books\learning
```

and replace it with the folder where you placed you SAS data sets and text files. If you are storing your SAS data sets and text files in separate places you will need to search separately for libname and infile statements and make changes appropriately:

```
libname learn 'c:\books\learning';
options fmtsearch=(learn);
```

Chapter 1 Solutions

```
*1-1;
/* Invalid variable names are:
   Wt-Kg (contains a dash)
   76Trombones (starts with a number)
*/

*1-3;
/* Number of variables is 5
   Number of observatyions is 10
*/

*1-5;
/* Default length for numerics is 8 */
```

Chapter 2 Solutions

```
*2-1;
*--------------------------------------------------*
 | Program name: stocks.sas  in c:\books\learning   |
 | Purpose: Read in raw data on stock prices and    |
 |    compute values                                |
 | Programmer: Ron Cody                             |
 | Date: June 23, 2006                              |
*--------------------------------------------------*;
*a;
data portfolio;
   infile 'c:\books\learning\stocks.txt';
   input Symbol $ Price Number;
   Value = Number*Price;
run;

title "Listing of Portfolio";
proc print data=portfolio noobs;
run;
```

```
*b;
title "Means and Sums of Portfolio Variables";
proc means data=portfolio n mean sum maxdec=0;
   var Price Number;
run;

*2-3;
/*
   EMF = 1.45*V + (R/E)*v**3 - 125;
*/

*2-5;
 /*need $ after Gender*/
data XYZ;
   infile "c:\books\learning\DataXYZ.txt";
   input Gender $ X Y Z;
   Sum = X + y + Z;
datalines;
Male 1 2 3
Female 4 5 6
Male 7 8 9
run;
```

Chapter 3 Solutions

```
*3-1;
*a - c;
data scores;
   infile 'c:\books\learning\scores.txt';
   input Gender : $1.
         English
         History
         Math
         Science;
   Average = (English + History + Math + Science) / 4;
 run;

 title "Listing of SCORES";
 proc print data=scores noobs;
 run;

*3-3;
data company;
   infile 'c:\books\learning\company.txt' dsd dlm='$';
   input LastName $ EmpNo $ Salary;
   format Salary dollar10.; /* optional statement */
run;

title "Listing of COMPANY";
proc print data=company noobs;
run;

*3-5;
data testdata;
   input X Y;
   Z = 100 + 50*X + 2*X**2 - 25*Y + Y**2;
datalines;
1 2
3 5
5 9
9 11
```

```
;

title "Listing of TESTDATA";
proc print data=testdata noobs;
run;

*3-7;
data cache;
   infile 'c:\books\learning\geocaching.txt' pad;
   ***Note: PAD not necessary but a good idea
      See Chapter 21 for a discussion of this;
   input GeoName  $  1-20
         LongDeg    21-22
         LongMin    23-28
         LatDeg     29-30
         LatMin     31-36;
run;

title "Listing of CACHE";
proc print data=cache noobs;
run;

*3-9;
data cache;
   infile 'c:\books\learning\geocaching.txt' pad;
   input @1  GeoName  $20.
         @21 LongDeg    2.
         @23 LongMin    6.
         @29 LatDeg     2.
         @31 LatMin     6.;
run;

title "Listing of CACHE";
proc print data=cache noobs;
run;

*3-11;
data employ;
   infile 'c:\books\learning\employee.csv' dsd missover;
   ***Note: missover is not needed but a good idea.
      truncover will also work
      See Chapter 21 for an explanation of missover
      and truncover infile options;
   informat ID $3. Name $20. Depart $8.
            DateHire mmddyy10. Salary dollar8.;
   input ID Name Depart DateHire Salary;
   format DateHire date9.;
run;

title "Listing of EMPLOY";
proc print data=employ noobs;
run;
```

Chapter 4 Solutions

```
*4-1;
*You will need to modify this libname statement;
libname learn 'c:\books\learning';

data learn.perm;
   input ID : $3. Gender : $1. DOB : mmddyy10.
```

```
         Height Weight;
   label DOB = 'Date of Birth'
         Height = 'Height in inches'
         Weight = 'Weight in pounds';
   format DOB date9.;
datalines;
001 M 10/21/1946 68 150
002 F 5/26/1950 63 122
003 M 5/11/1981 72 175
004 M 7/4/1983 70 128
005 F 12/25/2005 30 40
;

title "Contents of data set PERM";
proc contents data=learn.perm varnum;
run;

*4-3;
*Modify this libname statement;
libname perm 'c:\books\learning';

data perm.Survey2018;
   input Age Gender $ (Ques1-Ques5)($1.);
datalines;
23 M 15243
30 F 11123
42 M 23555
48 F 55541
55 F 42232
62 F 33333
68 M 44122
;

***Opening up a new session, you need to reissue
   a libname statement;
*Modify this libname statement;
libname perm 'c:\books\learning';
title "Computing Average Age";
proc means data=perm.survey2018;
   var Age;
run;
```

Chapter 5 Solutions

```
*5-1;
proc format;
   value agegrp 0 - 30 = '0 to 30'
               31 - 50 = '31 to 50'
               51 - 70 = '50 to 70'
               71 - high = '71 and older';
   value $party 'D' = 'Democrat'
                'R' = 'Republican';
   value $likert '1' = 'Strongly Disagree'
                 '2' = 'Disagree'
                 '3' = 'No Opinion'
                 '4' = 'Agree'
                 '5' = 'Strongly Agree';
run;

data voter;
   input Age Party : $1. (Ques1-Ques4)($1. + 1);
```

```
   label Ques1 = 'The president is doing a good job'
         Ques2 = 'Congress is doing a good job'
         Ques3 = 'Taxes are too high'
         Ques4 = 'Government should cut spending';
   format Age agegrp.
          Party $party.
          Ques1-Ques4 $likert.;
datalines;
23 D 1 1 2 2
45 R 5 5 4 1
67 D 2 4 3 3
39 R 4 4 4 4
19 D 2 1 2 1
75 D 3 3 2 3
57 R 4 3 4 4
;

title "Listing of Voter";
proc print data=voter;
***Add the option LABEL if you want to use the
   labels as column headings;
run;

title "Frequencies on the Four Questions";
proc freq data=voter;
   tables Ques1-Ques4;
run;

*5-3;
data colors;
   input Color : $1. @@;
datalines;
R R B G Y Y . . B G R B G Y P O O V V B
;
proc format;
   value $color 'R','B','G' = 'Group 1'
                'Y','O' = 'Group 2'
                ' '     = 'Not Given'
                Other   = 'Group 3';
run;

title "Color Frequencies (Grouped)";
proc freq data=colors;
   tables color / nocum missing;
   *The MISSING option places the frequency
    of missing values in the body of the
    table and causes the percentages to be
    computed on the number of observations,
    missing or non-missing;
   format color $color.;
run;

*5-5;
*Modify this libname statement;
libname learn 'c:\books\learning';
options fmtsearch=(learn);
proc format library=learn fmtlib;
   value yesno 1='Yes' 2='No';
   value $yesno 'Y'='Yes' 'N'='No';
   value $gender 'M'='Male' 'F'='Female';
   value age20yr
      low-20 = '1'
```

```
      21-40   = '2'
      41-60   = '3'
      61-80   = '4'
      81-high = '5';
run;
```

Chapter 6 Solutions

```
*6-1;
/*
Select File --> Import Data
Choose Excel and select Drugtest.xls.
*/

*6-3;
*Modify this libname statement;
libname readit 'c:\books\learning\soccer.xls';
title "Using the Excel Engine to read data";
proc print data=readit.'soccer$'n noobs;
run;
```

Chapter 7 Solutions

```
*7-1;
data school;
   input Age Quiz : $1. Midterm Final;
   if Age = 12 then Grade = 6;
   else if Age = 13 then Grade = 9;
   if Quiz = 'A' then QuizGrade = 95;
   else if Quiz = 'B' then QuizGrade = 85;
   else if Quiz = 'C' then QuizGrade = 75;
   else if Quiz = 'D' then QuizGrade = 70;
   else if Quiz = 'F' then QuizGrade = 65;
   CourseGrade = .2*QuizGrade + .3*Midterm + .5*Final;
datalines;
12 A 92 95
12 B 88 88
13 C 78 75
13 A 92 93
12 F 55 62
13 B 88 82
;

title "Listing of SCHOOL";
proc print data=school noobs;
run;

*7-3;
title "Selected Employees from SALES";
proc print data=learn.sales;
   where EmpID = '9888' or EmpID = '0177';
run;

proc print data=learn.sales;
   where EmpID in ('9888' '0177');
run;

*7-5;
data blood;
   set learn.blood;
```

```
   length CholGroup $ 6;
   select;
      when (missing(Chol)) CholGroup = ' ';
      when (Chol le 110) CholGroup = 'Low';
      when (Chol le 140) CholGroup = 'Medium';
      otherwise CholGroup = 'High';
   end;
run;

title "Listing of BLOOD";
proc print data=blood noobs;
run;

*7-7;
title "Selected Observations from BIBYCLES";
proc print data=learn.bicycles noobs;
   where Model eq "Road Bike" and UnitCost gt 2500 or
         Model eq "Hybrid" and UnitCost gt 660;
   *Note: parentheses are not needed since the AND
    operation is performed before OR.  You may inclue
    them if you wish;
run;
```

Chapter 8 Solutions

```
*8-1;
data vitals;
   input ID    : $3.
         Age
         Pulse
         SBP
         DBP;
   label SBP = "Systolic Blood Pressure"
         DBP = "Diastolic Blood Pressure";
datalines;
001 23 68 120 80
002 55 72 188 96
003 78 82 200 100
004 18 58 110 70
005 43 52 120 82
006 37 74 150 98
007  . 82 140 100
;

***Note: this program assumes there are no
   missing values for Pulse or SBP;
data newvitals;
   set vitals;
   if Age lt 50 and not missing(Age) then do;
      if Pulse lt 70 then PulseGroup = 'Low ';
      else PulseGroup = 'High';
      if SBP lt 140 then SBPGroup = 'Low ';
      else SBPGroup = 'High';
   end;
   else if Age ge 50 then do;
      if Pulse lt 74 then PulseGroup = 'Low';
      else PulseGroup = 'High';
      if SBP lt 140 then SBPGroup = 'Low';
      else SBPGroup = 'High';
   end;
run;
```

```
title "Listing of NEWVITALS";
proc print data=newvitals noobs;
run;

*8-3;
data test;
   input Score1-Score3;
   Subj + 1;
datalines;
90 88 92
75 76 88
88 82 91
72 68 70
;

title "Listing of TEST";
proc print data=test noobs;
run;

*8-5;
data logs;
   do N = 1 to 20;
      LogN = log(N);
      output;
   end;
run;

title "Listing of LOGS";
proc print data=logs noobs;
run;

*8-7;
data plotit;
   do x = 0 to 10 by .1;
      y = 3*x**2 - 5*x + 10;
      output;
   end;
run;

title "Problem 7";
proc sgplot data=plotit;
   series x=x y=y;
run;

*8-9;
data temperatures;
   do Day = 'Mon','Tues','Wed','Thu','Fri','Sat','Sun';
      input Temp @;
      output;
   end;
datalines;
70 72 74 76 77 78 85
;

title "Listing of TEMPERATURES";
proc print data=temperatures noobs;
run;

*8-11;
data temperature;
   length City $ 7;
```

```
   do City = 'Dallas','Houston';
      do Hour = 1 to 24;
         input Temp @;
         output;
      end;
   end;
datalines;
80 81 82 83 84 84 87 88 89 89
91 93 93 95 96 97 99 95 92 90 88
86 84 80 78 76 77 78
80 81 82 82 86
88 90 92 92 93 96 94 92 90
88 84 82 78 76 74
;

title "Temperatures in Dallas and Houston";
proc print data=temperature;
run;

*8-13;
data money;
   do Year = 1 to 999 until (Amount ge 30000);
      Amount + 1000;
      do Quarter = 1 to 4;
         Amount + Amount*(.0425/4);
      output;
      end;
   end;
   format Amount dollar10.;
run;

title "Listing of MONEY";
proc print data=money;
run;
```

Chapter 9 Solutions

```
*9-1;
data dates;
   input @1  Subj  $3.
         @4  DOB   mmddyy10.
         @14 Visit date9.;
   Age = yrdif(DOB,Visit,'Actual');
   format DOB Visit date9.;
datalines;
00110/21/195011Nov2006
00201/02/195525May2005
00312/25/200525Dec2006
;
title "Listing of DATES";
proc print data=dates noobs;
run;

*9-3;
options yearcutoff=1910;
data year1910_2006;
   input @1 Date mmddyy8.;
   format Date date9.;
datalines;
01/01/11
```

```
02/23/05
03/15/15
05/09/06
;
options yearcutoff=1920;
/* Good idea to set yearcutoff back to
   the default after you change it */
title "Listing of YEAR1910_2006";
proc print data=year1910_2006 noobs;
run;

*9-5;
data freq;
   set learn.hosp(keep=AdmitDate);
   Day = weekday(AdmitDate);
   Month = month(AdmitDate);
   Year = year(AdmitDate);
run;

proc format;
   value days 1='Sun' 2='Mon' 3='Tue'
              4='Wed' 5='Thu' 6='Fri'
              7='Sat';
   value months 1='Jan' 2='Feb' 3='Mar'
                4='Apr' 5='May' 6='Jun'
                7='Jul' 8='Aug' 9='Sep'
                10='Oct' 11='Nov' 12='Dec';
run;

title "Frequencies for Hospital Admissions";
proc freq data=freq;
   tables Day Month Year / nocum nopercent;
   format Day days. Month months.;
run;

*9-7;
title "Admissions before July 15, 2002";
proc print data=learn.hosp;
   where AdmitDate le '15Jul2002'd and
      AdmitDate is not missing;
run;

*9-9;
data dates;
   input Day Month Year;
   if missing(Day) then Date = mdy(Month,15,Year);
   else Date = mdy(Month,Day,Year);
   format Date mmddyy10.;
datalines;
25 12 2005
.  5  2002
12 8     2006
;

title "Listing of DATES";
proc print data=dates noobs;
run;

*9-11;
data intervals;
   set learn.medical;
   Quarters = intck('qtr','01Jan2006'd,VisitDate);
```

```
run;

title "Listing of INTERVALS";
proc print data=intervals noobs;
run;

*9-13;
data return;
   set learn.medical(keep=Patno VisitDate);
   Return = intnx('month',VisitDate,6,'sameday');
   format VisitDate Return worddate.;
run;

title "Return Visits for Medical Patients";
proc print data=return noobs;
run;
```

Chapter 10 Solutions

```
*10-1;
data subset_a;
   set learn.blood;
   where Gender eq 'Female' and BloodType='AB';
   Combined = .001*WBC + RBC;
run;

title "Listing of SUBSET_A";
proc print data=subset_a noobs;
run;

data subset_b;
   set learn.blood;
   Combined = .001*WBC + RBC;
   if Gender eq 'Female' and BloodType='AB' and Combined ge 14;
run;

title "Listing of SUBSET_B";
proc print data=subset_b noobs;
run;

*10-3;
data lowmale lowfemale;
   set learn.blood;
   where Chol lt 100 and Chol is not missing;
   /* alternative statement
   where Chol lt 100 and not missing(Chol);
   */
   if Gender = 'Female' then output lowfemale;
   else if Gender = 'Male' then output lowmale;
run;

title "Listing of LOWMALE";
proc print data=lowmale noobs;
run;

title "Listing of LOWFEMALE";
proc print data=lowfemale noobs;
run;

*10-5;
title "Listing of INVENTORY";
```

```
proc print data=learn.inventory noobs;
run;

title "Listing of NEWPRODUCTS";
proc print data=learn.newproducts noobs;
run;

data updated;
   set learn.inventory learn.newproducts;
run;

proc sort data=updated;
   by Model;
run;

title "Listing of updated";
proc print data=updated;
run;

*10-7;
proc means data=learn.gym noprint;
   var fee;
   output out=Meanfee(drop=_type_ _freq_)
          Mean= / autoname;
run;

data percent;
   set learn.gym;
   if _n_ = 1 then set Meanfee;
   FeePercent = round(100*fee / Fee_Mean);
   drop Fee_Mean;
run;

title "Listing of PERCENT";
proc print data=PERCENT;
run;

*10-9;
proc sort data=learn.inventory out=inventory;
   by Model;
run;

proc sort data=learn.purchase out=purchase;
   by Model;
run;

data pur_price;
   merge inventory
         purchase(in=InPurchase);
   by Model;
   if InPurchase;
   TotalPrice = Quantity*Price;
   format TotalPrice dollar8.2;
run;

title "Listing of PUR_PRICE";
proc print data=pur_price noobs;
run;

*10-11;
options mergenoby=nowarn;
data try1;
```

```
   merge learn.inventory learn.purchase;
run;

title "Listing of TRY1";
proc print data=try1;
run;

options mergenoby=warn;
data try2;
   merge learn.inventory learn.purchase;
run;

title "Listing of TRY2";
proc print data=try2;
run;

options mergenoby=error;
data try3;
   merge learn.inventory learn.purchase;
run;

title "Listing of TRY3";
proc print data=try3;
run;

*10-13;
/* Solution where the numeric identifier is converted
   to a character value */
proc sort data=learn.demographic_ID out=demographic_ID;
   by ID;
run;

data survey2;
   set learn.survey2(rename=(ID = NumID));
   ID = put(NumID,z3.);
   drop NumID;
run;

proc sort data=survey2;
   by ID;
run;

data combine;
   merge demographic_ID
         survey2;
   by ID;
run;

title "Listing of COMBINE";
proc print data=combine noobs;
run;

/* Solution where the character identifier is converted
   to a numeric value */
data demographic_ID;
   set learn.demographic_ID(rename=(ID = CharID));
   ID = input(CharID,3.);
   drop CharID;
run;

proc sort data=demographic_ID;
   by ID;
```

```
run;

proc sort data=learn.survey2 out=survey2;
   by ID;
run;

data combine;
   merge demographic_ID
         survey2;
   by ID;
run;

title "Listing of COMBINE";
proc print data=combine noobs;
run;
```

Chapter 11 Solutions

```
*11-1;
data health;
   set learn.health;
   BMI = (Weight/2.2) / (Height*.0254)**2;
   BMIRound = round(BMI);
   BMIRound_tenth = round(BMI,.1);
   BMIGroup = round(BMI,5);
   BMITrunc = int(BMI);
run;

title "Listing of HEALTH";
proc print data=health noobs;
run;

*11-3;
data miss_blood;
   set learn.blood;
   if missing(WBC) then call missing(Gender,RBC, Chol);
run;

title "Listing of MISS_BLOOD";
proc print data=miss_blood noobs;
run;

*11-5;
data psychscore;
   set learn.psych;
   if n(of Score1-Score5) ge 3 then
   ScoreAve = mean(largest(1,of Score1-Score5),
                   largest(2,of Score1-Score5),
                   largest(3,of Score1-Score5));
   if n(of Ques1-Ques10) ge 7 then
      QuesAve = mean(of Ques1-Ques10);
   Composit = ScoreAve + 10*QuesAve;
   keep ID ScoreAve QuesAve Composit;
run;

title "Listing of PSYCHSCORE";
proc print data=psychscore noobs;
run;

*11-7;
data _null_;
```

```
   x = 10; y = 20; z = -30;
   AbsZ = abs(z);
   ExpX = round(exp(x),.001);
   Circumference = round(2*constant('pi')*y,.001);
   put _all_;
run;

*11-9;
 data fake;
    do Subj = 1 to 100;
       if rand('uniform') le .4 then Gender = 'Female';
       else Gender = 'Male';
       Age = 9 + ceil(rand('uniform')*51);
       output;
   end;
run;

title "Frequencies";
proc freq data=fake;
   tables Gender / nocum;
run;

title "First 10 Observations of FAKE";
proc print data=fake(obs=10);
run;

*11-11;
data convert;
   set learn.char_num(rename=
     (Age = Char_Age
      Weight = Char_Weight
      Zip = Num_Zip
      SS = Num_ss));
   Age = input(Char_Age,8.);
   Weight = input(Char_Weight,8.);
   SS = put(Num_SS,ssn11.);
   Zip = put(Num_Zip,z5.);
   drop Char_: Num_:;
run;

title "Listing of CONVERT";
proc print data=convert noobs;
run;

*11-13;
data smooth;
   set learn.stocks;
   Price1 = lag(Price);
   Price2 = lag2(Price);
   Average = mean(Price, Price1, Price2);
run;

title "Plot of Price and Moving Average";
proc sgplot data=smooth;
   series x=Date y=Price;
   series x=Date y=Average;
run;
```

Chapter 12 Solutions

```
*12-1;
*One way to test the storage lengths is to use
 the LENGTHC function that returns storage lengths
 compared to the LENGTH function that returns the
 length of a character string, not counting
 trailing blanks;
data storage;
   length A $ 4 B $ 4;
   Name = 'Goldstein';
   AandB = A || B;
   Cat = cats(A,B);
   if Name = 'Smith' then Match = 'No';
      else Match = 'Yes';
   Substring = substr(Name,5,2);
   L_A = lengthc(A);
   L_B = lengthc(B);
   L_Name = lengthc(Name);
   L_AandB = lengthc(AandB);
   L_Cat = lengthc(Cat);
   L_Match = lengthc(Match);
   L_Substring = lengthc(Substring);
run;

title "Lengths of Character Variables";
proc print data=storage noobs;
   var L_:;
   *All variables starting with L_;
run;
/*
Variable   Storage Length
A                4
B                4
Name             9
AandB            8
Cat            200
Match            2
Substring        9
*/

*12-3;
data names_and_more;
   set learn.names_and_more;
   Name = compbl(Name);
   Phone = compress(Phone,,'kd');
run;

title "Listing of NAMES_AND_MORE";
proc print data=names_and_more noobs;
run;

*12-5;
data convert;
   set learn.names_and_more(keep=Mixed);
   Integer = input(scan(Mixed,1,' /'),8.);
   Numerator = input(scan(Mixed,2,' /'),8.);
   Denominator = input(scan(Mixed,3,' /'),8.);
   if missing(Numerator) then Price = Integer;
   else Price = Integer + Numerator / Denominator;
   Price = round(Price,.001);
```

```
   drop Numerator Denominator Integer;
run;

title "Listing of CONVERT";
proc print data=convert noobs;
run;

*12-7;
*Using one of the CAT functions;
data concat;
   set learn.study(keep=Group Subgroup);
   length Combined $ 3;
   Combined = catx('-',Group,Subgroup);
run;

title "Listing of CONCAT";
proc print data=concat noobs;
run;

*Without using CAT functions;
data concat;
   set learn.study(keep=Group Subgroup);
   length Combined $ 3;
   Combined = trim(Group) || '-' || put(Subgroup,1.);
run;

title "Listing of CONCAT";
proc print data=concat noobs;
run;

*12-9;
data spirited;
   set learn.sales;
   where find(Customer,'spirit','i');
run;

title "Listing of SPIRITED";
proc print data=spirited noobs;
run;

*12-11;
title "Subjects from ERRORS with Digits in the Name";
proc print data=learn.errors noobs;
   where anydigit(Name);
   var Subj Name;
run;

*12-13;
data exact within25;
   set learn.social;
   if SS1 eq SS2 then output exact;
   else if spedis(SS1,SS2) le 25 and
      not missing(SS1) and
      not missing(SS2) then output
      within25;
run;

title "Listing of EXACT";
proc print data=exact noobs;
run;
```

```
title "Listing of WITHIN25";
proc print data=within25 noobs;
run;

*12-15;
data numbers;
   set learn.names_and_more(keep=phone);
   length AreaCode $ 3;
   AreaCode = substr(Phone,2,3);
run;

title "Listing of NUMBERS";
proc print data=numbers;
run;

*12-17;
data personal;
   set learn.personal(drop=Food1-Food8);
   substr(SS,1,7) = '******';
   substr(AcctNum,5,1) = '-';
run;

title "Listing of PERSONAL (with masked values)";
proc print data=personal noobs;
run;
```

Chapter 13 Solutions

```
*13-1;
data survey1;
   set learn.survey1;
   array Ques{5} $ Q1-Q5;
   do i = 1 to 5;
      Ques{i} = translate(Ques{i},'54321','12345');
   end;
   drop i;
run;

title "List of SURVEY1 (rescaled)";
proc print data=survey1;
run;

*13-3;
data nonines;
   set learn.nines;
   array nums{*} _numeric_;
   do i = 1 to dim(nums);
      if nums{i} = 999 then
      call missing(nums{i});
   end;
   drop i;
run;

title "Listing of NONINES";
proc print data=nonines;
run;

*13-5;
data passing;
   array pass_score{5} _temporary_
      (65,70,60,62,68);
```

```
   array Score{5};
   input ID : $3. Score1-Score5;
   NumberPassed = 0;
   do Test = 1 to 5;
      NumberPassed + (Score{Test} ge pass_score{Test});
   end;
   drop Test;
datalines;
001   90 88 92 95 90
002   64 64 77 72 71
003   68 69 80 75 70
004   88 77 66 77 67
;
title "Listing of PASSING";
proc print data=passing;
   id ID;
run;
```

Chapter 14 Solutions

```
*14-1;
title "First 10 Observations in BLOOD";
proc print data=learn.blood(obs=10) label;
   id Subject;
   var WBC RBC Chol;
   label WBC = 'White Blood Cells'
         RBC = 'Red Blood Cells'
         Chol = 'Cholesterol';
run;

*14-3;
title "Selected Patients from HOSP Data Set";
title2 "Admitted in September of 2004";
title3 "Older than 83 years of age";
title4 "--------------------------------------";
proc print data=learn.hosp
           n='Number of Patients = '
           label
           double;
   where Year(AdmitDate) eq 2004 and
         Month(AdmitDate) eq 9 and
         yrdif(DOB,AdmitDate,'Actual') ge 83;
   id Subject;
   var DOB AdmitDate DischrDate;
   label AdmitDate = 'Admission Date'
         DischrDate = 'Discharge Date'
         DOB = 'Date of Birth';
run;
```

Chapter 15 Solutions

```
*15-1;
title "First 5 Observations from Blood Data Set";
proc report data=learn.blood(obs=5);
   column Subject WBC RBC;
   define Subject / display "Subject Number" width=7;
   define WBC / "White Blood Cells" width=6 format=comma6.0;
   define RBC / "Red Blood Cells" width=5 format=5.2;
run;
quit;
```

```
*15-3;
title "Demonstrating a Compute Block";
proc report data=learn.hosp(obs=5);
   column Subject AdmitDate DOB Age;
   define AdmitDate / display "Admission Date" width=10;
   define DOB / display;
   define Subject / display width=7;
   define Age / computed "Age at Admission" ;
   compute Age;
      Age = round(yrdif(DOB,AdmitDate,'Actual'));
   endcomp;
run;
quit;

*15-5;
title "Patient Age Groups";
proc report data=learn.bloodpressure;
   column Gender Age AgeGroup;
   define Gender / width=6;
   define Age / display width=5;
   define AgeGroup / computed "Age Group";
   compute AgeGroup / character length=5;
      if Age gt 50 then AgeGroup = '> 50';
      else if not missing(Age) then AgeGroup = '<= 50';
   endcomp;
run;
quit;

*15-7;
title "Mean Cholesterol by Gender and Blood Type";
proc report data=learn.blood;
   column Gender BloodType Chol;
   define Gender / group width=6;
   define BloodType / group "Blood Type" width=5;
   define Chol / analysis mean "Mean Cholesterol"
                         width=11 format=5.1;
run;
quit;

*15-9;
title "Report on the Survey Data Set";
proc report data=learn.survey split=' ';
   column ID Age Gender Salary Ques1-ques5 AveResponse;
   define ID / display width=4;
   define Age / display width=18;
   define Gender / display width=6;
   define Salary / display width=10 format=dollar10.;
   define Ques1 / display noprint;
   define Ques2 / display noprint;
   define Ques3 / display noprint;
   define Ques4 / display noprint;
   define Ques5 / display noprint;
   *Note: This solution will case an automatic
    character to numeric conversion;
   compute AveResponse;
      AveResponse = mean(of Ques1-Ques5);
   endcomp;
   /*********************************************
   To avoid the automatic conversion, substitute
   the code below for the compute block:
```

```
   compute AveResponse;
      Q1 = input(Ques1,1.);
      Q2 = input(Ques2,1.);
      Q3 = input(Ques3,1.);
      Q4 = input(Ques4,1.);
      Q5 = input(Ques5,1.);
      AveResponse = mean(of Q1-Q5);
   endcomp;
   ************************************************/

   define AveResponse / computed "Average Response" width=8 format=3.1;
run;
quit;
```

Chapter 16 Solutions

```
*16-1;
options fmtsearch=(learn);
***This is where the file formats.sas7bcat was
   placed;
title "Statistics on the College Data Set";
proc means data=learn.college
           n
           nmiss
           mean
           median
           min
           max
           maxdec=2;
   var ClassRank GPA;
run;

*16-3;
proc sort data=learn.college out=college;
   by SchoolSize;
run;

title "Statistics on the College Data Set - Using BY";
title2 "Broken down by School Size";
proc means data=college
           n
           mean
           median
           min
           max
           maxdec=2;
   by SchoolSize;
   var ClassRank GPA;run;

title "Statistics on the College Data Set - Using CLASS";
title2 "Broken down by School Size";
proc means data=learn.college
           n
           mean
           median
           min
           max
           maxdec=2;
   class SchoolSize;
   var ClassRank GPA;
run;
```

```
*16-5;
proc format;
   value rank 0-50 = 'Bottom Half'
             51-74 = 'Third Quartile'
            75-100 = 'Top Quarter';
run;

title "Statistics on the College Data Set";
title2 "Broken down by School Size";
proc means data=learn.college
           n
           mean
           maxdec=2;
   class ClassRank;
   var GPA;
   format ClassRank rank.;
run;

*16-7;
proc means data=learn.college noprint chartype;
   class Gender SchoolSize;
   var ClassRank GPA;
   output out=summary
          n= mean= median= min= max= / autoname;
run;

data grand(drop=Gender SchoolSize)
     bygender(drop=SchoolSize)
     bysize(drop=Gender)
     cell;
   drop _freq_;
   set summary;
   if _type_ = '00' then output grand;
   else if _type_ = '10' then output bygender;
   else if _type_ = '01' then output bysize;
   else if _type_ = '11' then output cell;
run;

title "Listing of GRAND";
proc print data=grand noobs;
run;

title "Listing of BYGENDER";
proc print data=bygender noobs;
run;

title "Listing of BYSIZE";
proc print data=bysize noobs;
run;

title "Listing of CELL";
proc print data=cell noobs;
run;
```

Chapter 17 Solutions

```
*17-1;
title "One-way Frequencies from BLOOD Data Set";
proc freq data=learn.blood;
   tables Gender BloodType AgeGroup / nocum nopercent;
```

```
run;

*17-3;
proc format;
   value cholgrp low-200  = 'Normal'
                 201-high = 'High'
                 .        = 'Missing';
run;

title "Demonstrating the MISSING Option";
title2 "Without MISSING Option";
proc freq data=learn.blood;
   tables Chol / nocum;
   format Chol cholgrp.;
run;

title "Demonstrating the MISSING Option";
title2 "With MISSING Option";
proc freq data=learn.blood;
   tables Chol / nocum missing;
   format Chol cholgrp.;
run;

*17-5;
proc format;
   value rank low-70  = 'Low to 70'
              71-high = '71 and higher';
run;

title "Scholarship by Class Rank";
proc freq data=learn.college;
   tables Scholarship*ClassRank;
   format ClassRank rank.;
run;

*17-7;
title "Blood Types in Decreasing Frequency Order";
proc freq data=learn.blood order=freq;
   tables BloodType / nocum nopercent;
run;
```

Chapter 18 Solutions

```
*18-1;
options fmtsearch=(learn);
title "Demographics from COLLEGE Data Set";
proc tabulate data=learn.college format=6.;
   class Gender Scholarship SchoolSize;
   tables Gender Scholarship all,
          SchoolSize / rts=15;
   keylabel n=' ';
run;

*18-3;
proc format;
   value $gender 'F' = 'Female'
                 'M' = 'Male';
run;
title "Demographics from COLLEGE Data Set";
proc tabulate data=learn.college format=6.;
   class Gender Scholarship SchoolSize;
```

```
   tables (Gender all)*(Scholarship all),
          SchoolSize all / rts=25;
   keylabel n=' '
            all = 'Total';
   format Gender $gender.;
run;

*18-5;
title "Descriptive Statistics";
proc tabulate data=learn.college format=6.1;
   class Gender;
   var GPA;
   tables GPA*(n*f=4.
               mean min max),
          Gender all;
   keylabel n = 'Number'
            all = 'Total'
            mean = 'Average'
            min = 'Minimum'
            max = 'Maximum';
run;

*18-7;
title "More Descriptive Statistics";
proc tabulate data=learn.college format=7.1 noseps;
   class Gender SchoolSize;
   var GPA ClassRank;
   tables SchoolSize all,
          GPA*(median min max)
          ClassRank*(median*f=7. min*f=7. max*f=7.)/ rts=15;
   keylabel all = 'Total'
            median = 'Median'
            min = 'Minimum'
            max = 'Maximum';
   label ClassRank = 'Class Rank'
         SchoolSize = 'School Size';
run;

*18-9;
title "Demonstrating Column Percents";
proc format;
   value $gender 'F' = 'Female'
                 'M' = 'Male';
run;
proc sort data=learn.college out=college;
   by descending Scholarship;
run;
proc tabulate data=college
              format=7.
              order=data
              noseps;
   class Gender Scholarship;
   tables (Gender all),
          (Scholarship all)*colpctn;
   keylabel colpctn = 'Percent'
            all = 'Total';
   format Gender $gender.;
run;
```

Chapter 19 Solutions

```
*19-1;
options fmtsearch=(learn);
ods listing close;
ods html path= 'c:\books\learning' file='prob19_1.html';

title "Sending Output to an HTML File";
proc print data=learn.college(obs=8) noobs;
run;

proc means data=learn.college n mean maxdec=2;
   var GPA ClassRank;
run;

ods html close;
ods listing;

*19-3;
ods listing close;
ods html file='prob19_3.html'
         style=journal;

title "Sending Output to an HTML File";
proc print data=learn.college(obs=8) noobs;
run;

proc means data=learn.college n mean maxdec=2;
   var GPA ClassRank;
run;

ods html close;
ods listing;

*19-5;
ods trace on;
proc univariate data=learn.survey;
   var Age Salary;
run;
ods trace off;

ods select quantiles;
proc univariate data=learn.survey;
   var Age Salary;
run;
```

Chapter 20 Solutions

```
*20-1;
title "Bar Chart for the Variable Status";
proc sgplot data=SASHelp.Heart;
   vbar Status;
run;

*20-3;
title "Mean Height by Sex";
proc sgplot data=SASHelp.Heart;
   hbar sex / response=Height stat=mean nofill
              barwidth=.25;
run;
```

```
*20-5;
title "Scatter Plot with Regression Line and Confidence Limits";
proc sgplot data=SASHelp.Heart;
   reg x=Height y=Weight / CLM CLI;
run;

*20-7;
title "Demonstrating PBSPLINE Smoothing";
proc sgplot data=SASHelp.Heart(obs=100);
   pbspline x=Height y=Weight;
run;

*20-9;
title "Histogram for Cholesterol";
proc sgplot data=SASHelp.Heart;
   histogram Cholesterol;
run;

*20-11;
title "Horizontal Box Plots";
proc sgplot data=SASHelp.Heart;
   hbox Cholesterol / group=Sex;
run;
```

Chapter 21 Solutions

```
*21-1;
data prob21_1;
   infile 'c:\books\learning\test_scores.txt' missover;
   /* or truncover */
   input Score1-Score3;
run;

title "Listing of PROB21_1";
proc print data=prob21_1 noobs;
run;

*21-3;
data prob21_3;
   infile 'c:\books\learning\scores_column.txt' pad;
   input Score1 1-2
         Score2 3-4
         Score3 5-6;
run;

title "Listing of PROB21_3";
proc print data=prob21_3 noobs;
run;

*21-5;
title "Summary Report from BICYCLES Data Set";
data prob21_5;
   set learn.bicycles end=lastrec;
   TotalUnits + units;
   Sum_of_Sales + TotalSales;
   file print;
   if lastrec then
      put "--------------------------------------"/
          "Total Units Sold is " TotalUnits comma10. /
          "Sales Total is " Sum_of_Sales dollar10.0;
run;
```

```
*21-7;
data prob21_7;
   infile 'c:\books\learning\file_A.txt'
          firstobs=2
          end=last_of_a;
   if last_of_a then infile 'c:\books\learning\file_B.txt'
          firstobs=2;
   input x y z;
run;

title "Listing of PROB21_7";
proc print data=prob21_7;
run;

*21-9;
data prob21_9;
   filename xyzfiles ('c:\books\learning\xyz1.txt'
                      'c:\books\learning\xyz2.txt');
   infile xyzfiles;
   input x y z;
run;

title "Listing of PROB21_9";
proc print data=prob21_9;
run;

*21-11;
data prob21_11;
   infile 'c:\books\learning\three_per_line.txt';
   input @1 (HR1-HR3)(3. +6)
         @4 (SBP1-SBP3)(3. +6)
         @7 (DBP1-DBP3)(3. +6);
run;

title "Listing of PROB21_11";
proc print data=prob21_11;
run;
```

Chapter 22 Solutions

```
*22-1;
proc format;
   value high_sbp low - <140 = 'Normal'
                  140 - high = 'High SBP';
   value high_dbp low - <90 = 'Normal'
                  90 - high ='High DBP';
run;

title "Frequencies on SBP and DBP";
proc freq data=learn.bloodpressure;
   tables SBP DBP / nocum nopercent;
   format SBP high_sbp.
          DBP high_dbp.;
run;

*22-3;
proc format;
   value high_sbp low - <140 = 'Normal'
                  140 - high = 'High SBP';
   value high_dbp low - <90 = 'Normal'
```

```
                    90 - high ='High DBP';
run;
data bloodpressure;
   set learn.bloodpressure;
   SBPGroup = put(SBP,high_sbp.);
   DBPGroup = put(DBP,high_dbp.);
run;

title "Listing of BLOODPRESSURE";
proc print data=bloodpressure noobs;
run;

*22-5;
proc format;
   invalue $convert
      0   - 65 = 'F'
      66  - 75 = 'C'
      76  - 85 = 'B'
      86  - high = 'A'
      other = ' ';
run;

data lettergrades;
   infile 'c:\books\learning\numgrades.txt';
   input ID $ LetterGrade $convert. @@;
run;

title "Listing of LETTERGRADES";
proc print data=lettergrades noobs;
run;

*22-7;
data control;
   set learn.dxcodes(rename=(Dx = Start
                             Description = Label));
   retain fmtname '$dxcodes' Type 'C';
run;

proc format cntlin=control;
   select $dxcodes;
run;

*22-9;
proc format;
   value muggle
      '01jan1990'd - '31dec2004'd = 'Too Early'
      '01jan2005'd - '31dec2005'd = [mmddyy10.]
      '01jan2007'd - high = 'Too Late';
run;

title "Listing of GYM";
proc print data=learn.gym noobs;
   format Date muggle.;
run;
```

Chapter 23 Solutions

```
*23-1;
data long;
   set learn.wide;
   array X_array[5] X1-X5;
```

```
   array Y_array[5] Y1-Y5;
   do Time = 1 to 5;
      X = X_array[Time];
      Y = y_array{Time];
      output;
   end;
   keep Subj Time X Y;
run;

title "Listing of LONG";
proc print data=long;
run;

*23-3;
proc transpose data=learn.wide
               out=long(rename=(col1=X)
                          drop=_name_);
   by Subj;
   var X1-X5;
run;

title "Listing of LONG";
proc print data=long;
run;
```

Chapter 24 Solutions

```
*24-1;
proc sort data=learn.dailyprices out=dailyprices;
   by Symbol Date;
run;
data lastprice;
   set dailyprices;
   by Symbol;
   if last.Symbol;
run;

title "Listing of LASTPRICE";
proc print data=lastprice noobs;
run;

*24-3;
proc sort data=learn.dailyprices out=dailyprices;
   by Symbol Date;
run;
data countit;
   set dailyprices;
   by Symbol;
   if first.Symbol then N_Days = 0;
   N_Days + 1;
   if last.Symbol;
   keep Symbol N_Days;
run;

title "Listing of COUNTIT";
proc print data=countit noobs;
run;

*24-5;
proc sort data=learn.dailyprices out=dailyprices;
   by Symbol Date;
```

```
run;
data first_last;
   set dailyprices;
   by Symbol;
   retain FirstPrice;
   if first.Symbol and last.Symbol then delete;
   if first.Symbol then FirstPrice = Price;
   if last.Symbol then do;
      Diff = Price - FirstPrice;
      output;
   end;
   keep Symbol Price Diff;
run;

title "Listing of FIRST_LAST";
proc print data=first_last noobs;
run;

*24.7;
proc sort data=learn.dailyprices out=dailyprices;
   by Symbol Date;
run;
data first_last;
   set dailyprices;
   by Symbol;
   if first.Symbol and last.Symbol then delete;
   Diff = dif(Price);
   if not first.Symbol then output;
   keep Symbol Price Diff;
run;

title "Listing of FIRST_LAST";
proc print data=first_last noobs;
run;
```

Chapter 25 Solutions

```
*25-1;
title "Listing produced on &sysday, &sysdate9 at &systime";
proc print data=learn.stocks(obs=5) noobs;
run;

*25-3;
%macro print_n(dsn, /* data set name */
               nobs /* number of observations to list */);
   title "Listing of the first &nobs Observations from "
         "Data set &dsn";
   proc print data=&dsn(obs=&nobs) noobs;
   run;
%mend print_n;

%print_n(learn.bicycles, 4)

*25-5;
proc means data=learn.fitness nway noprint;
   var TimeMile RestPulse MaxPulse;
   output out=summary
          mean= / autoname;
run;

data _null_;
```

```
   set summary;
   call symput('GrandTime',TimeMile_mean);
   call symput('GrandRest',RestPulse_mean);
   call symput('GrandMax',MaxPulse_mean);
run;

data compute_percents;
   set learn.fitness;
   P_TimeMile = round(100*TimeMile/&GrandTime);
   P_RestPulse = round(100*RestPulse/&GrandRest);
   R_MaxPulse = round(100*MaxPulse/&GrandMax);
run;

title "Fitness Stats as a Percent of Mean";
proc print data=compute_percents noobs;
run;
```

Chapter 26 Solutions

```
*26-1;
title "Observations from INVENTORY where Price > 20";
proc sql;
   select *
   from learn.inventory
   where Price gt 20;
quit;

*26-3;
proc sql;
   create table n_sales as
   select Name, TotalSales
   from learn.sales
   where Region eq 'North';
quit;

title "Listing of N_SALES";
proc print data=n_sales noobs;
run;

*26-5;
***Part 1;
proc sql;
   create table both as
   select l.Subj as LeftSubj,
          Height,
          Weight,
          r.Subj as RightSubj,
          Salary
   from learn.left as l inner join
        learn.right as r
   on left.Subj = right.Subj;
quit;

title "Listing of BOTH";
proc print data=both noobs;
run;

/* alternate code
proc sql;
   create table both as
   select l.Subj as LeftSubj,
```

```
          Height,
          Weight,
          r.Subj as RightSubj,
          Salary
   from learn.left as l,
        learn.right as r
   where left.Subj = right.Subj;
quit;

title "Listing of BOTH";
proc print data=both noobs;
run;
*/

***Part 2;
proc sql;
   create table both as
   select l.Subj as LeftSubj,
          Height,
          Weight,
          r.Subj as RightSubj,
          Salary
   from learn.left as l full join
        learn.right as r
   on left.Subj = right.Subj;
quit;

title "Listing of BOTH";
proc print data=both noobs;
run;

***Part 3;
proc sql;
   create table both as
   select l.Subj as LeftSubj,
          Height,
          Weight,
          r.Subj as RightSubj,
          Salary
   from learn.left as l left join
        learn.right as r
   on left.Subj = right.Subj;
quit;

title "Listing of LEFT";
proc print data=both noobs;
run;

*26-7;
proc sql;
   create table third as
   select *
   from learn.first union all corresponding
   Select *
   from learn.second;
quit;

title "Listing of THIRD";
proc print data=third;
run;
```

```
*26-9;
proc sql;
   create table percentages as
   select Subject,
          RBC,
          WBC,
          mean(RBC) as MeanRBC,
          mean(WBC) as MeanWBC,
          100*RBC / calculated MeanRBC as Percent_RBC,
          100*WBC / calculated MeanWBC as Percent_WBC
   from learn.blood(obs=10);
quit;

title "Listing of PERCENTAGES";
proc print data=percentages;
run;
```

Chapter 27 Solutions

```
*27-1;
title "List of Non-standard Phone Numbers";
data _null_;
   file print;
   input Plate_No $10.;
   if not prxmatch("/[A-Z]\d\d\d/",Plate_No) then
      put "Plate number " Plate_No "does not conform to pattern";
datalines;
ABC123
SASMAN
SASJEDI
345XYZ
low987
WWW999
;

*27-3;
title "List of Non-standard Phone Numbers";
data _null_;
   file print;
   retain Return;
   if _n_ = 1 then Return = prxparse("/\d\d\d\.\d\d\d\.\d\d\d\d/");
   input Phone $13.;
   if not prxmatch(Return,Phone) then
      put "Phone number " Phone "does not conform to pattern";
datalines;
(908)432-1234
800.343.1234
8882324444
(888)456-1324
;
```

Index

D

Q

R

CPSIA information can be obtained
at www.ICGtesting.com
Printed in the USA
BVHW092147040921
615825BV00003B/165